Fundamental Constants

1 unified atomic mass unit	$1\ u = 1.66 \times 10^{-27}$ kg $= 931$ MeV/c^2
Rest mass of proton	$m_p = 1.67 \times 10^{-27}$ kg
Rest mass of neutron	$m_n = 1.67 \times 10^{-27}$ kg
Rest mass of electron	$m_e = 9.11 \times 10^{-31}$ kg
Charge on an electron	$-e = -1.6 \times 10^{-19}$ C
Charge on a proton	$+e = +1.6 \times 10^{-19}$ C
Avogadro's number	$N_A = 6.02 \times 10^{23}$ molecules/mol
Universal gas constant	$R = 8.31$ J /mol/K
Boltzmann's constant	$k_B = 1.38 \times 10^{-23}$ J/K
Speed of light	$c = 3.00 \times 10^8$ m/s
Planck's constant	$h = 6.63 \times 10^{-34}$ J s
Electrical permittivity of vacuum	$\epsilon_o = 8.85 \times 10^{-12}$ C^2/N/m^2
Coulomb's constant	$k_e = 9.0 \times 10^9$ N m^2/C^2
Magnetic permeability of vacuum	$\mu_o = 4\pi \times 10^{-7}$ T m/A
Universal gravitational constant	$G = 6.67 \times 10^{-11}$ N m^2/kg^2
Atmosphere	1 atm $= 1.01 \times 10^5$ Pa
Electron volt	1 eV $= 1.6 \times 10^{-19}$ J

Earth - Moon System
Earth-Moon Distance $= 3.85 \times 10^8$ m

Property	Earth	Moon
Radius	6.38×10^6 m	1.74×10^6 m
Mass	5.98×10^{24} kg	7.35×10^{22} kg
Density	5.52 g/cm^3	3.34 g/cm^3
Gravity	9.8 m/s^2	1.63 m/s^2

SI Prefixes

Factor	Prefix	Symbol
10^{18}	exa	E
10^{15}	peta	P
10^{12}	tera	T
10^{9}	giga	G
10^{6}	mega	M
10^{3}	kilo	k
10^{2}	hecto	h
10^{1}	deka	da
10^{-1}	deci	d
10^{-2}	centi	c
10^{-3}	milli	m
10^{-6}	micro	μ
10^{-9}	nano	n
10^{-12}	pico	p
10^{-15}	femto	f
10^{-18}	atto	a

PLANETS

Planet	Distance from the sun (meters)	Mass (M_E)	Radius (R_E)	Temp. °C
Mercury	5.8×10^{10}	0.06	0.38	300
Venus	1.08×10^{11}	0.82	0.95	500
Earth	1.50×10^{11}	1.0	1.0	20
Mars	2.28×10^{11}	0.11	0.53	- 12
Jupiter	7.80×10^{11}	318	11	-120
Saturn	1.427×10^{12}	95	9.5	-130
Uranus	2.871×10^{12}	15	4.1	-180
Neptune	4.497×10^{12}	17	3.8	-200
Pluto	5.914×10^{12}	0.002	0.2	-250

SUN

Property	Value
Radius	6.96×10^8 m
Mass	1.99×10^{30} kg
Surface Temp	6400 K
Interior Temp	15.6×10^6 K
Rotation period	25 Days

PREFACE

I am very excited to present the 2010 Edition of the AP® Physics B Handbook and accompanying Instructor's Manual. I have been offering the two books for last 12 years.

For the 2010 addition, I have made substantial improvements in a variety of ways. I have expanded on the notes and information preceding the problems in every chapter. I have included many more illustrations to help with the grasp of the concepts. I have a few more problems and multiple choice questions. For the instructor's Manual, I have added complete solution to multiple choice questions instead of just the answers in the previous editions. Teacher is encouraged to supplement the material with problems and questions of greater difficulty if she/he deems it necessary.

This handbook is divided into two sections. The section for Fall semester includes two topics, namely, Newtonian Mechanics and Thermal Physics. The section for the Spring semester includes the remaining topics, namely, Fluids, Electricity & Magnetism, Wave & Optics, and Atomic & Nuclear Physics. Of course, teachers can follow the topics in any other sequence they are comfortable with.

Students and teachers of AP Physics B course will find this book very useful when going through the rigors of the course. For students, this will provide easy reference material to prepare for the quizzes and examinations. Students should refer to any algebra-based introductory college physics textbook for detailed explanation of the concepts and supplementary reading materials.

Any constructive criticism, suggestions, and comments about this handbook are most welcome. Please feel free to e-mail me.

Hasan Fakhruddin

Instructor of Physics
The Indiana Academy for
Science, Mathematics, and Humanities
BSU
Muncie, IN 47306

June 2010
E-mail: hfakhrud@bsu.edu

The problems marked ** are substantially above the level of AP Physics B and the teacher may assign them as homework with discretion.

CONTENTS

Chapter 1
Introduction - Units and Dimension

The Advanced Placement® Physics B Course. Abbreviated as AP® Physics B, this is a challenging, non-Calculus Physics course. One year each of algebra and geometry is required and some knowledge of trigonometry is desirable. The College Board, after surveying the need of a large number of colleges has designed the syllabus for this course. The AP Physics B examination of the College Board is normally held in the month of May. After passing this course, student may get credit for this course or may test out of an equivalent course in college.

What is Physics?

Physics is a branch of science that attempts to describe the fundamental nature of the Universe. In particular, Physics is the study of matter, energy, and their interaction.

All physical phenomena are parts of one or more of the following five areas of Physics:

1. **Mechanics** 2. **Thermodynamics** 3. **Electromagnetism** 4. **Relativity**
5. **Quantum Mechanics**

Physics is an experimental science. The laws and theories are either derived from experiments, or tested by the experiments or both. The experiments in Physics involve measurements of physical quantities. There are two systems of units for measurement commonly in use:

> 1. **British (English) System ;**   2. **Metric System**

British System is commonly used in everyday life in the United States. This system includes units such as inch, foot, mile, mph, pound, ounce, gallon etc.

In scientific field, we use the International System of Units abbreviated as **SI** following its French name, *Le Syste'me International d'Unite's*. This system is a subset of Metric System. In most part, we will be using SI units in this course.

In the International System of Units, there are **seven** physical quantities, which are defined to be fundamental quantities. In addition, there are two supplementary quantities. These quantities and their units are given in the table below:

FUNDAMENTAL OR BASE QUANTITIES AND THEIR UNITS

	Fundamental Quantities	**SI Units**
1	Length	meter
2	Mass	kilogram
3	Time	second
4	Electric Current	ampere
5	Temperature	Kelvin
6	Amount of Substance	mole
7	Luminous Intensity	candela

	Supplementary Fundamental Quantities	**SI Units**
1	Plane Angle	radians
2	Solid angle	steradians

Note: While stating the value of a quantity, it must be labeled with appropriate unit, if any.
An **atomic clock** uses precise microwave signal that is emitted when electrons undergo the transition between the two hyperfine levels of the ground state of the atom of the cesium-133 atom.

Definition of 'second': Since 1967, the second has been defined on the basis of cesium atomic clock. It is defined to be the duration of 9,192,631,770 periods of the radiation corresponding to the transition between the two hyperfine levels of the ground state of the cesium-133 atom.

Definition of 'meter': The definition of meter has gone through several refinements since its origination by French in 1790. At that time the meter was defined as one/ten-millionth of the distance from the equator to the north pole along a meridian through Paris. The Geneva Conference on Weights and Measures, in 1984, has defined the meter as *the distance light travels, in a vacuum, in 1/299,792,458 seconds with time measured by a cesium-133 atomic clock.*

Revised definition of 'inch': In 1959, an inch was defined to be exactly equal to 2.54 centimeters so that the International foot is exactly equal to 0.3048 meters.

Definition of 'kilogram': The kilogram is defined as being equal to the mass of the *International Prototype Kilogram* which is almost exactly equal to the mass of one liter of water.

Relationship between kilogram and pound: The weight of one kilogram approximately equals 2.2046 pounds.

DERIVED QUANTITIES:
Any physical quantity that is not one of the fundamental quantities described above is a derived quantity.
- Some of the physical quantities are actually the ratios of the same two quantities and hence have no units.
  - For example, index of refraction (n) and coefficient of friction (μ) which will be discussed later in the course.
- The unit of a derived quantity is a combination of one or more fundamental quantity.
  - For example, volume of a rectangular prism = length x breadth x height, speed = distance/time, and density = mass/volume.

SIGNIFICANT FIGURES:
The number of digits reported in a measurement is called its significant figure.

To determine the number of significant figures in a measurement, locate the
first non-zero digit on the left and count to the right extreme digit (including any zeros).
The number of digits counted is the number of significant figures in that measurement.

Measurement	Number of Significant figures
8.2	2
0.82	2
0.000082	2
0.820	3
820	2 or 3?
8.20×10^2	3
8.2×10^2	2

Rules for significant figures in mathematical operations:

Multiplication and Division: For multiplication and division, the product or quotient should contain the same number of significant figures as the number in the least precise factor.
e.g. 3.45 X 2.5 = 8.6 and 4.8/6.1523 =0.78

Addition and Subtraction: The result is not carried beyond the first column having a doubtful figure.
e.g. 2.185 +365.2 = 367.4 and 2.185 - 365.2 = -363.0

Scientific Notation: Any number can be represented in the form
$$p \times 10^q \text{ where } 1 \le p < 10 \text{ and } q \text{ is an integer.}$$
Examples:
2,451 = 2.451×10^3 (4 sig fig) = 2.45×10^3 (3 sig fig)
  = 2.5×10^3 (2 sig fig)   = 3×10^3 (1 sig fig)

.002451 = 2.451×10^{-3} (4 sig fig) = 2.45×10^{-3} (3 sig fig)
  = 2.5×10^{-3} (2 sig fig)   = 3×10^{-3} (1 sig fig)

Some commonly used prefixes for Powers of Ten

10^3	kilo (k)	10^{-3}	milli (m)
10^6	mega (M)	10^{-6}	micro (μ)
10^9	giga (G)	10^{-9}	nano (n)
10^{12}	tera (T)	10^{-12}	pico (p)

Examples for conversion of units:
Given:
1 ft = 0.3048 m   1 m = 100 cm   1 m = 39.37 in   1 mi = 1.609 km
1 h = 3600 s   1 km = 1000 m

Convert :
3 ft to cm
3 ft = 3 ft (0.3048 m/1 ft) (100 cm/ 1 m) = 91.44 cm

65 mph to m/s
65 mi / h = (65 mi / h) (1.609 km / 1 mi) (1000 m / 1 km) (1 h / 3600 s)= 29.06 m/s

1 g/cm^3 to kg/m^3
1 g/cm^3 = 1 g/cm^3 (1 kg / 1000 g) (100 cm / 1 m)^3 = 1000 kg / m^3

Does it make any cent?
$1 = 100¢
= 10¢ x 10¢
= $(1/10) x $(1/10)
= $(1/100)
= 1¢

Some Interesting facts & figures
<ul><li>One light-year: Distance traveled by light in one year.</li><li>One light-year = 9.461×10^{15} m = 5.879×10^{12} miles</li><li>Time taken by light to reach us from the sun = 8.33 minutes.</li><li>Time taken by light to reach us from the moon = 1.26 seconds.</li><li>Light travels 1 ft per nano seconds.</li><li>October 10 is National Metric Day.</li><li>π-day is celebrated on March 14 and π minute occurs twice on March 14 at 1:59 am, and 1:59 pm (3.14159)</li><li>If all the available gold in the world is molded in the form of a cube, it will be 54 ft x 54 ft x 54 ft</li><li>A *jiffy* is an actual unit of time equal to 1/100th of a second.</li><li>111,111,111 x 111,111,111 = ?</li></ul>

Problems:

1. Convert:

(a) 1 m^3 into cm^3 (b) 1 kg/m^3 into g/cm^3 (c) 1 ft^3 into in^3

[(a) 10^6 cm^3 (b) 0.001 g/cm^3 (c) 1728 in^3]

2. Give answers to correct significant figures:

(a) 2.5 x 3.09 (b) 4.758/9.41   (c) 3.467+5.4+8.2983481 (d) 7.987-2.55-3.4482

3. Calculate the average density of the earth. Assume earth to be a spherical with a radius of 6.38×10^3 km and mass of 6.0×10^{24} kg. Express your answer in g/cm^3 with correct number of significant figures.

[5.5 g/cm^3]

Multiple Choice: Choose the best alternative

1. The SI Unit for temperature is

(A) Fahrenheit   (B) Celsius   (C) Kelvin   (D) Roemer   (E) None of these

2. The prefix used for 10^9 is

(A) nano   (B) giga   (C) mega   (D) milli   (E) tera

3. The prefix micro is used for

(A) 10^3 (B) 10^6 (C) 10^{-6} (D) 10^{-3} (E) 10^{-9}

4. The number of significant figures in 9.180×10^5 is

(A) 1   (B) 2   (C) 3   (D) 4   (E) 5

5. When 10^6 is divided by 10^{-3} the resultant number can be expressed by the prefix

(A) nano   (B) giga   (C) kilo   (D) milli   (E) pico

Chapter 2
Scalars and Vectors

Most physical quantities can be classified as vectors and scalars.
Vector: A vector is a physical quantity that has magnitude and direction
Scalar: A scalar is a physical quantity that has magnitude but no direction.

A vector can be represented by an 'arrow' (——►) in a diagram. For example:

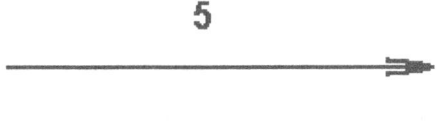

Represents a vector of magnitude 5 units directed due east or in the positive x direction

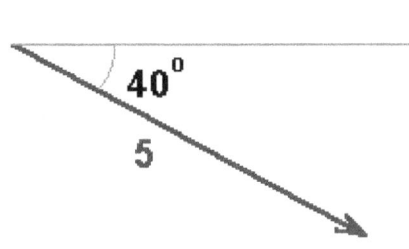

Represents a vector of magnitude 5 units 40° south of east or - 40° from the x-axis

Components of a Vector: A vector a lying in xy-plane can be resolved into two components along x and y axes as illustrated below.

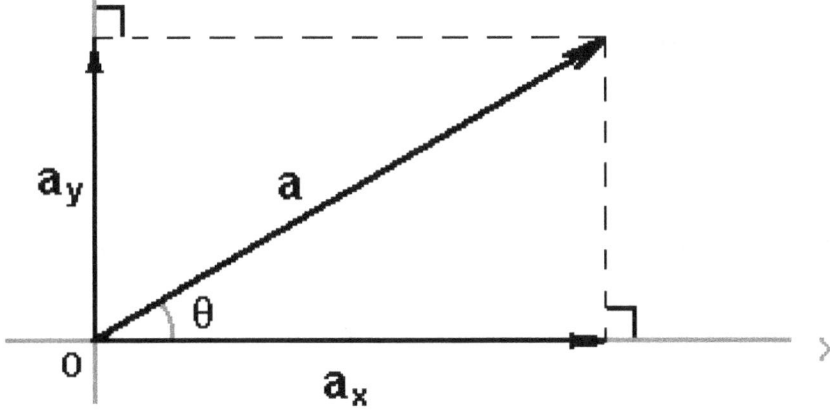

The x-component of a is $a_x = a\cos\theta$
The y-component of a is $a_y = a\sin\theta$

The components a_x and a_y can be positive or negative depending on the quadrant in which the angle θ lies.

The magnitude of a is $= \sqrt{(a_x^2 + a_y^2)}$

Note: If a vector **a** (↑) is vertical then $a_x = 0$ and $a_y \neq 0$
If a vector **a** (→) is horizontal then $a_x \neq 0$ and $a_y = 0$

Multiplication of a vector by a scalar:
A vector **a** can be multiplied by a scalar quantity s. The result is another vectors s**a**. The new vector s**a** has magnitude s times that of **a**. The direction of s**a** is same as **a** if s is positive and opposite if s is negative.

Addition and subtraction of scalar quantities:
Addition and subtraction of scalar quantities is trivial: it is same as the addition and subtraction of real numbers.

For example: In a large container one pours 4 liters of water and then 3 liters of water the total water poured in the container is 4 + 3 = 7 liters. Now if 5 liters of water is poured out from the container the amount of water remaining in the container is 7 liters – 5 liters = 2 liters.

Addition and subtraction of vector quantities:
Addition of two vectors can be performed by any of the following three methods:

1. Parallelogram Addition of Vectors: Let **a** and **b** be two vectors. Draw these two vectors from a common point as shown in the diagram below. Complete the parallelogram taking these two vectors as the adjacent sides. Then the diagonal passing through the common point represents the sum of the vectors **a** and **b**.

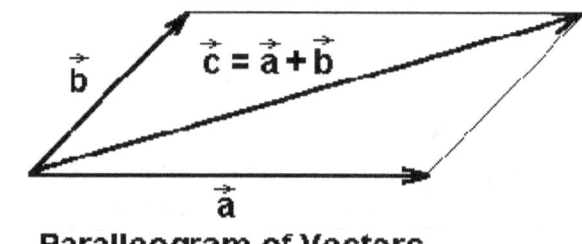

Paralleogram of Vectors

2. Triangle addition of Vectors: This method is geometrically equivalent to the method above. First, draw one of the vectors, say the vector **a**. Then draw the vector **b** from the head of the vector **a**. The vector from the tail of vector **a** to the head of the vector **b** represents the sum of the vectors **a** and **b**.

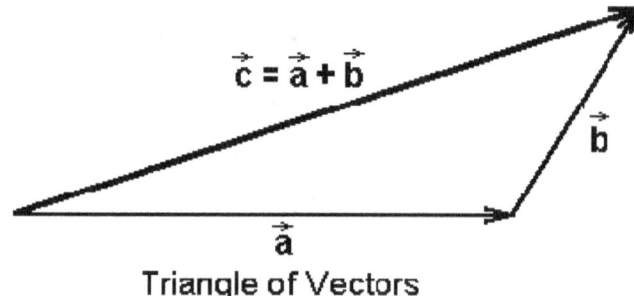

Triangle of Vectors

3. Component Method of Addition of Vectors:

Let **a**, **b**, and **c** be three vectors to be added by the Component Method. Let **R** be the sum of these vectors. The relationship between **R**, **a**, **b**, and **c** can be written as the vector equation

$$R = a + b + c$$

To determine the magnitude and the direction of the resultant vector **R**, first resolve every given vector into its x and y components. Thus

$a_x = a \cos \theta$
$a_y = a \sin \theta$

a is resloved into a_x and a_y

b is resloved into b_x and b_y

c is resloved into c_x and c_y (see the diagram above)

6

Then sum of all the x components of the vectors is the x component of the resultant and sum of all the y components of the vectors is the y component of the resultant. That is,

$$R_x = a_x + b_x + c_x; \qquad R_y = a_y + b_y + c_y$$

Now the _magnitude_ and _direction_ of the resultant vector R is given by

$$R = \sqrt{R_x^2 + R_y^2} \quad \text{and} \quad \theta = \tan^{-1}\left(\frac{R_y}{R_x}\right)$$

The algebra of the vectors thus developed can be applied for the addition of vector quantities such as displacement, velocity, linear momentum, force, and fields.

Problems:
1. For each vector in the diagram below obtain its x and y components:

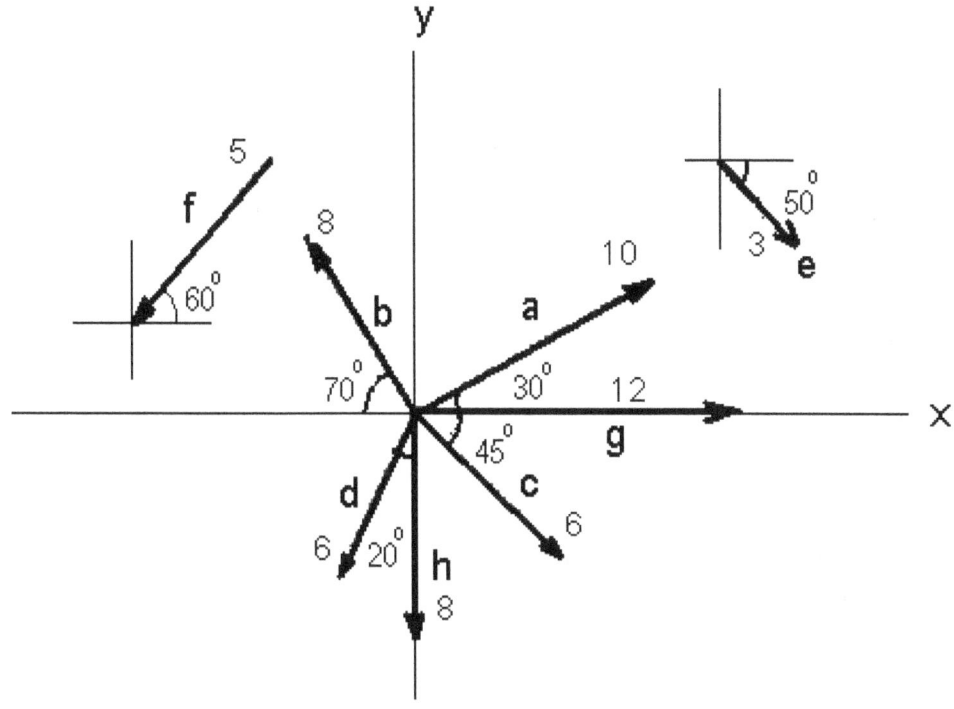

[$a_x = 8.7$, $a_y = 5.0$; $b_x = -2.7$, $b_y = 7.5$; $c_x = 4.2$, $c_y = -4.2$; $d_x = -2.1$, $d_y = -5.6$; $e_x = 1.9$, $e_y = -2.3$; $f_x = -2.5$, $f_y = -4.3$; $g_x = 12$, $g_y = 0$; $h_x = 0$, $h_y = -8$]

2. Below, add the given vectors in each case to obtain the magnitude and direction of the resultant.

(i) **a** = 8, +25° from x axis; **b** = 12, +70° from x axis
(ii) **a** = 50, -45° from x axis; **b** = 40, +30° from x axis
(iii) **a** = 20, -80° from negative x axis (2nd quad), **b** = 20, -30° from x axis (4th quad)
(iv) **a** = 10, +55° from x axis; **b** = 40, +120° from x axis (2nd quad), **c** = 25, -15° from x axis (4th quad).

3. Two force vectors F_1 and F_2 have equal magnitudes of 80 newtons each. F_1 is pointing $60°$ N of E and F_2 is pointing $60°$ S of E. Calculate the magnitude and direction of the vector $F_1 + F_2$.

[80 newtons, East]

4. Given vector **a**, magnitude 25, angle $+ 40°$ from x-axis, and **b**, magnitude 40, angle $160°$ from x axis, determine third vector c such that the sum $a + b + c = 0$

Some interesting facts & figures
• **Besides vectors and scalars there are other categories for physical quantities such as tensors, spinors, pseudoscalars, and pseudovectors.** • **Physical examples of pseudovectors include the magnetic field, torque, vorticity, and the angular momentum.** • **Physical examples for pseudoscalars include the magnetic charge (as it is mathematically defined, regardless of whether it exists physically), the magnetic flux, and the helicity.**

Multiple Choice:

1. A vector starts from the origin of a coordinate system. Its x-component is positive and the y-component is negative. The vector must be in the
(A) first quadrant   (B) second quadrant   (C) third quadrant
(D) fourth quadrant   (E) None of these

2. Two vectors have equal magnitude and cancel each other out. The angle between them must be
(A) $0°$ (B) $90°$ (C) $180°$ (D) $270°$ (E) $360°$

3. The positive x-component of a vector is half the magnitude of the vector. The vector makes an angle of _____ degrees with the x-axis.
(A) $45°$ (B) $90°$ (C) $30°$ (D) $60°$ or $- 60°$ (E) $60°$ only

4. The x and y-components of a vector are 0 and -5 respectively. The vector
(A) is horizontal and pointing in the positive x-direction
(B) is horizontal and pointing in the negative x-direction
(C) is vertical and pointing in the positive y-direction
(D) is vertical and pointing in the negative y-direction
(E) does not lie in the xy-plane

5. Two vectors of equal non-zero magnitudes are symmetric about the y-axis, one in the first quadrant and second in the second quadrant respectively. Which of the following is true about the x and y components of their resultant vector?
(A) x-component = 0; y-component $\neq$ 0
(B) x-component $\neq$ 0; y-component =0
(C) x-component$\neq$ 0; y-component $\neq$ 0
(D) x-component = 0; y-component = 0
(E) Information insufficient for a definite answer

Questions:
1. Can two vectors having different magnitudes be combined to give a zero vector?

2. If three vectors add up to zero resultant, must they all be in the same plane?

3. A person somewhere on the Earth travels 10 mi south, then 10 mi east, and then 10 mi north. He is back to starting point. Which place on the Earth is he?

4. Taking into account the orbital and rotational motion of the earth, does a book lying on a table move faster, slower, at the same speed during the night as compared to day?

Chapter 3
Motion in One Dimension

For a body moving along straight line in a given direction, its speed is same as the magnitude of its velocity and the distance traveled is same as the magnitude of its displacement.

If the acceleration is constant (uniform acceleration) one can derive following four basic equations involving displacement, velocity, acceleration, and time:

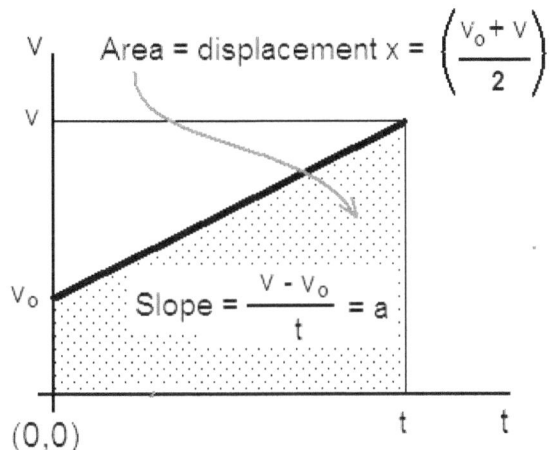

Area = displacement $x = \left(\dfrac{v_o + v}{2}\right) t$

Slope = $\dfrac{v - v_o}{t} = a$

$$x = v_{av}\, t\,, \quad [v_{av} = \tfrac{1}{2}(v_o + v)]$$
$$v = v_o + a\, t$$
$$x = v_o\, t + \tfrac{1}{2} a\, t^2$$
$$v^2 = v^2_{\,o} + 2\, a\, x$$

Motion Under Uniform Gravity:
For a body moving freely under gravity its acceleration is constant and acts vertically downward. The magnitude of this acceleration is **g = 9.8 m/s^2**.

If a body is falling vertically down or thrown vertically upward the above four equations can be applied when proper signs are used for the variables involved.

Sign conventions to be used for one-dimensional motion under gravity:

Take the magnitude of a vector quantity to be **positive** if it is pointing **upward** and **negative** if it is pointing **downward**.

Since the displacement is along **y-axis** replace the variable **x** by **y**. And, since **g** is always pointing downward (even if the body is moving upward) replace acceleration a by **-g** and then always use **g = + 9.8 m/s^2**. Thus the above 4 equations can be modified as follows:

$$y = v_{av}\, t$$
$$v = v_o - g\, t$$
$$y = v_o\, t - \tfrac{1}{2} g\, t^2$$
$$v^2 = v^2_{\,o} - 2\, g\, y$$
[Take g = +9.8 m/s^2]

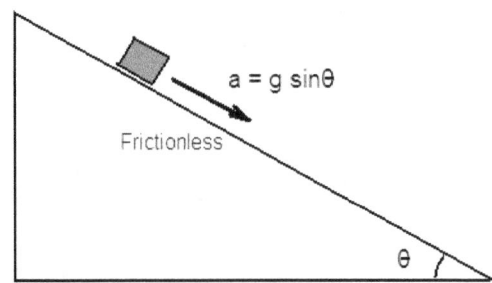

a = g sinθ

Frictionless

θ

Acceleration of a body down a frictionless inclined plane: A frictionless inclined plane makes an angle θ with the horizontal. If a particle is allowed to slide down this inclined plane it will have a uniform acceleration **a** given by $\quad a = g\,\sin\theta$

SOLVED PROBLEMS:

1. A plane's touchdown speed on a runway is 60 m/s. Its velocity reduces at a constant rate to 50 m/s after having traveled a distance of 200 m.
(a) Calculate the plane's acceleration.

$v^2 = v_o^2 + 2ax$ ➜ $(50 \text{ m/s})^2 = (60 \text{ m/s})^2 + 2a(200)$ ➜ $a = -2.75 \text{ m/s}^2$

(b) If the plane continues with the same acceleration

(i) How much longer will the plane take to come to a stop?

Here take $v_o' = 50$ m/s. $v' = 0$ m/s, and $a = -2.75$ m/s
$v' = v_o' + at$ ➜ $0 \text{ m/s} = 50 \text{ m/s} + (-2.75 \text{ m/s}^2)t$ ➜ 18.2 s

(ii) How much farther will the plane travel before coming to a stop?

$v'^2 = v_o^2 + 2ax'$ ➜ $0 = (50 \text{ m/s})^2 + 2(-2.75 \text{ m/s}^2)x'$ ➜ x' = 455 m

2. A ball is thrown vertically up from the ground. It reaches a maximum height of 30 m.

(a) Calculate the velocity with which the ball was thrown up.

Maximum height ➜ $v = 0$ m/s
$v_y^2 = v_{oy}^2 - 2gy$ ➜ $0 = v_o^2 - 2(9.8 \text{ m/s}^2)(30 \text{ m})$ ➜ $v_o = 24.2$ m/s

(b) Calculate the time the ball took to reach the maximum height.

$v_y = v_{oy} - gt$ ➜ $0 = 24.2 \text{ m/s} - (9.8 \text{ m/s}^2)t$ ➜ $t = 2.47$ s

(c) Calculate the time it takes to return to the ground.

When the ball returns to the ground its y-displacement = 0 m
$y = v_{oy}t - \frac{1}{2} gt^2$ ➜ $0 = (24.2 \text{ m/s})t - (4.9 \text{ m/s}^2)t^2$
➪ $t(24.2 - 4.9t) = 0$ ➜ $t = 0$ s or 4.94 s; choose 4.94 s

(d) Calculate the velocity just before the ball hits the ground.

When the ball returns to the ground its y-displacement = 0 m

$v_y^2 = v_{oy}^2 - 2gy$ ➜ $v_y^2 = (24.2 \text{ m/s})^2 - 2(9.8 \text{ m/s}^2)(0 \text{ m})$
➜ $v_y^2 = 585.64 \text{ m}^2/\text{s}^2$ ➜ $v_y = \pm 24.2$ m/s (choose minus sign).

3. An electron is traveling with a velocity of 0.8c. It enters an electric field which brings it to a stop in 8.0 ns.
[c = *speed of light* = 3×10^8 m/s; 1 ns = 1 nanosecond = 1×10^{-9} s]

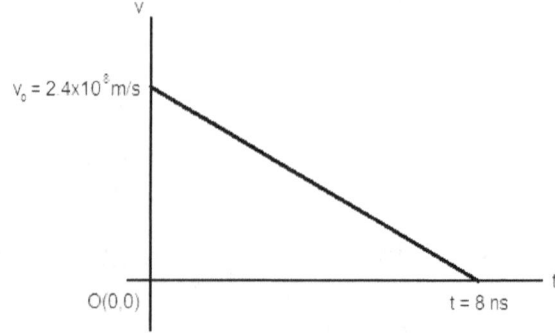

(a) Sketch a v vs. t graph for the motion.
(b) Calculate the acceleration of the electron.
$v = 0$ m/s; $v_o = 0.8$ c $= 0.8 \times 3 \times 10^8$ m/s $= 2.4 \times 10^8$ m/s; $t = 8.0 \times 10^{-9}$ s

$v = v_o + at$ ➜ $0 \text{ m/s} = 2.4 \times 10^8 \text{ m/s} + a(8.0 \times 10^{-9} \text{ s})$ ➜ $a = -3 \times 10^{16} \text{ m/s}^2$

(c) Calculate the distance in the electric field over which the electron will take to come to a stop.

10

$x = ½ (v_o + v)t = ½ (2.4 \times 10^8 \text{ m/s} + 0 \text{ m/s})(8 \times 10^{-9} \text{ s}) = 0.96 \text{ m}$

4. Two cars A and B are at x = 0 when t =0. Their v vs. t graphs are shown below:

(a) Calculate their accelerations over the 8-s interval.

For A: It has zero acceleration for the first 20 s.
From 20 s to 60 s, $a_A = (0 \text{ m/s} - 25 \text{ m/s})/(60 \text{ s} - 20 \text{ s}) = 0.625 \text{ m/s}^2$

For B: $a_B = (0 \text{ m/s} - (-25 \text{ m/s})) /(60 \text{ s} - 0 \text{ s}) = 0.417 \text{ m/s}^2$

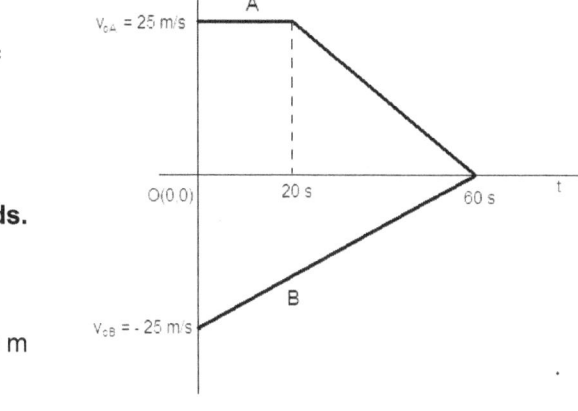

(b) Calculate the displacement of each car over 60 seconds.

For A: x_A = Area under the graph = (25 m/s x 20 s) + ½ (25 m/s)((60 s – 20 s) = 1000 m
For B: x_B = Area 'under/ the graph = ½ (-25 m/s)(60 s) = - 750 m

(c) Calculate the distance between the two cars at 60 s.

x_A is positive (to the right of the origin) and x_B is negative (to the left of the origin). The total distance = 1000 m + 750 m = 1750 m

Some interesting facts & figures
• While rate of change of displacement is velocity and the rate of change of velocity is acceleration, the rate of change of acceleration is called 'jerk'. Equivalent terms for this quantity are 'jolt', 'surge', and 'lurch'. This quantity is useful in engineering design of roller coasters, 'whiplash' suffered by an accident victim, and rapid changes in acceleration causing wear and tear of cuttint tools
• Flying from London to New York by a supersonic jet, due to the time zones crossed, you can arrive 2 hours before you leave.
• Cheetah's can accelerate from 0 to 70 km/h in 3 seconds.
• Hawaii is moving toward Japan 4 inches every year.
• The surface speed record on the moon is 10.56 miles per hour. It was set with the lunar rover.
• If you could drive to the sun -- at 55 miles per hour it would take about 193 years.
• When glass breaks, the cracks move at speeds of up to 3,000 miles per hour.
• Alexander Graham Bell, who invented the telephone, also set a world water-speed record of over seventy miles an hour at the age of seventy two.
• Everest is increasing it's height 10 cm per year
• The moon moves away from the Earth at a rate of 1.25 inches per year?
• Flea's can jump 130 times higher than their own height. In human terms this is equal to a 6ft. person jumping 780 ft. into the air.

PROBLEMS:

1. Two runners A and B initially 120 m apart on a straight track run toward each other with constant speeds of 5 m/s and 3 m/s respectively. Determine after how long and at what position on the track will they meet.   *[15 s, 75 m from A's end]*

2. Two runners A and B initially 200 m apart on a straight track run toward each other. The runner A starts with an initial velocity of 5 m/s and maintains a constant acceleration of 0.1 m/s². The runner B starts with an initial velocity of 2 m/s and maintains a constant acceleration of 0.3 m/s². Determine after how long and at what position on the track

11

will they meet.

3. A car starts with an initial velocity of 5 m/s and an acceleration of 0.1 m/s². Determine:

    (a) How fast is car moving after it has traveled 100 m.   *[6.7 m/s]*

    (b) How long does the car take to cover the distance of 100 m?   *[17.1 s]*

4. A particle starts with initial speed $v_o = 20$ m/s from the origin (0,0) and moves in the positive x-direction with an acceleration of a = -1.5 m/s².

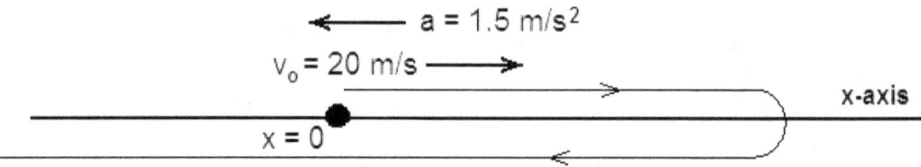

Determine

    (a) Time t when the particle stops momentarily (v = 0).

    (b) The distance x when the particle stops momentarily

    (c) Time when the particle returns to the origin (0,0).

    (d) The speed of the particle at x = 100 m

    (e) The time t when x = 100 m

    (f) The time t when x = -100 m

(g) On the
coordinate
systems below,
draw the graphs
for t = 0 to t = 30
second

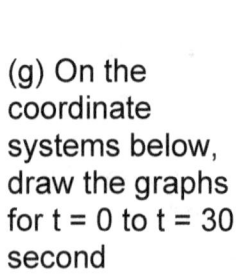
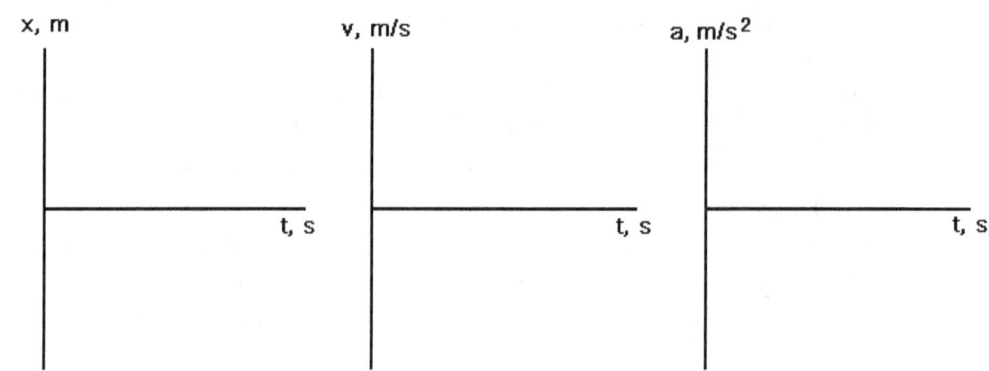

5. A coin is dropped from the top of a building. It reaches the ground after 2.3 s.

    (a) Determine the height of the building.   *[25.9 m]*

    (b) Determine the velocity of the coin just before it hits the ground.   *[-22.5 m/s]*

12

6. A student throws a ball vertically up. The ball leaves his hand and returns after 1.8 s.
   (a) With what velocity did the student launch the ball?

   (b) How high did the ball go?

7. A hot air balloon is rising vertically up with a constant velocity of 3.0 m/s.
 When the balloon is 200 m above the ground, a boy in the balloon releases
a coin, at rest with respect to himself, outside the basket.
(a) Determine the time the coin will take to hit the ground.   *[6.7 s]*

   (b) Determine the velocity of the coin just before it hits the ground. *[-62.7 m/s]*

   (c) Determine the maximum height reached by the coin.   *[200.46 m]*

**8. Two balls are released from the same height but 1.2 s apart. How long after the first ball is released will the two balls be 12 m apart?

9. A ball thrown vertically up reaches a height of 10.6 m in 1.1 s.
   (a) Determine the initial velocity of the ball.   *[15.0 m/s]*

   (b) Determine the maximum height reached by the ball.   *[11.5 m]*

10. A ball released from the top of a building hits the ground with a speed of 20 m/s. Determine the height of the building and the time it took the ball to reach the ground.

11. Two objects, 89 mi. apart are moving toward each other. Their speeds are 17 mph and 43 mph respectively. How far apart will they be 1 min before collision?
[1 mi]

12. A particle is placed at the top of a 3-m long frictionless inclined plane. It slides down to the bottom of the plane in 1.25 s. Determine the angle of the inclined plane with the horizontal.

13. A frictionless inclined plane makes an angle of 43° with the horizontal. A particle starts from rest from the top of the plane and slides to the bottom of the plane in 2.0 second.
   (a) Determine the acceleration of the particle down the inclined plane. *[6.68 m/s²]*

   b) Determine the length of the inclined plane.   *[13.4 m]*

(c) Determine the velocity of the particle at the bottom of the plane.   *[13.4 m/s]*
**14. Two joggers A and B start from a point on a 120-m circular jogging trail. They move with speeds of 3 m/s and 1 m/s respectively. When and where do they meet again if they are moving?
  (a) in the same direction?

  (b) in the opposite direction?

**15. A hiker starts to climb up a hill at 6:00 AM. On his way up either he is moving up or he may be resting at some place but never climbing down. He reaches the top at 6:00 PM. He stays at the top overnight and begins climbing down at 6:00 AM the next day. On his way down either he is moving down or resting somewhere. He reaches the bottom of the hill at 6:00 PM.
 At how many places is he there at the same time on the two days?

16. Mr. Fiz is returning home at a speed of 2 mi/h with his dog Ix. He unleashes Ix when they are still 3 miles from his house. Ix happily begins running back and forth between the house and his master with a constant speed of 3 mi/h. Ix does not waste anytime while turning around. By the time Mr. Fiz reaches home, how many miles has Ix run?

17. A particle starts from origin with zero initial velocity and moves in the positive x-direction at constant acceleration. It's acceleration verses time graph is shown below.

(a) Determine the distance covered by the particle in the 4 seconds.   *[48 m]*

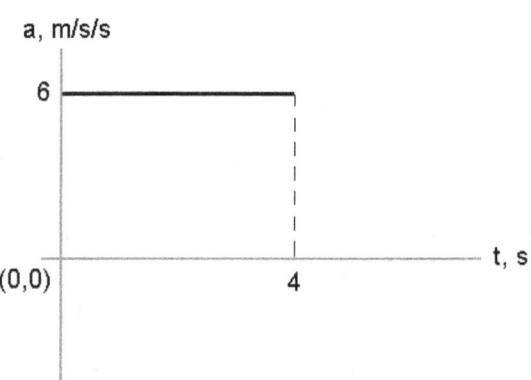

14

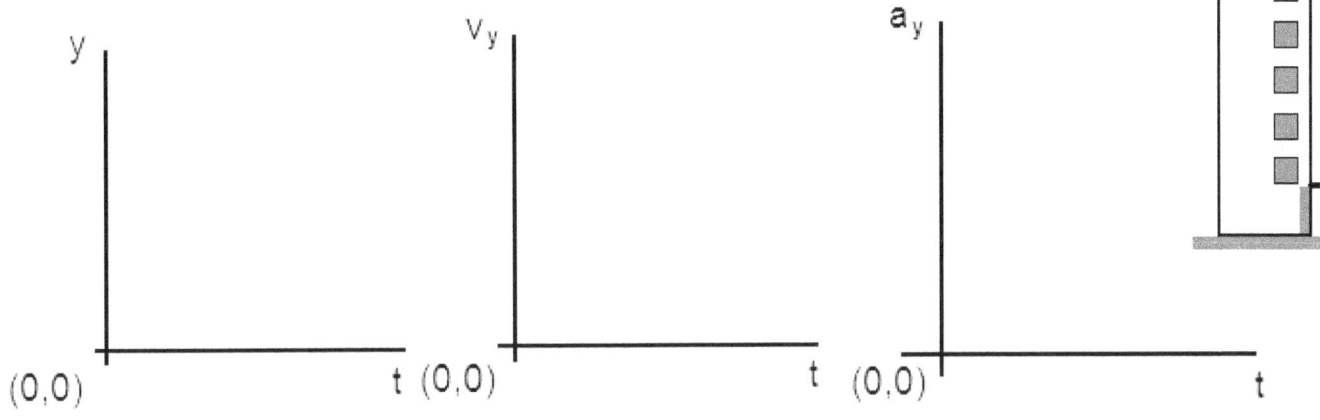

(b) Determine the velocity of the particle at t = 4 s. 24 m/s   *[24 m/s]*

18. From the edge of the top of a 68-m tall building a ball is thrown vertically up with an initial speed of v_{oy} = 24 m/s. (see the fig. below).

   (a) Calculate the time taken by the ball to reach the ground.

   (b) Calculate the maximum height reached by the ball above the ground

   (c) Calculate the velocity with which the ball hits the ground.

   (d) Sketch y vs. t, v_y vs. t, and a_y vs. t graphs for the ball while it is in the air.

   (e) Sketch v_y vs. y and a_y vs. y graphs for the ball while it is in the air

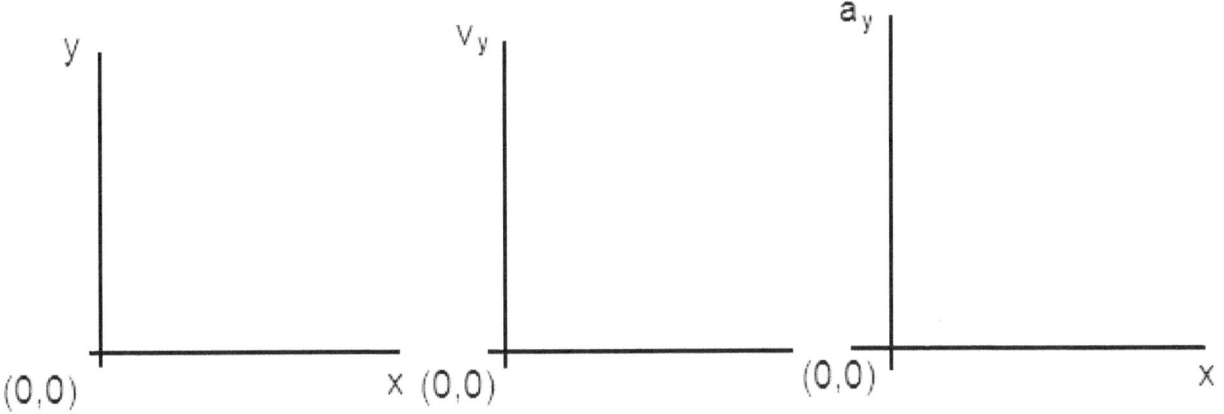

19. Chelsea beats Grace by 10 meters in a 100-m dash. Next, Chelsea starts 10 meters behind the start line and they race again. Who will win? or will there be a tie? Assume same constant speed in both races.

15

Multiple Choice:

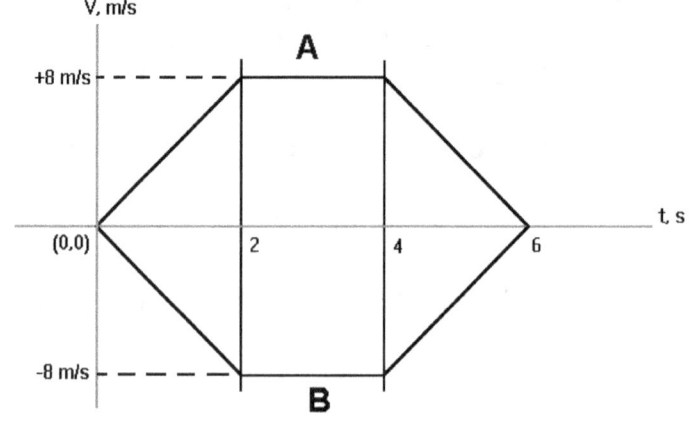

1. Two persons A and B start from origin and move along x-axis. Their velocity vs. time graphs for 6 seconds is shown below.
During these 6 seconds, when do they meet again?

(A) 2 seconds later
(B) 4 seconds later
(C) 6 seconds later
(D) 2 seconds, 4 seconds, and 6 seconds later
(E) never

2-3. A person starts from the origin and moves along the x-axis with constant acceleration for 8 seconds. His v vs. t graph is shown below.

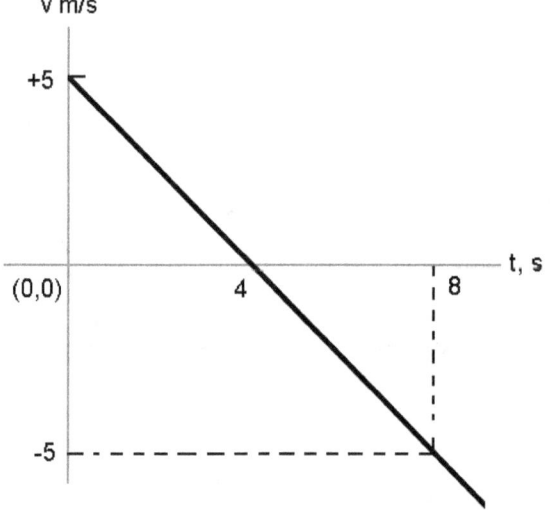

2. At what time is he farthest from the origin?

(A) 0 seconds   (B) 4 seconds
(C) 8 seconds

3. When does he return back to the origin?

(A) 4 seconds later
(B) 8 seconds later
(C) 6 seconds later
(D) Never
(E) Information insufficient

4 -6

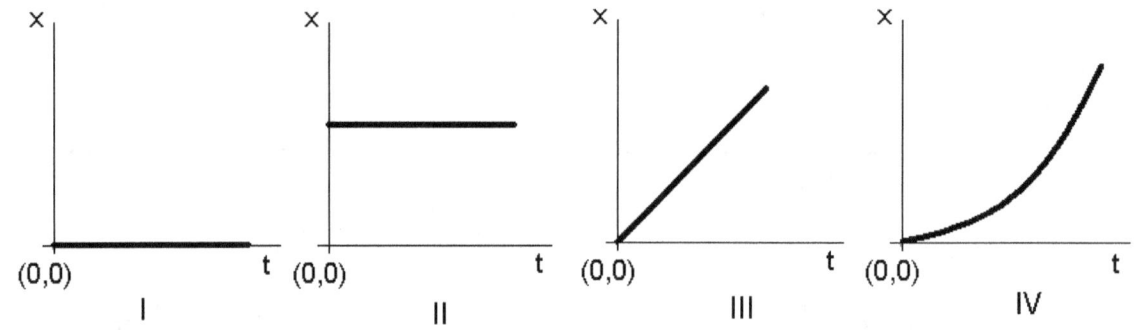

The x vs. t graphs above represent the motion of particles moving along x-axis.

4. Which graph(s) represent a stationary particle?

5. Which graph(s) represent a constant velocity?

6. Which graph(s) represent a variable velocity?

Chapter 4
Projectile Motion
(Without air resistance)

Motion in two dimensions: Motion in two dimensions can be resolved into two one-dimensional motions: along x and y-axes. The x-displacement is affected by the x-components of the particle's velocity and acceleration (v_x and a_x), and the y-displacement is affected by the y-components of its velocity and acceleration (v_y and a_y). However, these two one-dimensional motions are not totally independent. They are related by time (t) as a parameter.

Projectile Motion: The above approach can be applied to various cases of two-dimensional motion under gravity as described below:

1. A projectile thrown horizontally from a height
(for example, from the top of a building, tower, table, cliff, or a plane etc.)

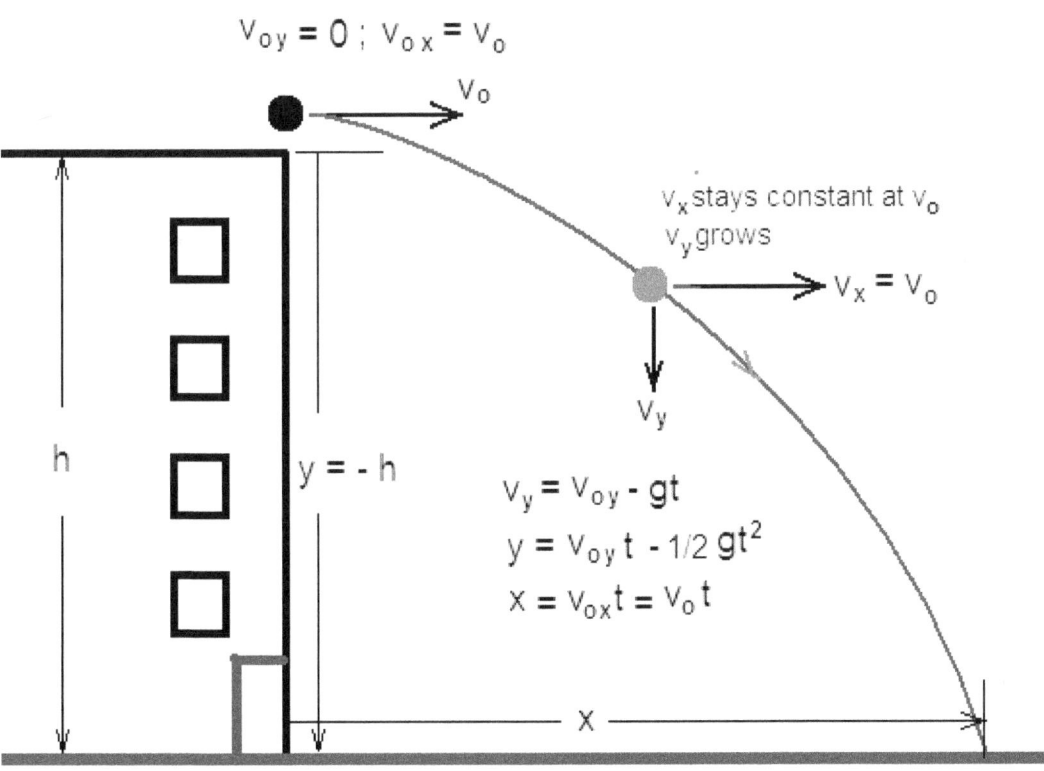

$$v_{oy} = 0 \; ; \; v_{ox} = v_o$$

$$v_o$$

v_x stays constant at v_o
v_y grows

$$v_x = v_o$$

$$v_y$$

$$h$$

$$y = -h$$

$$v_y = v_{oy} - gt$$
$$y = v_{oy}t - 1/2\, gt^2$$
$$x = v_{ox}t = v_o t$$

$$x$$

No acceleration in the x-direction $\longrightarrow$ $a_x = 0$
Acceleration in the y-direction $a_y = -g$

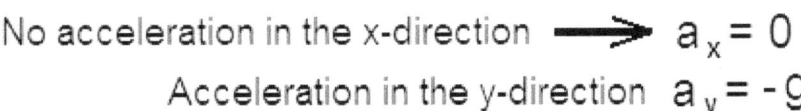

2. A projectile thrown at an angle from the horizontal from a height

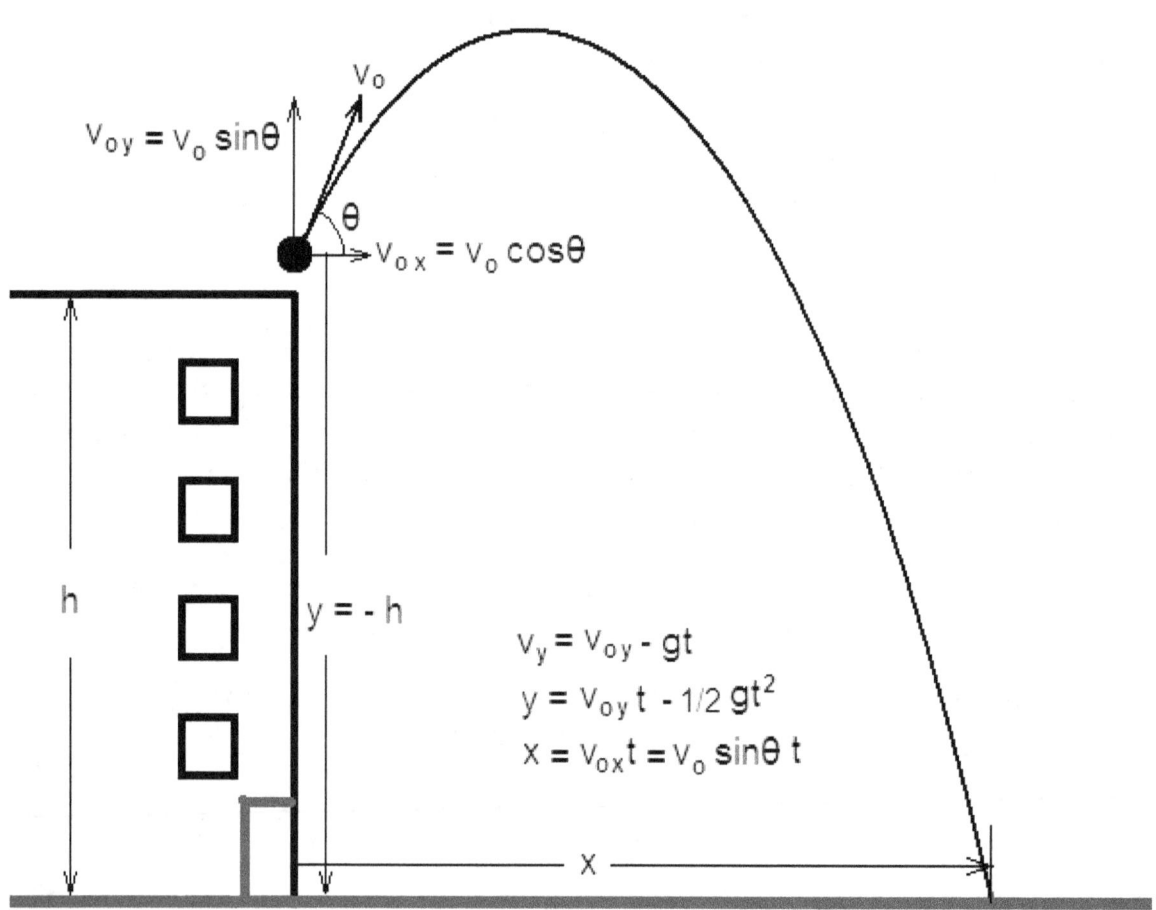

$$v_y = v_{oy} - gt$$
$$y = v_{oy} t - 1/2\, gt^2$$
$$x = v_{ox}t = v_o \sin\theta \; t$$

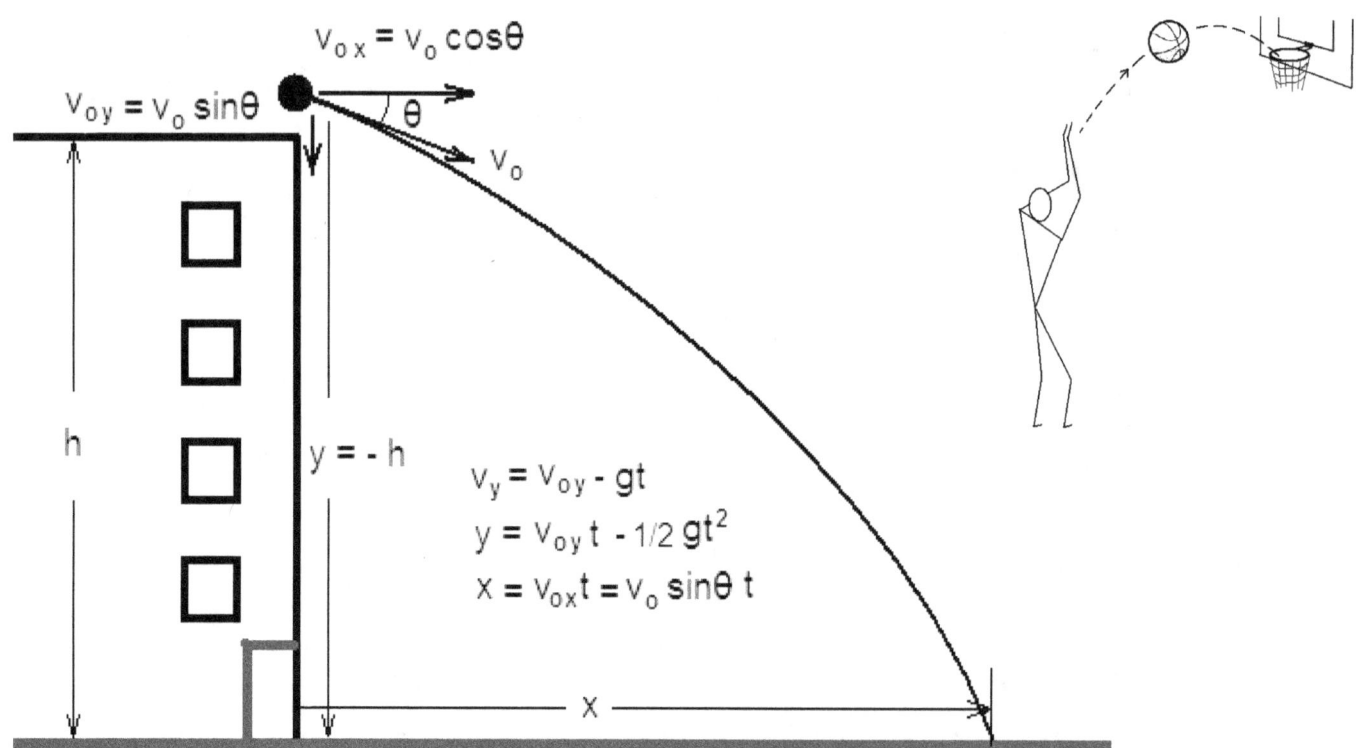

$$v_y = v_{oy} - gt$$
$$y = v_{oy} t - 1/2\, gt^2$$
$$x = v_{ox}t = v_o \sin\theta \; t$$

3. A projectile thrown at an angle from a level ground and returned to the ground

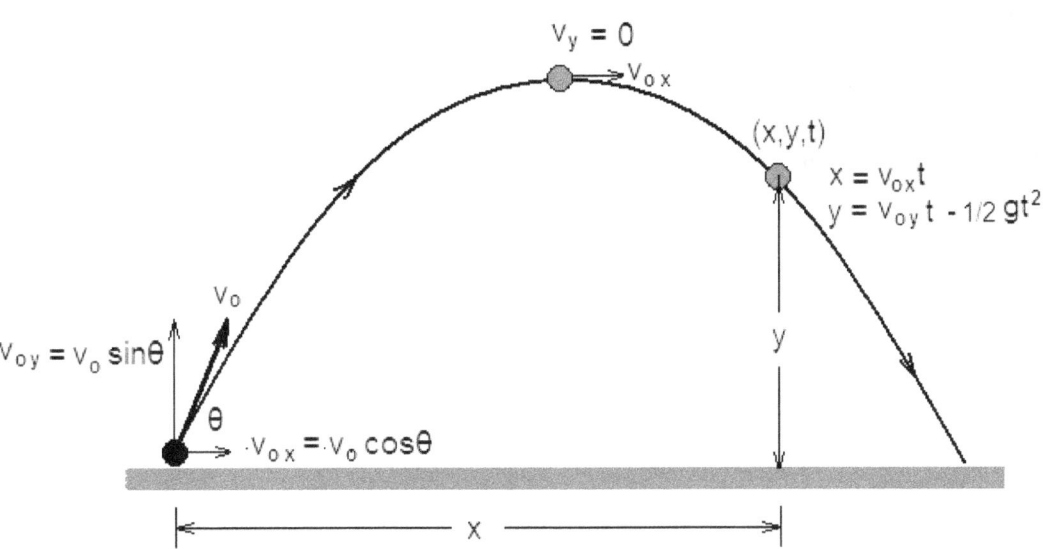

In all of the above cases of projectile motion, there is no acceleration in the x direction. Hence the x component of the velocity stays constant through out the motion. The acceleration in the y direction is same as the acceleration due to gravity.

Thus common for all the above three cases are the following equations:

In the x-direction, $x = v_{0x}t$

$v_x = v_{ox}$

In the y-direction, $v_y = v_{0y} - gt$ $y = \frac{1}{2}(v_{0y} + v_y)t$

$y = v_{0y}t - \frac{1}{2}gt^2$ $v_y^2 = v_{0y}^2 - 2gy$

Let a projectile be launched from the origin of a coordinate system, with an initial velocity v_0 at an angle θ from the x-axis. If the projectile returns to x-axis then its range R, maximum height reached H, and the time of flight T are given by:

$R = v_0^2 \sin(2\theta)/g$

$H = (v_0^2 \sin^2\theta)/2g$

$T = 2(v_0 \sin\theta)/g$

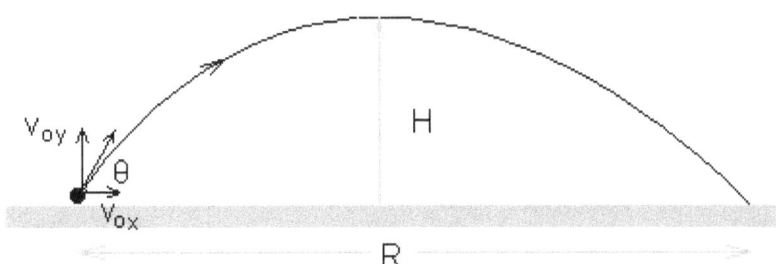

In this case the horizontal component of the velocity, **v_ox** stays constant.
The vertical component **v_y** decreases from **v_oy** at the initial point of launch to **zero** at the top of the trajectory to **−v_oy** when back to the ground.

Some interesting facts & figures
•   The longest recorded flight of a chicken is thirteen seconds.
•   If a cannon is backed up against a tree and fired, the shot will go farther.
•   If two projectiles are launched with the same speed at launch angles α and β, their ranges will be

same if α + β = 90°

Problems:
{Note: Ignore air resistance in all of the following problems on projectile motion.}

1. From the top of a 30-m tall building a ball is thrown <u>horizontally</u> with a speed of 8.0 m/s. Determine
(a) The time it takes for the ball to hit the ground.   *[2.5 s]*
(b) The distance from the building at which the ball hits the ground.   *[20 m]*
(c) The magnitude and direction of the velocity of the ball just before it hits the ground.
                                        [25.8 m/s, 71.9° below horizontal]
(d) The speed of the ball after 1.0 second.   *[12.7 m/s]*
(e) Sketch the graphs of the horizontal and vertical components of the velocity of the ball vs. time.

(f) How far is the ball from its initial position 2 seconds after it is launched?   *[25.3 m]*

2. A stone thrown horizontally from the top of a cliff hits water after 4.3 seconds at a distance of 52 m from the cliff. Determine:
(a) The height of the cliff
(b) Initial velocity of the stone
(c) The magnitude and direction of the velocity of the stone just before it hits the water

3. From the top of a 1.8-m high table, a cat jumps to the floor with initial velocity horizontal. The cat lands 3.0 m from the table. Determine the initial velocity of the cat.   *[4.9 m]*

4. A coin is thrown horizontally with an initial speed of 25 m/s from the top of a vertical cliff. The velocity of coin just before it hits water makes an angle of 75° from the horizontal.
(a) Determine the vertical component v_y of the velocity of the coin just before it hits water.

(b) Determine the time taken by the coin to reach water.

(c) Determine the height of the cliff.

(d) Determine how far from the cliff the coin hits the water.

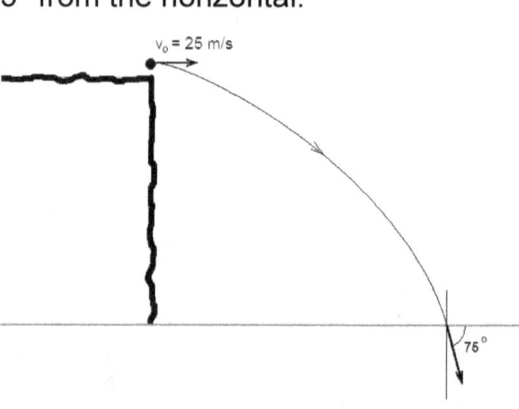

**(e)After how much time from the start is the coin

traveling at 35° from the horizontal?

5. A ball is thrown horizontally from the top of a building. The ball reaches the ground in 2.4 s. The velocity of the ball just before the impact makes an angle of 60° from the horizontal.

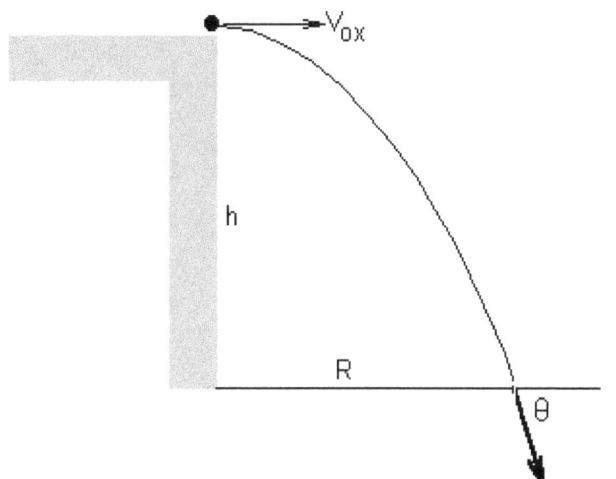

(a) Determine the height of the building.   [28.2 m]

(b) Determine the vertical component of the velocity of the ball just before the impact.
[-23.5 m/s]

(c) Determine the horizontal component of the velocity of the ball just before the impact.
[13.6 m/s]

(d) How far from the building does the ball hit the ground?   [32.6 m]

(e) Sketch graphs of (i) v_x vs. t   (ii) v_y vs. t   (iii) x vs. t   (iv) y vs. t

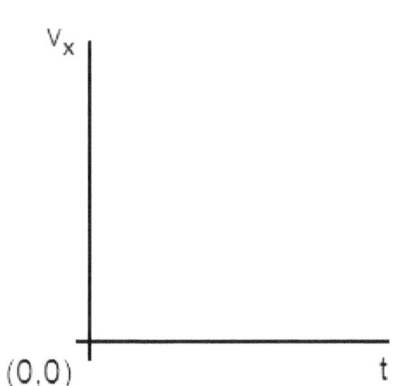

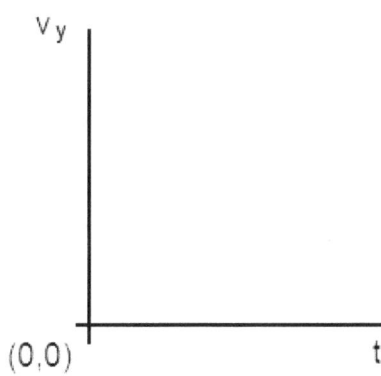

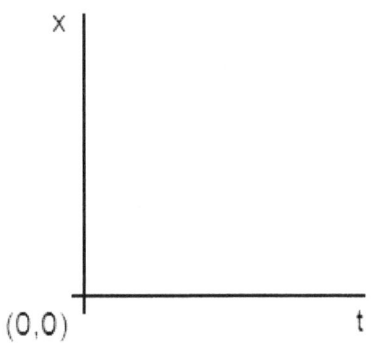

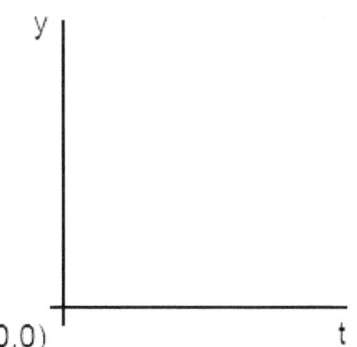

6. A football is kicked from the ground at an angle of 55°. It has a 'hang time' of 4.4 seconds.
(a) Determine the initial velocity of the ball.

(b) Determine the horizontal distance traveled by the ball.
(c) Determine the maximum height reached by the ball.

(d) Determine the time at which the ball is at 1/2 the maximum height.

7. A projectile is launched from ground at an angle of 34°. Its initial velocity is 25 m/s. There is a 6 -m tall wall 40 m from the initial position of launch.

$v_0 = 25$ m/s

34°

40 m

6 m

Determine whether the projectile will clear the wall or hit it. *[It will clear the wall by 2.72 m]*

8. A relief plane is flying horizontally toward a target on the ground. The plane is flying at a speed of 250 m/s at an altitude of 1000 m. It is supposed to drop a package to the target. How far horizontally before the target should the package be released?

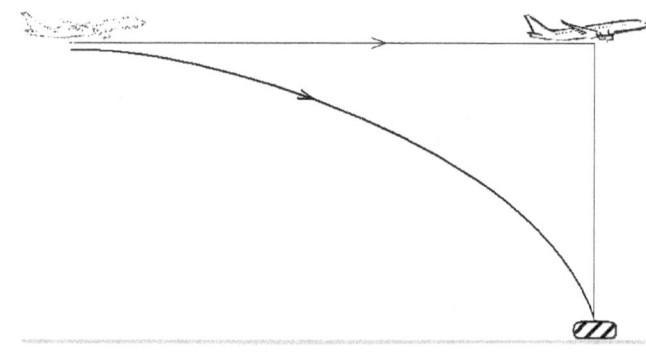

9. A ball is launched at 40° from the ground. It reaches a maximum height of 80 m.
(a) Determine its horizontal range.   *[381 m]*
(b) Determine the time of flight.   *[8.08 s]*

10. The range of a projectile on earth is 100 m. What will be its range on the moon if launched with the same initial conditions? g (moon) = 1.6 m/s^2

**11. A projectile is launched with an initial velocity of 30 m/s so that it covers a horizontal range of 45 m. Determine the maximum height reached by the projectile. (There are two answers).
[2.96 m; 43.0 m]

**12. A projectile is launched with initial speed v_0 over a level ground to achieve a maximum range. The corresponding maximum height is H. Now the projectile is launched with the same initial speed but vertically up. It reaches a maximum height of H'. Show that H' = 2H

13. A coin is launched from the top of a cliff, 30 m above water, at an angle of 75° above horizontal with a speed of 30 m/s.

(a) Calculate the time it takes to hit water.   *[6.81 s]*

(b) Calculate the horizontal displacement of the coin before it hits the water.   *[52.9m]*

(c) Calculate the speed of the coin just before it hits water.   *[38.6 m/s]*

14. A stunt motorcyclist wants to jump with his motorcycle across a crocodile infested 50-m wide river, between two vertical cliffs. He places a 45° ramp at the edge of one cliff. He speeds up and takes off from the ramp with a velocity of 70 km/h (? m/s)?   Will he be able to make it across the river?

Multiple Choice:
1- 3
As shown in the diagram above, a baseball is thrown from a level ground. The ball follows a parabolic trajectory before it hits the ground.

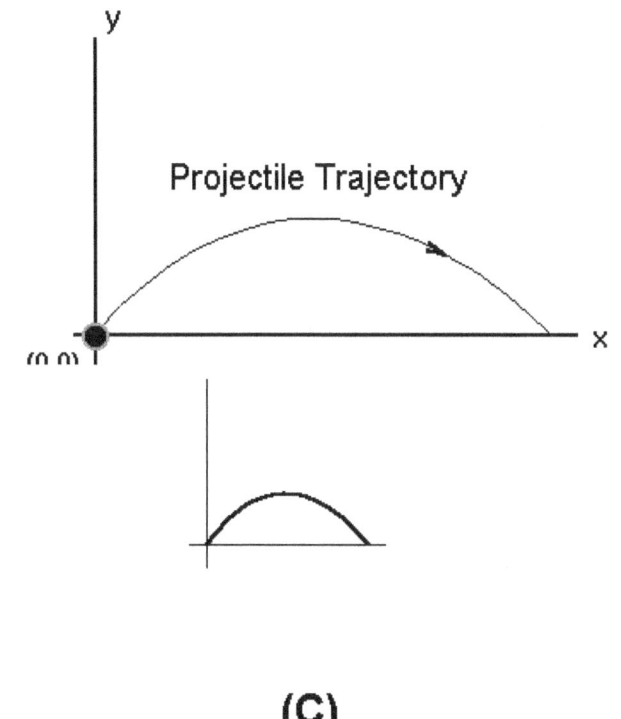

Projectile Trajectory

(0,0)

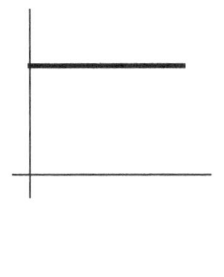

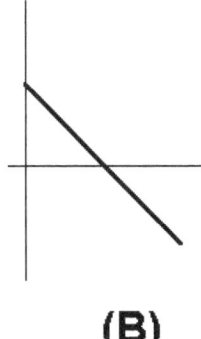

(A)   **(B)**   **(C)**

1. Consider the following three graphs:

Which, if any, graph best represents the relationship?
(a) v_y vs. t
(b) v_x vs. t
(c) y vs. t
(d) x vs. t

2. The speed of the ball is
(A) constant along the trajectory (B) maximum at the top (C) minimum at the top
(D) minimum at the end (E) minimum in the beginning

3. The acceleration of the ball is
(A) constant along the trajectory (B) maximum at the top (C) minimum at the top
(D) minimum at the end (E) minimum in the beginning

Chapter 5
Newton's Laws of Motion

Sir Isaac Newton [1642-1727]
Kinematics is the study motion without inclusion of the causes for motion. Newton's three laws of motion were formulated to include the causes of motion and thus gave rise to a very important branch of physics called 'Dynamics'. In fact 'Mechanics' (now, classical mechanics) is based on these three laws.

NEWTON'S FIRST LAW OF MOTION: *Every body continues to be in a state of rest or in a state of uniform motion along a straight line unless acted upon by an impressed force to act otherwise.*

Consider a body on which net force is zero. If this body is at rest it will continue to be at rest. If it is moving at constant velocity it will continue in that state. Hence the Newton's first law defines two physical quantities called **inertia** and **force**.

Sir Isaac Newton

Inertia is the tendency of a body to stay in the state of rest or the state of constant velocity.
Force is that which changes or tries to change the state of rest or the state of constant velocity of a body.

NEWTON'S SECOND LAW OF MOTION: *Net force acting on a body equals mass times the acceleration of the body. The direction of the net force is same as that of the acceleration of the body.*
Mathematically, this law states that,
$\Sigma F = m\ a$ (vector equation)
$\Sigma F_x = m\ a_x$, $\Sigma F_y = m\ a_y$ (scalar equation)
NEWTON'S THIRD LAW OF MOTION: *To every action there is an equal and opposite reaction.*

Action and reaction are two equal and opposite forces acting on two interacting bodies.
Rocket launching, recoil on firing a rifle, ability to walk on ground, and a ball thrown to a wall bouncing back are some of the examples illustrating this law.

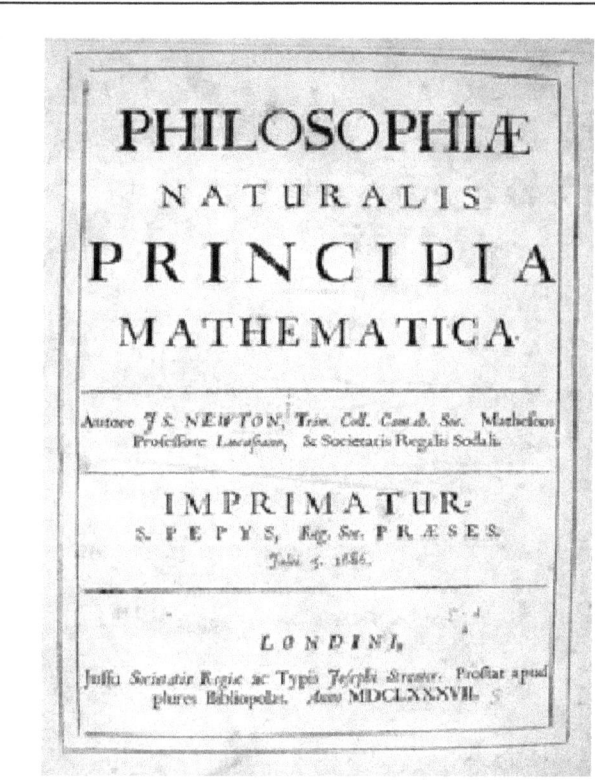

This is Newton's Three-volume wor
It contains the explanations for his
Laws of Motion and the
Law of Universal Gravitation

Caution regarding a misconception! A body does not necessarily move in the direction in which a force F is applied on it, i.e. given the direction of only the acceleration a, it is impossible to tell in what direction the body is moving.

However, a body will move in the direction of F and a under the following conditions:
- *The body was initially stationary.*
- *The body was initially moving in the same direction as F and a*
- *The body was initially moving opposite to the F and a and over a period of time its velocity slows down to zero and then reverses its direction.*

Some interesting facts & figures
• **Newton is considered by many scholars and members of general public as one of the most influential persons in the human history.** • **Newton built the first reflecting telescope.** • **Newton practiced alchemy.**

Multiple Choice: Choose the best alternative.

1. According to the Newton's Third Law, action and reaction are equal and opposite. The reason they don't cancel out is because
    (A) they act at different times
    (B) they act on two different bodies
    (C) the interacting bodies may not have equal masses
    (D) the two bodies may not be in the same state of motion before they interact.
    (E) none of these

2. According to the Newton's Second Law, acceleration of a body is given by:
    (A) a = Fm   (B) a = F/m   (C) a = m/F
    (D) a = ½ Fm^2 (E) none of these

3. A property called inertia is defined by Newton's
    (A) First Law of Motion
    (B) Second Law of Motion
    (C) Third Law of Motion
    (D) Law of Universal Gravitation
    (E) Law of Cooling

4. According to Newton's Second Law, if the net force on a particle is zero the particle
    (A) must be at rest
    (B) must be at rest or moving with constant speed along a straight line
    (C) must be moving at constant acceleration along a straight line
    (D) must be moving in uniform circular motion
    (E) must be performing simple harmonic motion

5. At a particular moment an object is moving due east and a force is acting on the object due north. The acceleration of the object at this moment must be
    (A) due east
    (B) due north
    (C) between east and north
    (D) west
    (E) south

Chapter 6
Newton's Laws applications

Newton's laws can be applied to a variety of situation in which one particle or a system of particles is(are) acted upon by one or more forces. There are various ways in which forces can be applied, namely: friction, tension in the string, reaction force from a surface, spring, gravitational force, electric force, magnetic force, nuclear forces etc.

Body or Particle?: The terms *body* and *particle* are interchangeably used to describe the object on which force acts. It should be understood that if <u>rotational motion is ignored</u> an object of any size (electron, baseball, car, planet, galaxy...) is a particle.

Weight (W): It is the gravitational force acting on a particle. $\qquad$ **W = mg**

Weight on Moon: The gravity on the moon is $1/6^{th}$ that on the earth. Hence

$$\textbf{W (on the moon) = } 1/6^{th} \textbf{ W (on the earth)}$$

Normal Reaction (F_N or N): It is the force from a surface on a particle. This force is always perpendicular to the surface. The normal force is equal and opposite to the force with which the particle <u>pushes</u> on the surface.

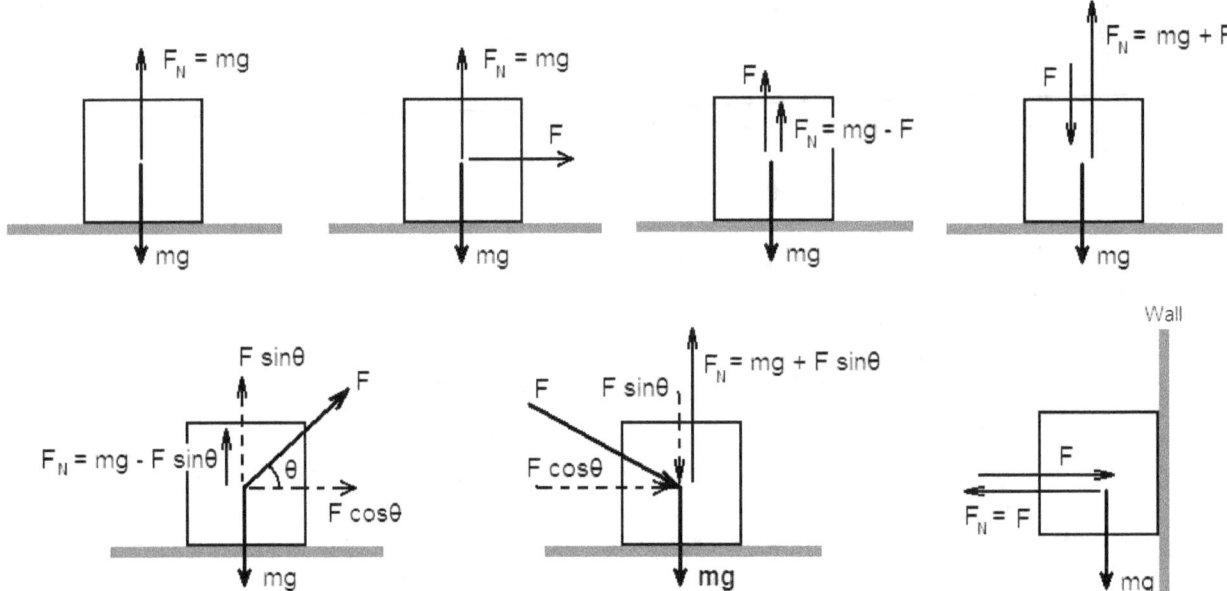

In all cases, except the first, above F is an external force impressed on the block.
When F is horizontal (parallel to the surface below the block) it has no effect on the normal force F_N

Significance of <u>Normal Force</u> (F_N):

- *Normal force gives a person feeling of weight!*
- *This feeling of weight may be different from the person's real weigh mg.*
- *The reading in the scale on which a person stands is the same as the (upward) normal force the scale needs to apply to support the person.*
- *The reading in the scale may be different from the real weight mg if they are in a state of acceleration.*

Thus, | F_N = Reading in the weighing scale = Apparent weight = The weight a person feels |

FREE BODY DIAGRAM:

A free body diagram of for every object involved in the problem helps simplify procedure to apply Newton's second law as illustrated in figure below in which a strong man is hauling a small car up:

Once all the forces acting on a particle of mass **m** are identified, these forces can be added by vector method to determine the resultant force (ΣF) acting on that particle. Then, applying Newton's Second Law of Motion we get,

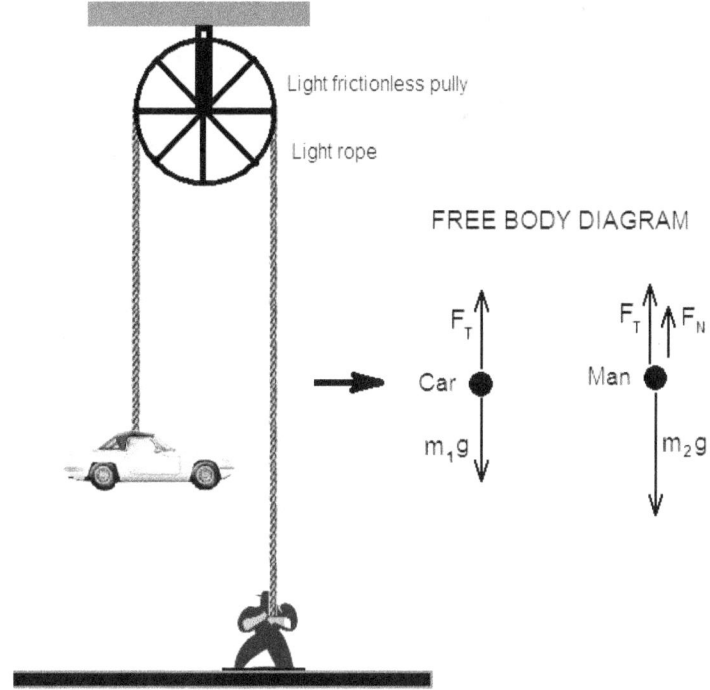

FREE BODY DIAGRAM

$$\Sigma F = ma \qquad \text{(vector equation)}$$

{Note: This equation can also be written as

$$F_{net} = ma\}$$

This vector equation should be written in scalar form before solving. Hence we have,

$$\Sigma F_x = ma_x, \quad \Sigma F_y = ma_y.$$
(scalar equations)

This procedure is repeated for other particles if present. These equations can then be solved to obtain the unknown quantities.

APPLICATION:

A load is attached to a cable of negligible mass.
The relationship among the quantities m, a, g, and F_T is given by

$$\boxed{F_T - mg = ma}$$

In this equation,

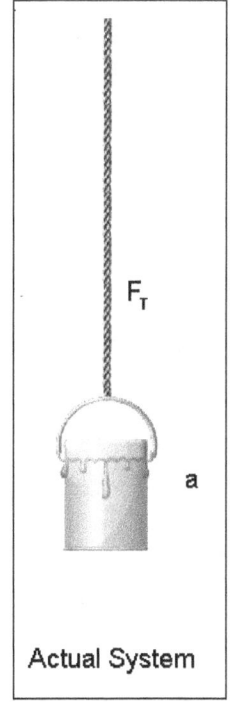

Actual System

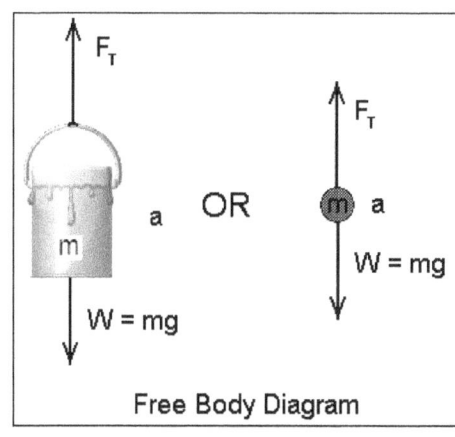

Free Body Diagram

a is + if it is ↑ i.e.

- The body is moving upward with increasing speed
- The body is moving downward with decreasing speed.

a is - if it is ↓ i.e.,

- The body is moving upward with decreasing speed
- The body is moving downward with increasing speed.

a is zero i.e.

- if the body is stationary
- moving with constant velocity of any magnitude.

Hooke's Law and Ideal Spring

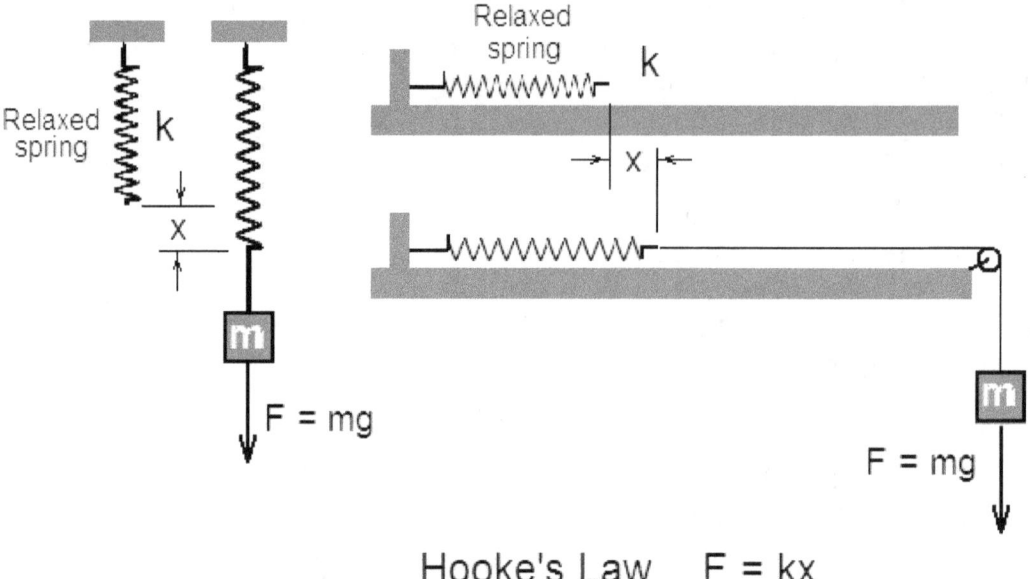

Hooke's Law   F = kx

A spring that obeys Hooke's Law is called an *ideal* spring; its mass is negligible.

According to the Hooke's Law: A force (F_{ext}) applied to extend or compress a spring is proportional to the change in length (x) of the spring.

Thus, $F_{external} = kx,$

k is called the elastic constant of the spring or just spring constant.

The restoring force (F) in the spring is thus, $F_{restoring} = -kx$

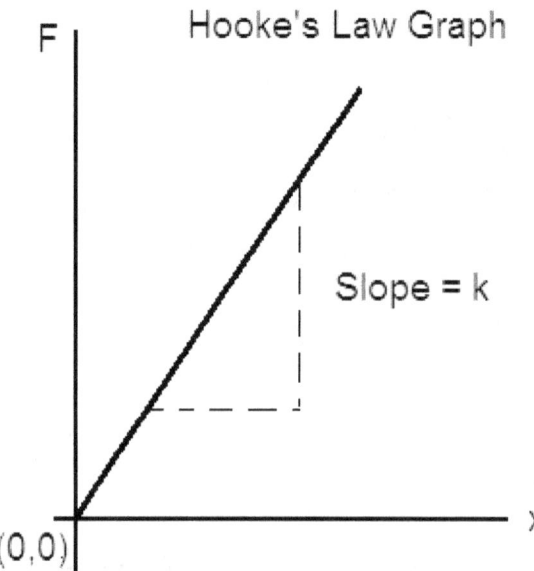

Some interesting facts & figures
• **Human jaw muscles can generate a force of 200 pounds (890 newtons) on the molars.**
• **Significance of 'Inertial Force': Nonsense! There is no such entity!**
• **Gravity, Electric Force, Magnetic force, Nuclear strong force, Nuclear weak force are fundamental forces of nature**
• **Elastic force, Friction, Normal force, Tension, Adhesive force, Cohesive force, and Intermolecular force are manifestations of electric and magnetic forces; hence they are non-fundamental forces.**

PROBLEMS:

1. A block of mass m is suspended by a string. Determine the tension in the string in each of the cases shown. *[49 N, 38 N, 60 N, 49 N]*

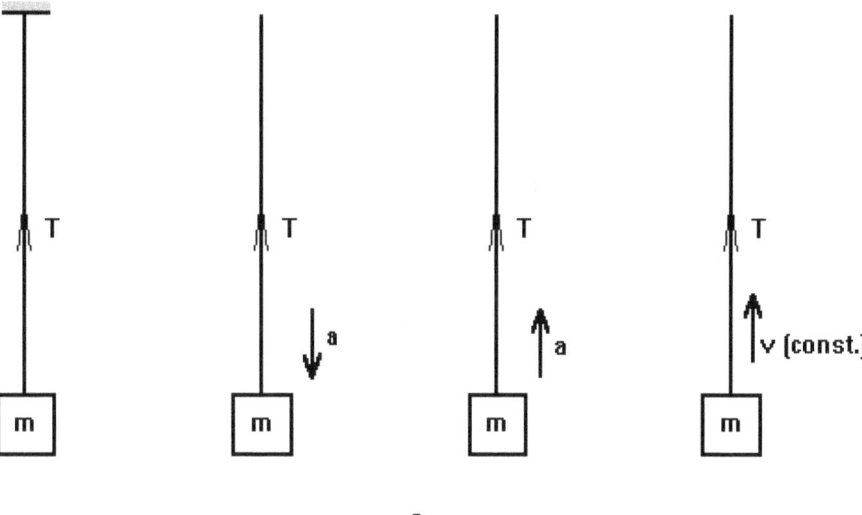

$m = 5.0 \text{ kg}, \quad a = 2.2 \text{ m/s}^2, \quad v = 5.2 \text{ m/s}$

2. Two blocks of masses M and m respectively are suspended from two strings as shown in the diagram. Determine the tensions T_1 and T_2 in each case.

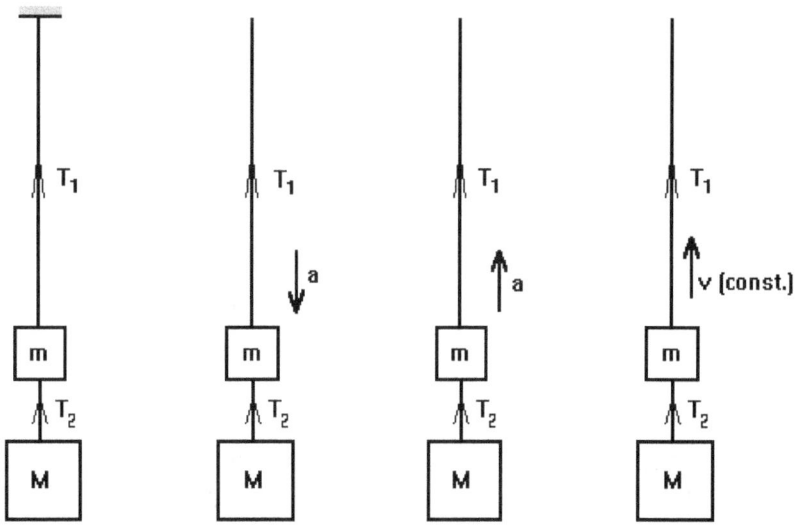

$M = 8.0 \text{ kg}, \quad m = 5.0 \text{ kg}, \quad a = 2.2 \text{ m/s}^2, \quad v = 5.2 \text{ m/s}$

3. As shown in the diagram below, three blocks are being pulled to the right on a horizontal frictionless surface. Determine the acceleration of the system and the tensions T_1 and T_2 in the strings. *[a=5 m/s², T₁=30 N, T₂=20 N]*

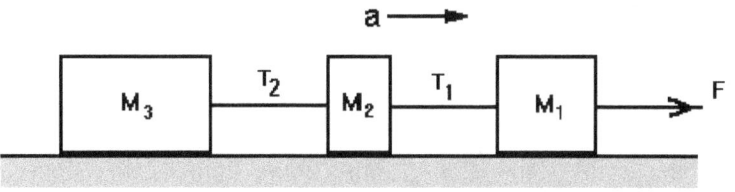

Frictionless surfaces

$M_1 = 10 \text{ kg}, \quad M_2 = 2 \text{ kg}, \quad M_3 = 4 \text{ kg}, \quad F = 80 \text{ N}$

4. In each of the two cases show below, two blocks are kept in contact on a horizontal frictionless surface and pushed by a force horizontal force F. Determine the magnitude of the contact force F' in each case.

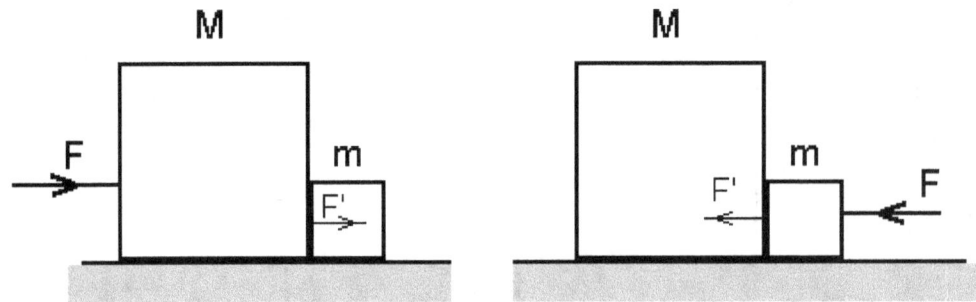

Frictionless surfaces

$$M = 20 \text{ kg}, \quad m = 1 \text{ kg}, \quad F = 84 \text{ N}$$

5. For the 'Atwood's Machine' shown in the diagram determine the tension in the string and the acceleration of the system in terms of m, M, and g.   [a=(M-m)g/(M+m), T=2mMg/(M+m)]

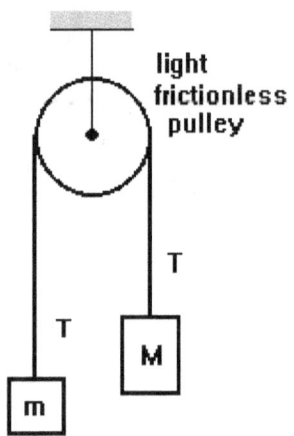

6. Three blocks each of mass m are suspended from a pulley as shown in the diagram. Determine the tensions T_1, T_2, and acceleration a in terms of m and g.

Also calculate the same quantities for m = 8 kg.

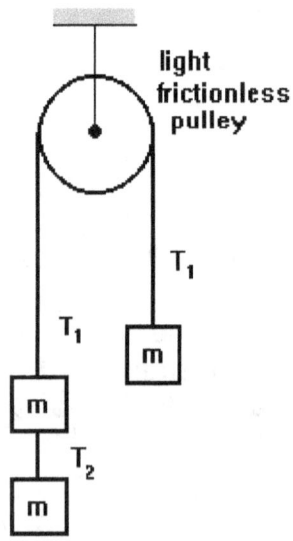

7. For the system of blocks and strings shown in the diagram, determine the tension T in the string and the acceleration a of the system in terms of m, M, and g.

Determine the same quantities if
M = 10 kg and m = 2 kg.

[a=mg/(m+M) = 1.63 m/s², T=mMg/(m+M)=16.3 N]

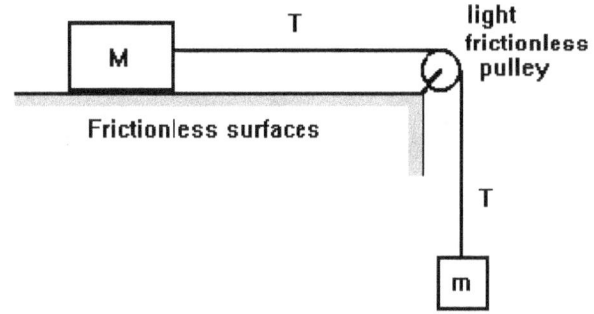

8. For the system of blocks and strings shown in the diagram, determine the tensions T_1 and T_2 in the strings and the acceleration of the system in terms of m, M, and g.

Determine the same quantities if
M = 10 kg and m = 1 kg.

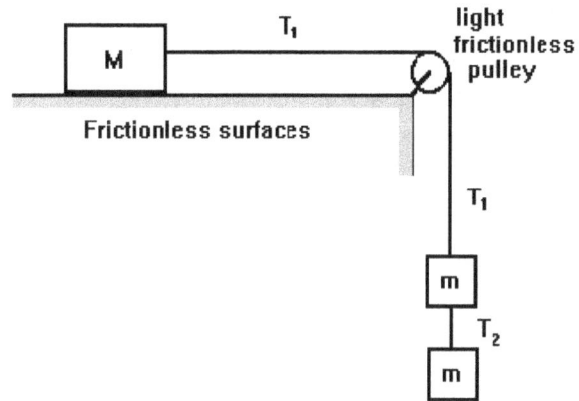

9. For the system of blocks and strings shown here, determine the force F applied on the system and the tensions T_1 and T_2 in the strings.

Take M = 10 kg, m_1 = 5 kg, m_2 = 3 kg, and a = 4 m/s².

[F=150.4 N, T_1=110.4 N, T_2=41.4 N]

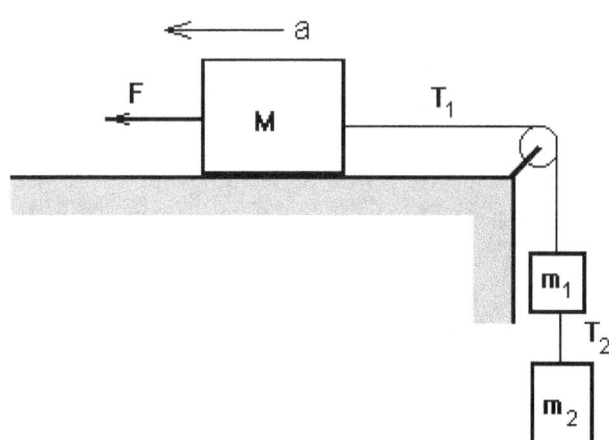

**10. For the system of blocks and strings shown here, determine the tension T in he string and the accelerations of m and M in terms of m, M, and g. Calculate the numerical values of the same quantities if m = 3 kg and M = 7 kg.

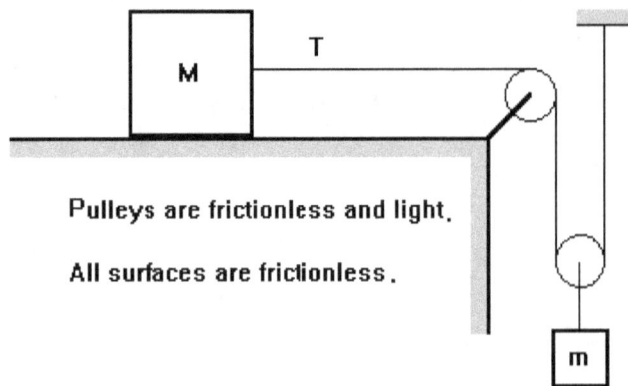

Pulleys are frictionless and light.

All surfaces are frictionless.

11. In the diagram below, a mass is suspended from three strings as shown. Determine the tension in each string.
[T_1 = 36.8 N, T_2 = 47.1 N, T_3 = 49 N]

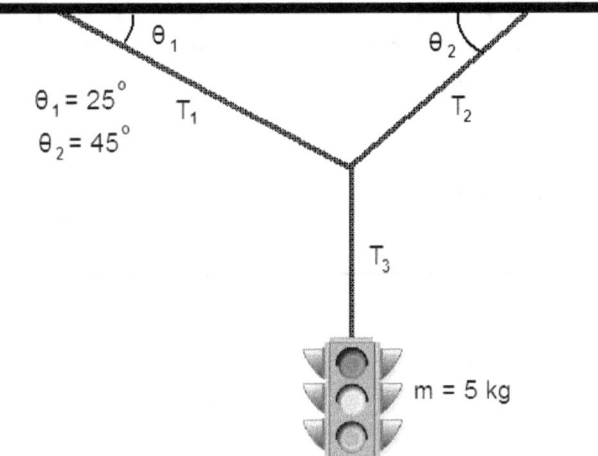

$\theta_1 = 25°$

$\theta_2 = 45°$

m = 5 kg

12. A junk car of mass 2000 kg is attached to the cable of a crane by a hook. The car is initially at rest on ground. The car is then lifted up with uniform acceleration.
(a) If the tension in the cable is 20,000 N calculate the acceleration of the car.

After 5 seconds, the car accidentally breaks free from the hook.
(b) Calculate the time after the hook breaks, in which the car hits the ground.

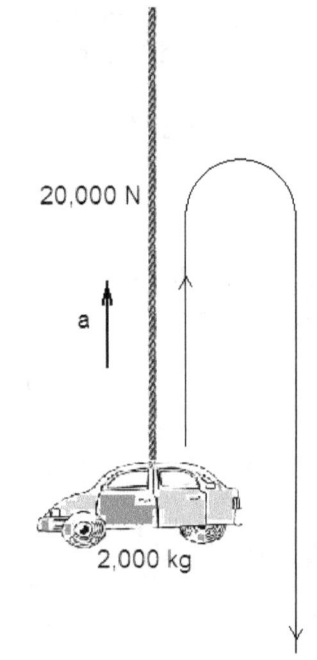

20,000 N

a

2,000 kg

13. A spring is suspended vertically from a ceiling. It is stretched by 0.05 m when a 2.5-kg block is suspended from it.

(a) Calculate the elastic constant, k, of the spring. *[490 N/m]*

(b) How much <u>additional</u> mass be used to give the spring an <u>additional</u> extension of 0.01 m?

[0.5 kg]

Multiple Choice:

1. A crate of mass m is attached to a cable and is being lowered with an acceleration of a = -¼ g. The tension in the cable must be
(A) ¼ mg   (B) ¾ mg   (C) 5/4 mg   (D) mg   (E) 0

2. A crate of mass m is attached to a cable and is being raised with an acceleration of a = +¼ g. The tension in the cable must be
(A) ¼ mg   (B) ¾ mg   (C) 5/4 mg   (D) mg
(E) 0

3.
A block of weight W is suspended by two strings as shown in the above. T_1 and T_2 are the tensions in the two strings. The tension T_2 is equal to

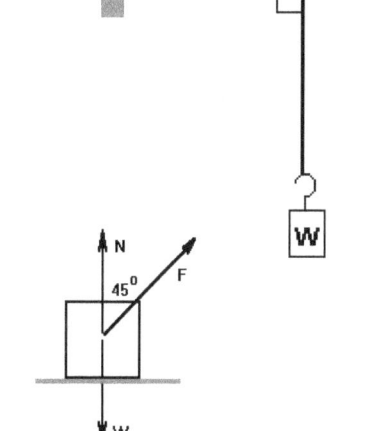

(A) W   (B) 0   (C) √2 W
(D) W / √2   (E) 2 W

4.

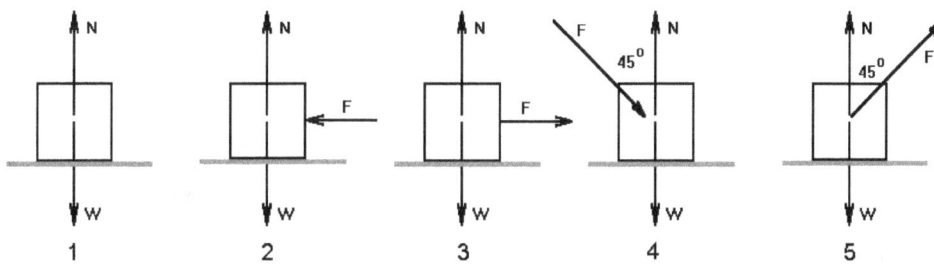

In which of the cases shown above is the normal force N equal to W?
(A) All of them   (B) 1 only   (C) 1, 2, and 3 only   (D) 4 and 5 only
(E) The answer depends on whether the surfaces in contact have friction or are frictionless.

5. A physics book of weight 20 N is placed on top of a chemistry book of weight 15 N on a horizontal table. The reaction force from the chemistry book on the physics book is
(A) 35N   (B) 20 N   (C) 15 N   (D) 17.5 N   (E) 5 N

6. As shown in the adjacent diagram, a man of mass M is using a light rope to hold a crate of mass m stationary. The pulley is frictionless. The normal force from the floor on the person must be
(A) Mg   (B) mg   (C) Mg + mg   (D) Mg – mg   (E) zero

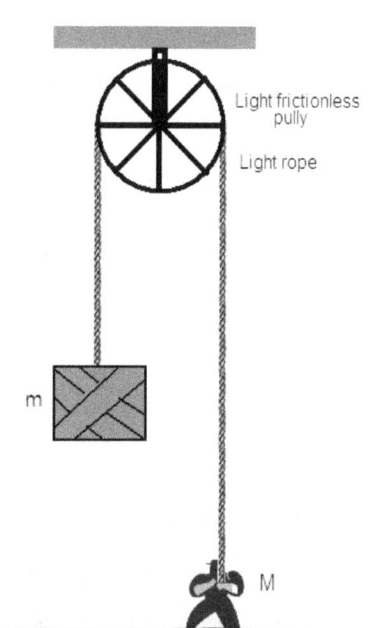

7. A bucket is suspended from an ideal spring. When 4 kg of water is added to the bucket the spring extends by 0.08 m. The spring constant of the spring must be
(A) 50 N/m   (B) 0.02 N/m   (C) 0.32 N/m   (D) 490 N/m   (E) 0.00204 N/m

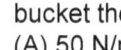

Chapter 7
Newton's Laws: Weight and Weightlessness

For an object of mass m, its weight is defined as $W = mg$ where g is the acceleration due to gravity at the point where the object is located.

It is one thing to have weight but another to <u>feel</u> the weight. In a gravitational field, whether a person feels his true weight, less weight, more weight, or weightless depends on the acceleration of the person.

The feeling of weight is caused by the normal force (F_N) acting on the person from another object, say floor of a room. Also if the person is standing on a weighing scale, <u>the reading shown by the scale is same as the normal force N</u> and that is the weight the person feels.

To understand this let us consider the elevator problem:

Let a person of mass m stand on a weighing scale inside an elevator. Consider various cases of acceleration of the elevator:

Case 1: $a = 0$

(*Either the elevator is stationary or it is moving up or down at constant velocity*)

Then applying Newton's Second Law for the components in the y-direction, we get

$$F_N - mg = m(0) \rightarrow F_N = mg$$

Thus the scale reads and the person feels the true weight mg.

Case 2: a is downward (negative)

(*Either the elevator is moving down with increasing speed or moving up with decreasing speed*)

Then applying Newton's Second Law for the components in the y-direction, we get

$$F_N - mg = m(-a) \rightarrow F_N = mg - ma$$

Thus the scale reads and the person feels weight, less by an amount ma.

Case 3: a is upward (positive)

(*Either the elevator is moving up with increasing speed or moving down with decreasing speed*)

Then applying Newton's Second Law for the components in the y-direction, we get,

$$F_N - mg = m(+a) \rightarrow F_N = mg + ma$$

Thus the scale reads and the person feels weight, more by an amount ma.

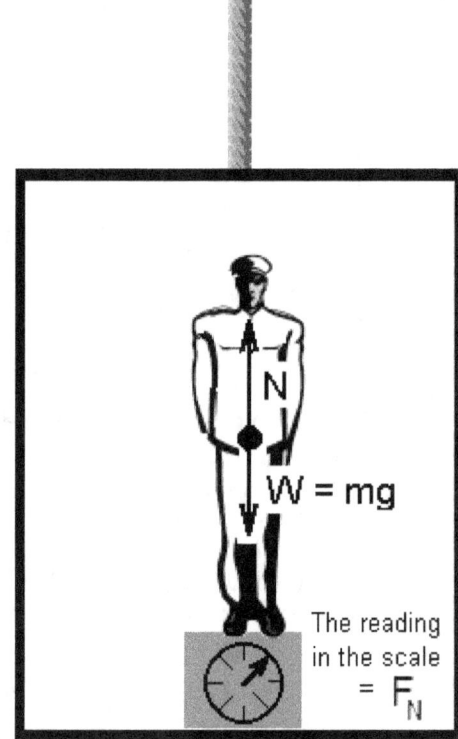

N

W = mg

The reading in the scale = F_N

$$F_N - mg = ma$$

Case 4:   a is downward = -g
(the elevator is in free fall)

Then applying Newton's Second Law for the components in the y-direction, we get

$F_N - mg = m(-g)$ ➜ $F_N = mg - mg = zero$

Thus the scale reads zero weight and the person feels weightless. The person can float around inside the elevator just as astronauts do inside the space shuttle.

Case 5: a is downward and its magnitude is >g

(The elevator is forced to accelerate downward)

The world turns upside down for the person inside the elevator. He can walk on the ceiling of the elevator.

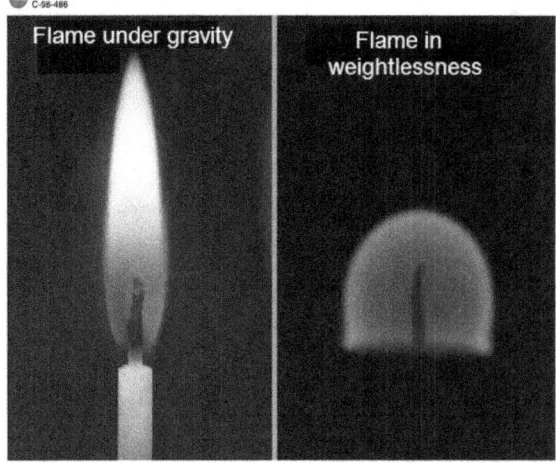

Flame under gravity   Flame in weightlessness

National Aeronautics and Space Administration
Lewis Research Center

Some interesting facts & figures
• The tallest building in the world, Bhuj Dubai' in UAE has 58 Otis elevators. The elevator for the deck travels at 20 mph (32 km/h) at about a floor per second.
• The world's two fastest elevators are in Taipie 101 building in Taiwan. They travel at 36 mph (60 km/h)
• Astronaut cannot weigh themselves while in orbit around the earth. They can determine their mass using a machine called Body Mass Measuring Device (BMMD)

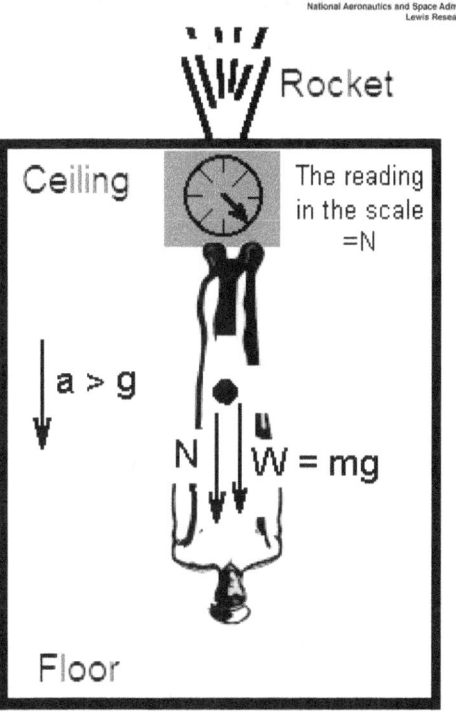

PROBLEMS:

1. A person of mass 50 kg stands on a weighing scale in an elevator. The scale reads in kilogram as well as in newtons. Determine the reading in the scale (in kg and N) for the following cases:

(a) The elevator is stationary.   *[50 kg, 490 N]*

(b) The elevator is accelerating upward at 4 m/s².   *[70.4 kg, 690 N]*

(c) The elevator is accelerating downward at 4 m/s².   *[29.6 kg, 290 N]*

(d) The elevator is moving up at a constant velocity of 4 m/s.   *[50 kg, 490 N]*

2. A person is holding a 5-kg briefcase stationary while traveling in an elevator. Determine the force applied by the person to support the briefcase in the following situations:
(a) The elevator is stationary.

(b) The elevator is accelerating upward at 2 m/s^2.

(c) The elevator is accelerating downward at 2 m/s^2.

(d) The elevator is moving up at a constant velocity of 5 m/s.

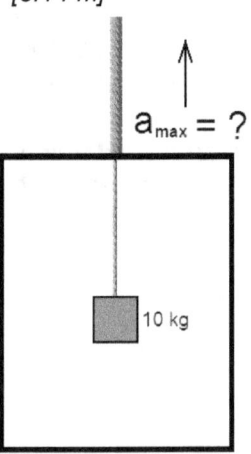

**3. An elevator is accelerating downward at 3.5 m/s^2. A person standing inside releases a coin from 2 m above the floor.
(a) How long will the coin take to reach the floor?   [0.8 s]

(b) From the point of view of a person on a floor outside the elevator how much distance did the coin fall?   [3.14 m]

4. A block of mass 10 kg is suspended from the ceiling of an elevator with a string whose breaking tension is 120 N. What is the maximum upward acceleration the elevator can have without breaking the string?

Multiple Choice:

1. A boy of weight 600 N stands on a scale inside an elevator that is moving <u>upward</u>. The scale reads 570 N. Which of the following is correct description of the speed of the elevator?

(A) Its speed is increasing.
(B) Its speed is decreasing.
(C) Its speed is constant but not zero.
(D) Its speed is constant and zero.
(E) It is in free fall.

2. An astronaut is in circular orbit in a satellite around the earth. The astronaut feels weightless because
(A) Centrifugal and centripetal forces acting on her cancel each other.
(B) She is in vacuum.
(D) She is wearing an antigravity suit.
(E) She and the satellite are both accelerating at the same rate toward the earth.
(F) At that distance, the earth's gravity is zero.

3. A stunt man is clinging to a cable of fixed length attached to a helicopter that is moving in a vertical direction. The tension in the cable is less than the weight of the man. The helicopter must be
(A) moving upward
(B) moving downward
(C) accelerating upward
(D) accelerating downward
(E) stationary

Chapter 8
Friction

When surface of a body slides over the surface of another body, frictional forces oppose the motion. Frictional forces act on both the bodies and are equal in magnitude and opposite in directions.

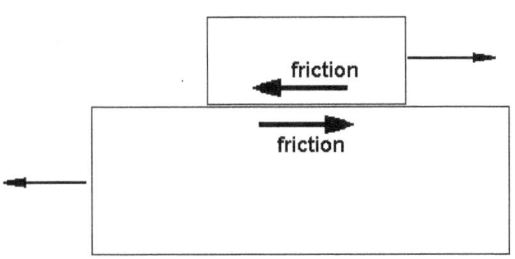

Frictional force is proportional to the normal force (F_N) acting between the two surfaces. The normal force indicates how hard the surfaces are pushing against each other. Also the fictional force depends on the nature of surfaces in contact.

STATIC FRICTION (f_s):

When a body tries to slide over another body but the motion has <u>not</u> yet begun the force of friction acting between the bodies is called static friction. This force increases up to certain maximum value, $f_s(max)$, in response to the external force trying to move the body. The body will begin moving only when the applied external force exceeds $f_s(max)$ which is given by

$$f_s(max) = \mu_s \, F_N$$

where μ_s is called coefficient of static friction and depends on the nature of surfaces in contact.

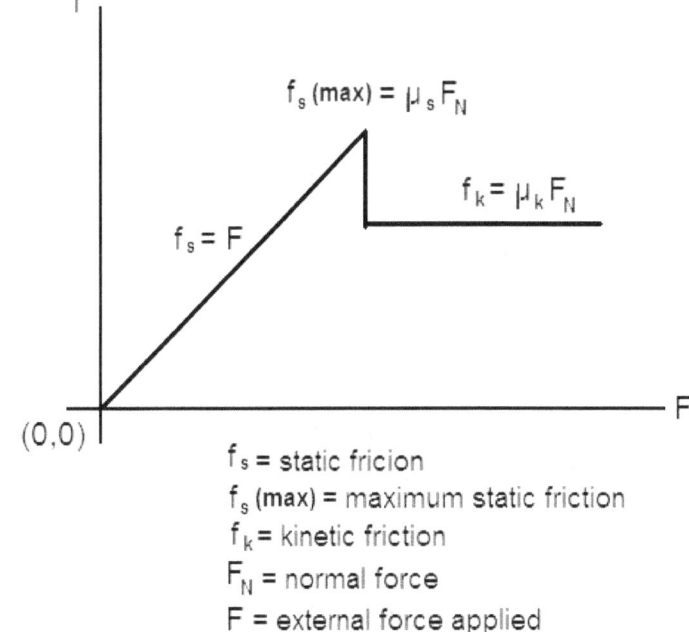

f_s = static friction
f_s (max) = maximum static friction
f_k = kinetic friction
F_N = normal force
F = external force applied
on the object

KINETIC FRICTION (f_k):

The force of kinetic friction acts between two surfaces in <u>motion</u>. The magnitude of this force is given by

$$f_k = \mu_k \, F_N$$

For any two given surfaces, $\mu_k < \mu_s$.

Normal Force (F_N): The normal force indicates how hard a surface is pushing against the other. In the diagrams below, the normal force is indicated in various different situations:

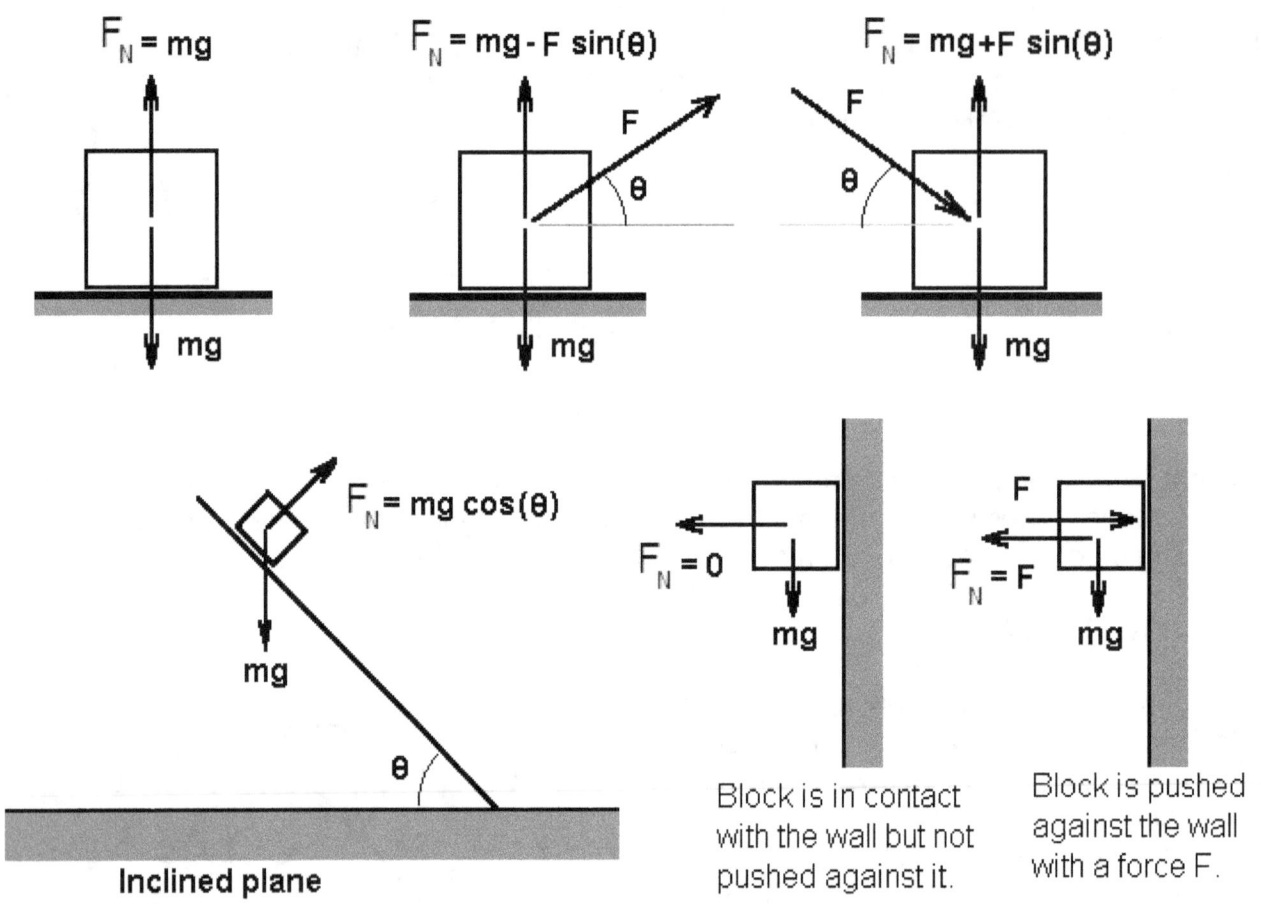

$$F_N = mg$$

$$F_N = mg - F\sin(\theta)$$

$$F_N = mg + F\sin(\theta)$$

$$F_N = mg\cos(\theta)$$

Inclined plane

$F_N = 0$

Block is in contact with the wall but not pushed against it.

$F_N = F$

Block is pushed against the wall with a force F.

Some interesting facts & figures
• Teflon has a very low coefficient of friction of 0.05.
• Argonne National Laboratory has developed a hard and nearly-frictionless-carbon coating of amorphou carbon that has coefficient of friction of around 0.004.

PROBLEMS:

1. A block of mass m=4.0 kg is placed on a horizontal surface. For the two surfaces in contact, μ_s = 0.8 and μ_k = 0.3.

(a) Determine the normal force N acting on the block.   *[39.2 N]*

(b) Determine the minimum horizontal force F required to set the block in motion. *[31.36 N]*

(c) If the same force F is maintained after the motion has begun determine the acceleration of the block.   *[4.9 m/s²]*

2. A block of mass m = 5.8 kg is placed on a horizontal surface (μ_s = 0.7). A force F is applied at an angle of 35° <u>above</u> horizontal as in the diagram below. Determine the minimum value of F required to set the block in motion.

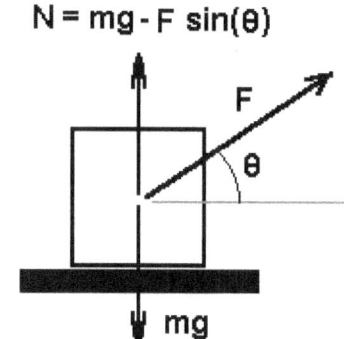

$$N = mg - F \sin(\theta)$$

3. A block of mass m = 5.8 kg is placed on a horizontal surface (μ_s = 0.7). A force F is applied at an angle of 35° <u>below</u> horizontal as in the diagram below. Determine the minimum value of F required to set the block in motion.   *[95.3 N]*

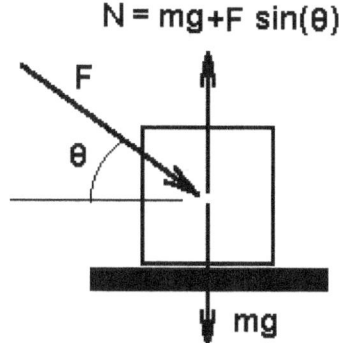

$$N = mg + F \sin(\theta)$$

3. For the system shown below, the block on the table is moving to the right and μ_k = 0.25.
   (a) Draw Free-body diagrams for m and M.
   (b) Determine the acceleration a of the system and the tension T in the string.

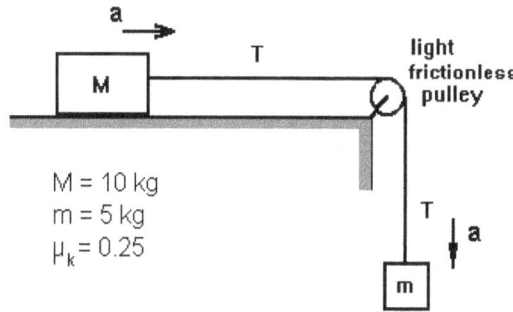

M = 10 kg
m = 5 kg
μ_k = 0.25

5. A block of mass m = 12 kg is placed on a plane inclined at an angle of 33°. The coefficient of kinetic friction is μ_k = 0.3. The block is given a gentle push to begin motion so that the kinetic friction is effective. The block slides down the inclined plane starting from zero speed.
(a) Determine the normal force acting on the block.   *[98.6 N]*

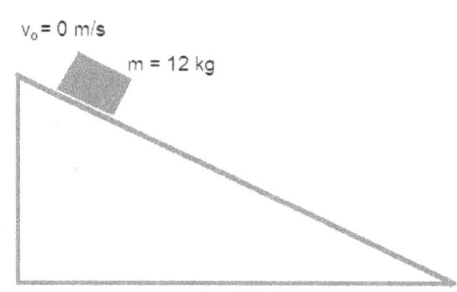

(b) Determine the frictional force acting on the block.   *[29.6 N]*

(c) Determine the net force acting down the inclined plane on the block.   *[34.4 N]*

(d) Determine the acceleration of the block down the inclined plane.   *[2.9 m/s²]*

(e) Determine the time taken by the block to travel 1 m down the plane.   *[0.83 s]*

6. A block of mass m = 1.2 kg is placed in contact with a vertical wall. The value of μ_s between the surfaces is 0.45. Determine the horizontal push F required to prevent the block from sliding down.

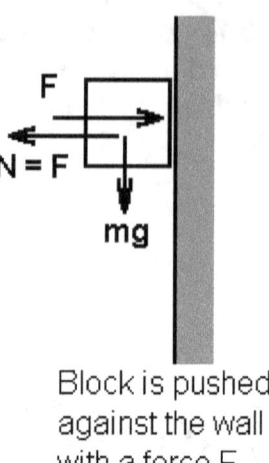

Block is pushed against the wall with a force F.

7. A student has to push a 3-kg book against a wall with a minimum force of 42 N to prevent it from sliding down. Determine the coefficient of static friction between the book and the wall. Ignore the force between the hand and the book.
[0.7]

8. A book is placed on a plane that is initially horizontal. For this system μ_s = 0.68. Now one end of the plane is lifted up and the angle of inclination is slowly increased.
   At what angle will the book begin to slide down?

9. A book is placed on a plane that is initially horizontal. For this system μ_k = 0.68. Now one end of the plane is lifted up and the angle of inclination is slowly increased through small steps. At each step the book is given a gentle push down the plane to overcome the static friction. At what angle will the book begin to slide down?
[34.2°]

10. A 2.0-kg book is launched horizontally along a table with an initial speed of 0.48 m/s. The book comes to rest after travelling a distance of 1.18 m.

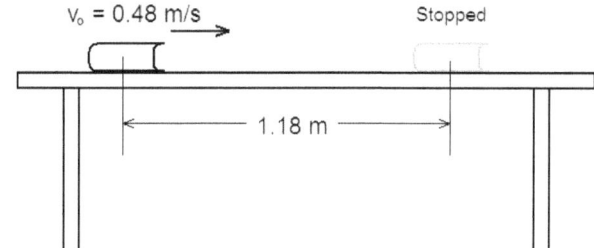

    (a) Determine the acceleration of the book.

    (b) Determine the force of friction (f_k) acting on the book.

    (c) Determine the coefficient of friction μ_k between the book and the table.

Multiple Choice:

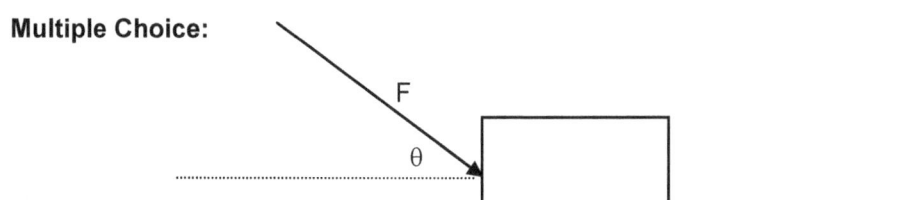

1. A crate of mass m on a horizontal floor is pushed by a force F as shown in the figure above. If the acceleration due to gravity is g, the normal force from the floor on the crate is given by
(A) mg   (B) mg - Fcosθ   (C) mg + Fcos θ   (D) mg – Fsinθ   (E) mg + Fsinθ

2. A block placed on the inclined plane is stationary. What is true about frictional force acting on the block?
(A) It is zero.   (B) It is acting up the plane   (C) It is acting down the plane.
(D) Its direction depends on the angle of the inclined plane.
(E) It is up the plane if the angle is less than 45° and down the plane if the angle is greater than 45°

3. A block of weight of weight 100 N is placed on a horizontal surface. The coefficient of static friction between the surfaces is 0.7. A horizontal force of 50 N is not able to move the block along the surface. Which of the following is the magnitude of the frictional force acting on the block?
(A) 0 N   (B) 35 N   (C) 50 N   (D) 70 N   (E) 100 N

4. A boy of mass m is sitting at the center of the flat bed of a pickup truck. The truck begins to accelerate at a and the boy does <u>not</u> slip away. The force of friction acting on the boy must be
(A) ma
(B) mg
(C) mg – ma
(D) mg + ma
(E) depends on the coefficient of static friction

5. A book of weight W is placed on a horizontal surface of coefficients of friction μ_k and μ_s. A horizontal force F is applied to keep moving the book at constant velocity v. Thus
(A) F = 0
(B) F = μ_sW
(C) F = μ_kW
(D) F = μ_sW – μ_kW
(E) F = Wv

Chapter 9
Inclined Plane

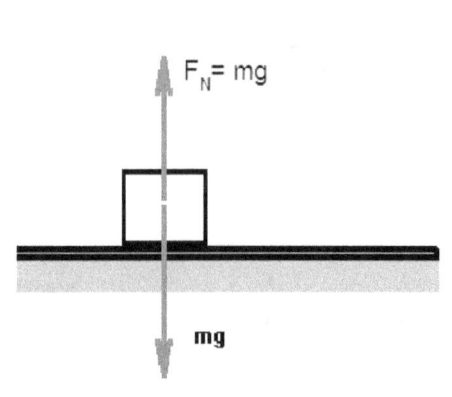

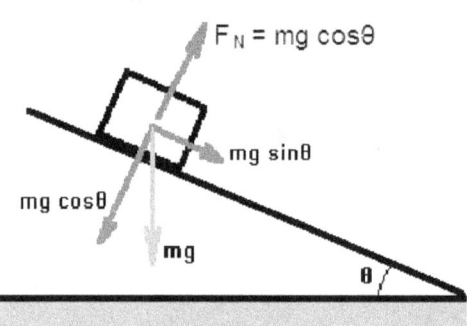

$F_N = mg\cos\theta$

$mg\sin\theta$

$mg\cos\theta$

mg

θ

Horizontal plane   Inclined plane

Frictionless inclined plane: W = mg

The component of the weight W on the block acting down the plane is $W_{||} = mg\sin\theta$.
The component of the weight W on the block perpendicular to the plane is $W_\perp = mg\cos\theta$. This component is the normal force F_N.
If the plane is frictionless, then $W_\perp = F_N$ has no effect on the motion down the plane. The acceleration down the plane is $a = W_{||}/m = mg\sin\theta/m = g\sin\theta$.

a = g sinθ only for frictionless plane

Inclined Plane with friction:

$f_k = \mu_k F_N = \mu_k mg\cos\theta$

1. If the block is moving down the plane with constant velocity, i.e. if a = 0 then

$mg\sin\theta = f_k$

2. If the block is moving down the plane with an acceleration a then

$mg\sin\theta - f_k = ma$

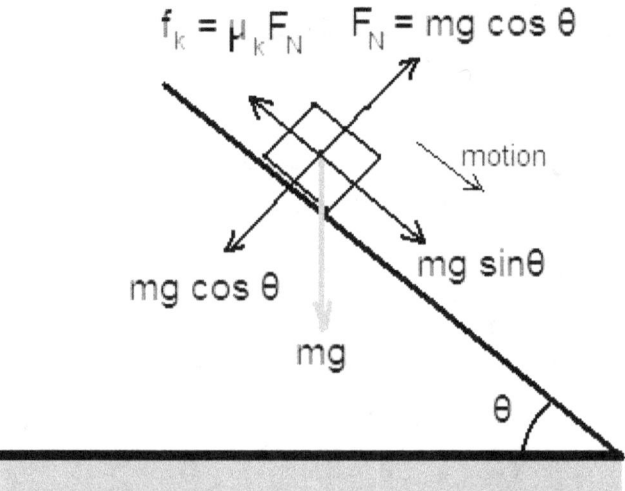

$f_k = \mu_k F_N$ $F_N = mg\cos\theta$

motion

$mg\sin\theta$

$mg\cos\theta$

mg

θ

Some interesting facts & figures
• The steepest street in the world - Baldwin Street at 35% grade– is in Dunedin, New Zealand
• Italy's 'Spacemaker' at 42 meters (137.8 feet) height and 60° slope, is the tallest plummet slide in the world. It can bring riders up to a speed of 100 km/h (62.1 mp/h).

PROBLEMS:

1. A block is placed at the top of a 3-m long frictionless plane inclined at an angle of 47°.
 (a) Determine the time taken by the block to reach the bottom of the inclined plane. *[0.91 s]*

 (b) Calculate the speed of the block at the bottom of the plane.   *[6.56 m/s]*

2. For the frictionless system of inclined plane and blocks as shown in the diagram below, determine the acceleration of the blocks and the tension in the string.

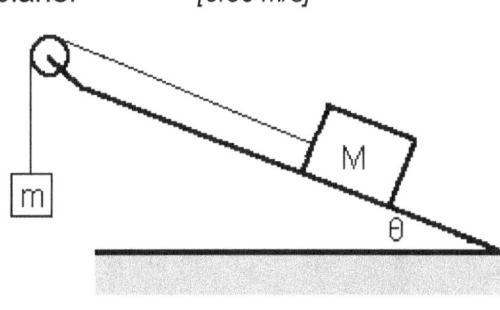

3. Do problem 2 for M = 15 kg.
   [a = 0.35 m/s², T = 20.3 N]

M = 8 kg, m = 2 kg, θ = 10°

4. A goose slides down an ice covered 18° sloping <u>frictionless</u> roof. The length of the roof is 8 m.

(a) Calculate the time taken by the goose to reach the edge of the room.

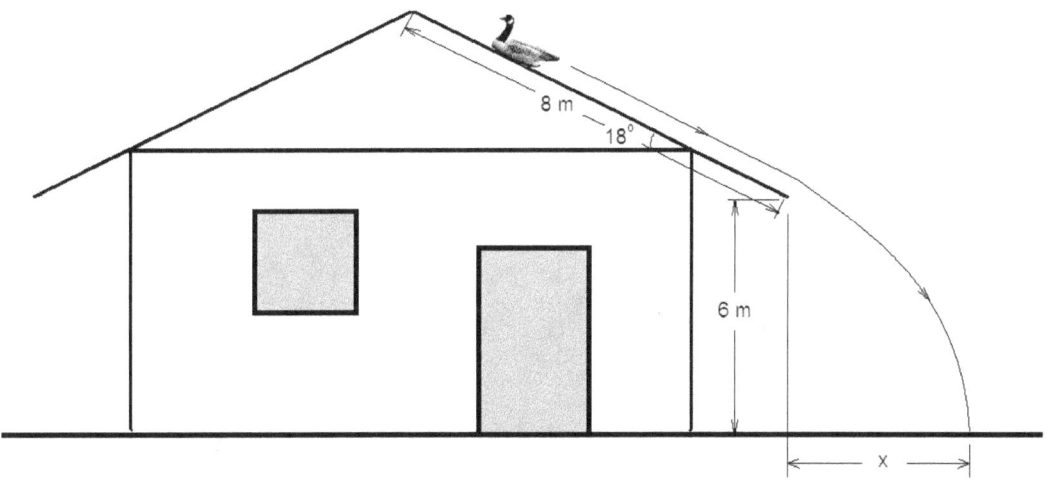

(b) Calculate the speed of the goose at the edge of the roof.

The edge of the roof is 6 m above the ground.

(c) Calculate how far horizontally from the edge (x) of the roof will the goose land on the ground. (Assume it does not try to fly off.)

(d) Calculate the speed with which the goose hit the ground.

5. A 42-kg boy climbs up and sits at the top of a straight playground slide at a height of 5.2 m above the ground. The coefficient of kinetic friction between him and the slide is $\mu_k = 0.23$. He begins to slide down the slide starting from $v_0 = 0$ m/s. The slide makes an angle of 52° with the horizontal.

(a) Draw a free-body diagram for the boy.

(b) Calculate the components $W_\perp$, and $W_\parallel$ of the boy's weight.   *[253 N; 324 N]*

(c) Calculate the acceleration of the boy down the slide.   *[6.33 m/s²]*

Multiple Choice:

1. A block is placed on a plane and the angle of the plane is gradually increased. At a certain angle α the block breaks into motion down the plane. The coefficient of static friction between the block and the plane must be
sin α
(B) cos α
(C) tan α
(D) cot α
(E) sec α

2-3. A plane is inclined at an angle α. A box of mass m placed on the plane is at rest. The coefficient of static friction between the surfaces is μ_s.

2. The normal force on the box from the plane is
(A) mg cos α
(B) mg cos $\alpha - \mu_s mg$
(C) mg cos $\alpha + \mu_s mg$
(D) $\mu_s mg -$ mg cos α
(E) $\mu_s mg$

3. The box is staying at rest on the plane. Which of the following is the correct relationship between the static friction f_s and $W_\parallel$?
(A) $f_s = W_\parallel$
(B) $f_s = 0$
(C) $f_s = W_\perp$
(D) $f_s > W_\parallel$
(E) $f_s < W_\parallel$

Chapter 10
Uniform Circular Motion

Uniform Circular Motion (UCM): It is motion of a particle along a circular path at constant speed **v**.

Period (T): The period of the UCM is the time taken to complete one revolution or rotation.

Frequency (f): It is the number of revolutions or rotations a particle makes per unit time. **f = 1/T** Its SI unit is Hertz (Hz) which is one revolution (rotation) per second.

Centripetal acceleration (a_c): A particle of mass m undergoing uniform circular motion experiences an acceleration, a_c, which is always pointing toward the center of the circle; hence the name, *centripetal* acceleration. It is constant in magnitude but continuously varying in direction.

Centripetal Force (F_c): The net force, $\Sigma F = F_c$ acting on the particle is also acting toward the center; hence the name *centripetal* force. It is constant in magnitude but continuously varying in direction.

Velocity (v): The velocity of the particle is always tangential to the circle. It is constant in magnitude but continuously varying in direction.

Important formulae:
- For the centripetal acceleration:

$$a_c = \frac{v^2}{r}$$

- For centripetal force:

$$F_c = \frac{mv^2}{r}$$

- The time period T, the frequency of rotation f, the radius of the circular path, and the speed of the particle undergoing UCM are related by the equations

$$T = \frac{2\pi r}{v} = \frac{1}{f}$$

Significance of 'Centrifugal Force':
'Centrifugal force' is felt by a person or object if they are a part of the system that is rotating. <u>It is not a real force and does not play a part in solving physics problems</u>. For example, people inside a car feel this force when the car is taking a turn or a person standing on the floor of merry-go-round maybe thrown off away from the center if he is not holding on to a support.

Various Kinds of Forces that act as Centripetal Forces:

1. Tension in a string ($F_c = F_T$):

For a rock tied to a string and whirled in a horizontal circle at constant speed, tension in the string is the centripetal force T. Thus,

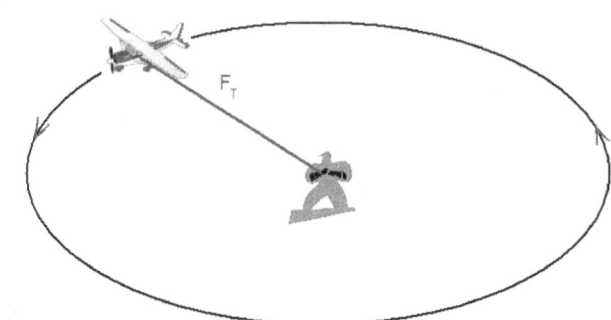

$$F_T = \frac{mv^2}{r}$$

2. Static friction ($F_c = f_s$ or $F_c = f_s(max)$:

For a car taking a turn around a street corner or a coin on a turntable static friction acts as the centripetal force F_c
Thus,

$$f_s = \frac{mv^2}{r}$$

or for max value $f_s(max)$

$$\mu_s mg = \frac{mv^2}{r}$$

3. Normal force ($F_c = F_N$):

For a particle undergoing UCM in contact with the inside of a circular surface(see Problem 9 below). , the normal force applied by the surface is the centripetal force. Thus,

$$F_N = \frac{mv^2}{r}$$

4. Force of gravity ($F_c = F_g$):

For a satellite in circular orbit (discussed in Chapter 11 ahead) around earth or any massive body, the gravitational force on the satellite is the centripetal force. Thus,

$$\frac{Gm_1 m_2}{r^2} = \frac{mv^2}{r}$$

5. Weight of another object ($F_c = m'g$):

Weight of another object m' can create tension in a string which in turn acts like a centripetal force for a particle m in UCM (see Problem 4 below).
Thus,

$$m'g = \frac{mv^2}{r}$$

Banking of the Road:

If a road has a curve and friction may not be sufficient to provide the necessary centripetal force the road needs to be banked. The banking angle θ depends on the speed of the vehicles and the radius of curvature of the road. It is given by:

$$\tan\theta = \frac{v^2}{rg}$$

At this angle, a vehicle can take a safe turn at speed v even in the absence of friction. Note that the angle does not depend on the mass of the vehicle.

Note: *Centripetal force being always perpendicular to the displacement of the particle, does not do any work* (discussed in Chapter 12 ahead) *on the particle. Hence it does not change the speed or kinetic energy of the particle.*

Artificial Gravity in Spinning Space Station:

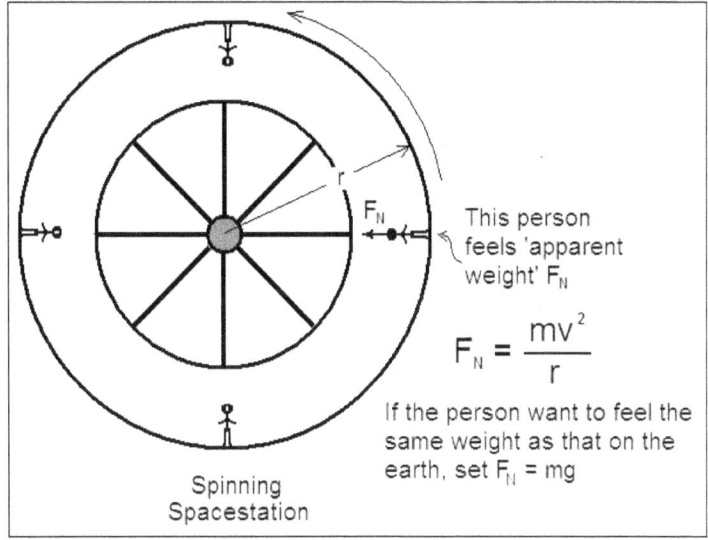

This person feels 'apparent weight' F_N

$$F_N = \frac{mv^2}{r}$$

If the person want to feel the same weight as that on the earth, set $F_N = mg$

Spinning Spacestation

Normal force (apparent weight) at the bottom of a roller coaster

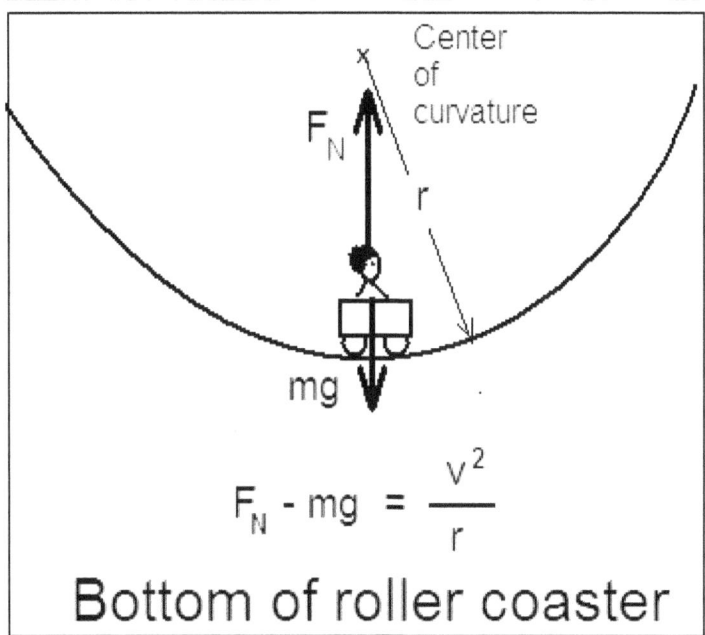

Center of curvature

F_N

r

mg

$$F_N - mg = \frac{v^2}{r}$$

Bottom of roller coaster

Normal force (Apparent weight) at the top of a roller coaster, or arched bridge or airplane trajectory

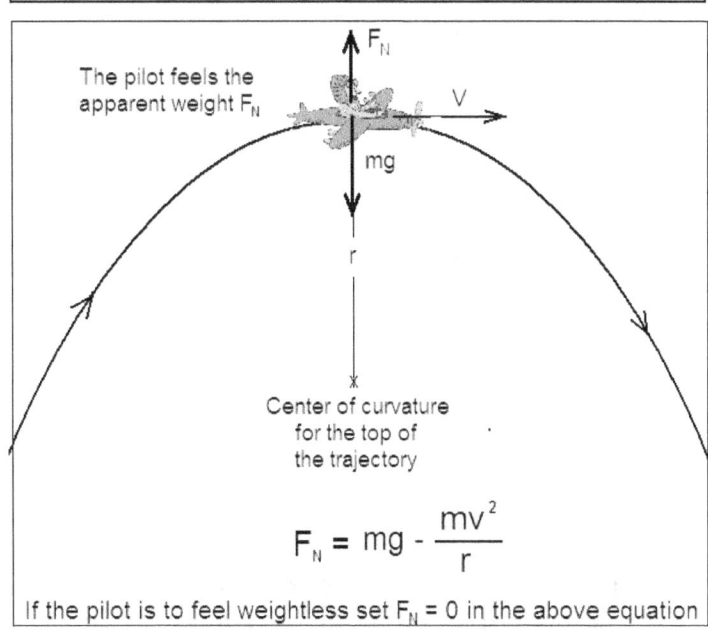

The pilot feels the apparent weight F_N

F_N

V

mg

r

Center of curvature for the top of the trajectory

$$F_N = mg - \frac{mv^2}{r}$$

If the pilot is to feel weightless set $F_N = 0$ in the above equation

Some interesting facts & figures
• The Earth's period of rotation (about its N-S axis) is 24 hours.
• The Earth's period of revolution around the sun is 1 year.
• The Earth's average velocity orbiting the sun is 107,220 km per hour.
• The average speed of the moon around the earth is 1.022 km/s or 3,680 km/h, or 2,200 mi/h
• George Washington Gale Ferris originally designed and constructed Giant Wheel later came to be known as Ferris Wheel. This wheel served as a landmark for the 1893 World's Columbian Exposition in Chicago

PROBLEMS:

1. A 1.8-kg stone tied to one end of a string is whirled at a constant speed of 2.5 m/s along a horizontal circle of radius 1.3 m. Determine

     a. the time period for the circular motion　　　　*[3.27s]*

     b. the frequency of the circular motion　　　　*[0.31 Hz]*

     c. the centripetal acceleration　　　　*[4.81 m/s²]*

     d. the tension in the string　　　　*[8.67 N]*

2. A car is taking a 20-m radius turn around a corner. If the coefficient of static friction between the tires and the surface of the road is μ_s= 0.6, determine the maximum possible safe speed of the car to avoid slipping.

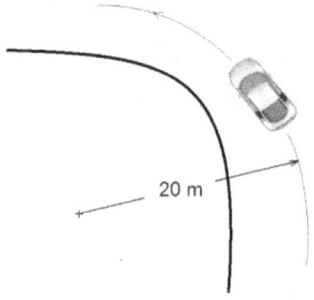

20 m

3. A highway has a curve of radius 350 m at a place. Determine the angle of banking at the curve for safe driving at 100 mph (44.7 m/s) without any assistance from the friction.　　　　*[30.2°]*

4. A string passing through a hole in a table carries a mass m at one end and a mass M at the other end. Mass m is placed on a horizontal frictionless table and mass M is hanging freely. The length of the string from the hole to the mass m is r. Determine the frequency with which the mass m must move in a circle of radius r so that the mass M stays at rest.

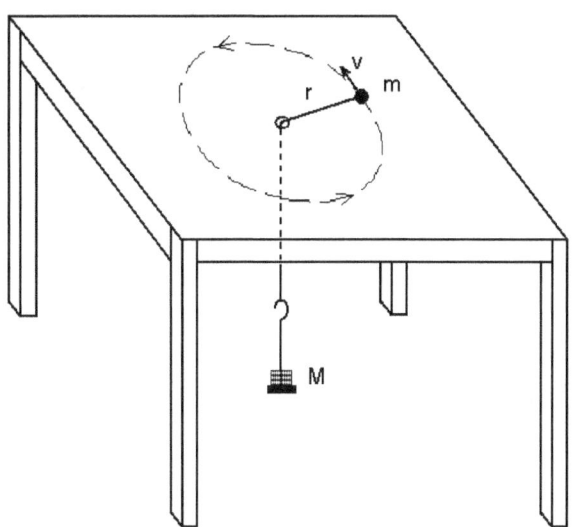

5. The vertical cross section of a bridge is a circular arc of radius 300 m. What is the maximum speed at which the car will not leave the surface while passing at the top of the bridge?

[54.2 m/s]

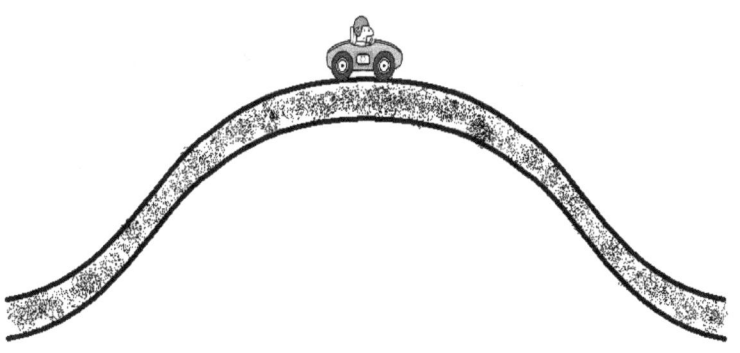

**6. Determine the apparent loss of weight of a person standing at the equator, due to the rotation of the earth. If the earth begins to rotate faster, what would be the length of the day at the rotational frequency at which the person would feel weightless?

7. A conical pendulum has a string of length 1.8 m at an angle of $30°$ from the vertical. The bob has a mass of 0.3 kg.
(a) Determine the tension F in the string.   [3.40 N]

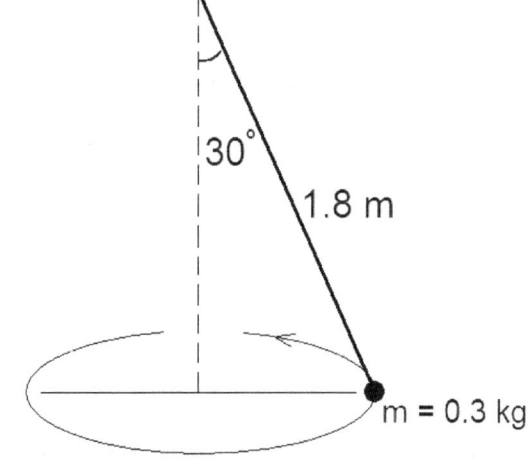

(b) Determine the linear speed of the bob.
          [2.25 m/s]

(c) Determine the period of this conical pendulum.   [2.51 s]

**8. A conical pendulum has a bob of mass 0.1 kg and a string of length 1.25 m. Its time period is 2.00 s. Determine

      a. the angle made by the string with the vertical direction

      b. the tension in the string

9. A rotor (gravitron) in an amusement park has a radius of 2.3 m. The coefficient of static friction between the rider and the rotor wall is 0.5. Determine the minimum frequency of the rotor above which the rider will not slip down when the floor of the rotor drops.   *[0.46 Hz]*

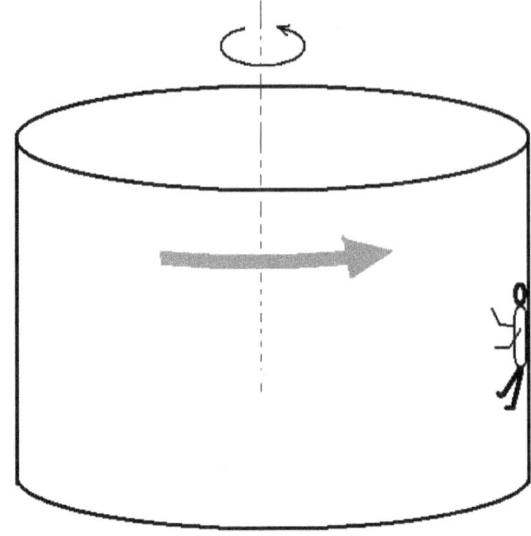

10. A coin is placed on a turn-table at .12 m from the axis of rotation. The coefficient of static friction between the coin and the turn-table is 0.1. When the turn-table is turned on, at what frequency will the coin begin to slip away?

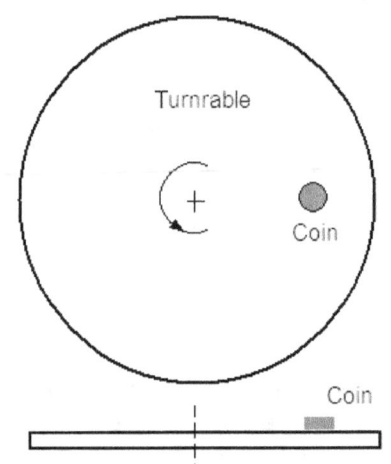

11. A certain space station has a radius of 50 m. Determine the time period of the rotation of the space station such that the astronaut near the edge of the space-station feels the same weight as that on the surface of the earth. How much weight will another astronaut at half the distance from the axis to the edge will feel compared to his weight at the surface of the earth.

         [14.2 s, ½ W_{earth}]

Circular Motion in Vertical Plane

12. An airplane flies at the top of a vertical circular trajectory of radius 5 km. What should be the speed of the plane that the persons inside feel weightless?

         [221.4 m/s]

13. The bottom of a roller coaster is a vertical circular arc of radius 25 m. The roller coaster passes the bottom at a speed of 30 m/s. Determine the apparent weight a rider of weight 600 N would feel.

14. The top of a roller coaster is a vertical circular arc of radius 25 m. A roller coaster rolls at the top with a speed 10 m/s. Determine the apparent weight of a 600-N person at the top.

         [355 N]

15. A Ferris wheel of radius 12 m is spinning with a time period of 1 min. Determine the apparent weight of a rider as a multiple of his actual weight (i.e. F_N/W) at

     a. the bottom   [1.01mg]

     b. the top   [0.99mg]

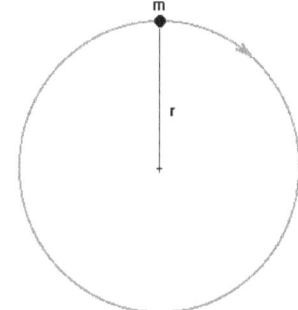

MULTIPLE CHOICE:

1. A quantity which remains constant in uniform circular motion is
(A) Centripetal force   (B) Acceleration   (C) Speed
(D) momentum   (E) None of these

2. A coin placed on a horizontal turntable is revolving without slipping with the turntable. The centripetal force on the coin is
(A) Its own weight   (B) static friction   (C) kinetic friction
(D) normal force   (E) nuclear strong

3. As shown in the diagram here, a particle m is moving clockwise along a circular path with constant speed. Which of the following correctly represent the directions of the centripetal force F_c, centripetal acceleration a_c, and the velocity v for this particle?

(A)    (B)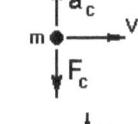

(C)     (D)

(E)

4. The centripetal force in uniform circular motion acts
(A) toward the center   (B) away from the center   (C) tangential to the circle
(D) perpendicular to the circle given by right hand thumb rule
(E) there is no such force

5. For a satellite in circular orbit around the earth, the centripetal force is provided by
(A) the gravitational pull of the moon
(B) the gravitational pull of the sun
(C) the Coriolis force
(D) the electromagnetic force from the launch site
(E) the gravitational pull from the earth

6. The period of the moon around the earth is approximately equal to
(A) 1 hour
(B) 1 day
(C) 1 month
(D) 1 year
(E) 4 years

7. The period of rotation of the earth is
(A) 1 hour
(B) about 1½ hours
(C) 12 hours
(D) 24 hours
(E) 1 year

Chapter 11
Newton's Law of Gravitation

If two particles of masses **m₁** and **m₂** are separated by a distance **r**, the gravitational force of attraction between the two particles is given by

$$F = G \frac{m_1 m_2}{r^2}$$

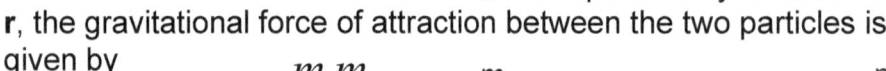

G is the Universal gravitational constant
$G = 6.67 \times 10^{-11} \ N \ m^2/kg^2$

Gravity is the weakest of the 5 fundamental forces: gravitational, electromagnetic, nuclear weak, and nuclear strong

Acceleration due to gravity at a point above a planet depends on the distance, r, of the point from the center of the planet

The acceleration due to gravity **g** a distance **r** from the center of any planet of mass **M** is given by

$$g = G \frac{M}{r^2}$$

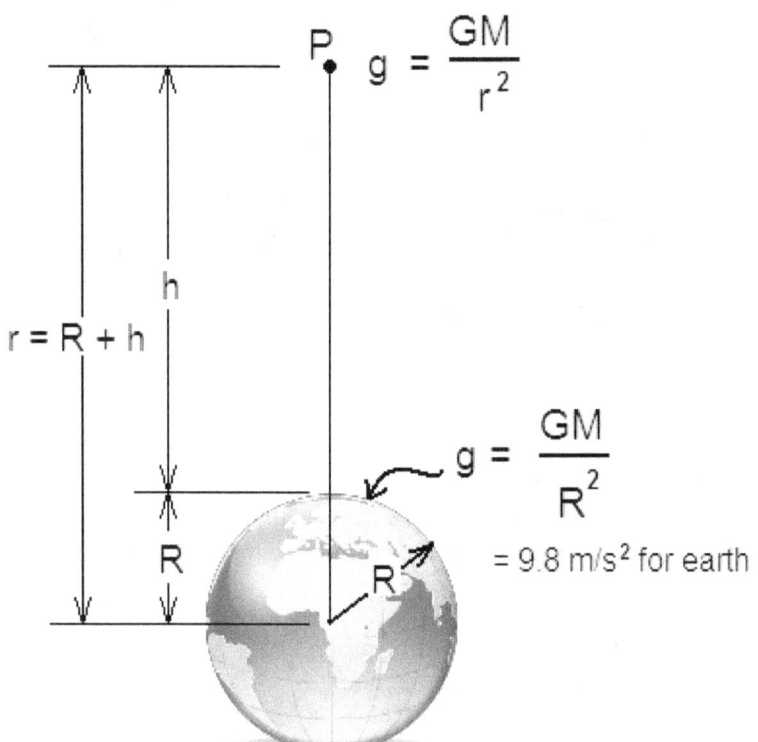

$g = \dfrac{GM}{r^2}$

$r = R + h$

$g = \dfrac{GM}{R^2}$

$= 9.8 \ m/s^2$ for earth

Earth-Moon Distance = 3.85×10^8 m
Earth-Sun Distance = 1.49×10^{11} m

Body	Mass (kg)	Radius (m)
Moon	7.35×10^{22}	1.74×10^6
Earth	5.98×10^{24}	6.38×10^6
Jupiter	1.90×10^{27}	6.98×10^7
Sun	1.99×10^{30}	6.96×10^8

Some interesting facts & figures
• An Astronaut can be up to 2 inches taller returning from space.
• The cartilage disks in the spine expand in the absence of gravity.
• Due to gravitational effects, you weigh slightly less when the moon is directly overhead.
• Since 1959, more than 6,000 pieces of 'space junk' (abandoned rocket and satellite parts) have fallen out of orbit - many of these have hit the earth's surface.
• There is a high and low tide because of our moon and the Sun.
• The Earth's equatorial circumference (40,075 km) is greater than its polar circumference (40,008 km).
• The greatest tide change on earth occurs in the Bay of Fundy. The difference between low tide and high tide can be as great as 54 ft. 6 in. (16.6 meters).
• Skylab, the first American space station, fell to the earth in thousands of pieces in 1979. Thankfully most over the ocean.
• The Skylab astronauts grew 1.5 - 2.25 inches (3.8 - 5.7 centimeters) due to spinal lengthening and straightening as a result of zero gravity.
• The largest meteorite crater in the world is in Winslow, Arizona. It is 4,150 feet across and 150 fee deep.
• 1978 Pluto's moon, Charon, discovered.
• The size of Earth is roughly the geometric mean of the size of the universe and the size of an Atom.
• The mass of a human is roughly the geometric mean of the mass of Earth and the mass of the proton.
• February 1865 is the only month in recorded history not to have a full moon.

PROBLEMS

1. How far apart must two particles each of mass 1 kg be placed so that the force between them is 1 N. *[8.2x10^{-6} m]*

2. Determine the mass of the earth (M_e) given
   R_e= 6.38 X 10^6 m,   g = 9.8 m/s^2

3. Determine the force of attraction between the earth and the moon. *[1.98x10^{20} kg]*

4. Determine the value of 'g_m' on moon and show that it is about 1/6th of g on the surface of the earth.

5. Determine the <u>distance above the earth</u> surface at which the value of 'g' drops to 90% of its value at the surface of the earth. *[3.36x10^5 m/s]*

6. The mass and the radius of a planet are twice those of the Earth's radius and mass respectively. A person weighs 100 lb on Earth. How much will he weigh on the other planet?

7. Four identical masses, each 3.0 kg are placed at the corners of a square 10.0 m on a side. Determine the net force on any mass due to the remaining three masses.
   [1.15x10^{-11} N, 45°]

**8. A mass M is split into two particles, one of mass m and the other of mass M-m and the two particles are placed a distance r apart. What ratio M/m maximizes the gravitational force between the two particle.

9. Below are the masses of three particles and the coordinates at which they are placed:
m_1 = 3.0 kg at (0.0 m, 0.0 m);
m_2 = 5.0 kg at (4.0 m, 0.0 m) ;
m_3 = 7.0 kg at (4.0 m, 3.0 m)
Determine the net force on m_1 due to the other two particles. $[1.12x10^{-10} N, 17.5°]$

10. A neutron star has a radius of 10 km and mass same as the mass of the sun. Determine the value of 'g' on the surface of the neutron star. How long will it take for an object to fall a distance of 2.0 m (not taking into account the effect of the rotation of the neutron star).

Satellites in Circular Orbits

Gravitational pull is the centripetal force on the satellite:

$$G\frac{mM}{r^2} = \frac{mv^2}{r}$$

The speed of satellite in circular orbit about a body of mass M:

$$v = \sqrt{\frac{GM}{r}}$$

The time period (T) of the satellite is proportional to $r^{3/2}$ (Kepler's Third law)

$$T^2 = \frac{4\pi^2}{GM}r^3$$

PROBLEMS:

11. A satellite is placed into a circular orbit 1000 km above the surface of the earth (r = 1000 km + 6400 km = 7400 km) . Determine
     a. the time period (T) of the satellite $[6.33x10^3 s]$
     b. the speed (v) of the satellite $[7.34x10^3 m/s]$

12. For a geosynchronous communications satellite (time period 24 hours) determine,
     a. the distance from the surface of the earth
     b. the speed of the satellite

13. The moon's orbit around the earth is approximately a circle of radius 3.84 x 10⁸ m. Its time period is about 27.3 days. With this information determine the mass of the earth.

4

[6.03x10^{24} kg]

**14. The mean distance of the planet Mars from the sun is about 1.52 times that of the earth from the sun. Calculate the length of the year on Mars in terms of the length of the year on the earth.

15. Calculate the mass of the planet Jupiter from the following information:
Io, one of the moons of Jupiter has a circular orbit of radius
4.2 X10^8 m and its period is 42.5 hours.   *[1.87x10^{27} kg]*

**16. An object at rest on the surface of the earth completes 'one orbit' around the earth in 24 hours. But this object is not a satellite. What would be the time period of a hypothetical satellite at the level of the earth's surface?

**17. A satellite in a low orbit around the earth can lose energy due to a very thin atmosphere. As it loses energy it falls to a lower orbit. How does its speed change?
[Its speed increases.]

Multiple Choice:

1. A satellite A of mass of m and a satellite B of mass of 2m are both in the same circular orbit around the earth. Which of the following is true about these satellites?
    I. Both have same speed.
    II. Both have same acceleration
    III. The KE of B is twice that of A

(A) All three   (B) I and II only   (C) III only   (D) I and III only
(E) None of these

2. Time period of a satellite in circular orbit around a planet does not depends on
(A) its distance from the planet   (B) mass of the satellite
(C) gravitational field of the planet   (D) mass of the planet   (E) all of these

3. Which of the following is responsible for astronauts in orbit around the earth feeling weightless?
(A) Very weak or no gravity at that distance
(B) They are in vacuum.
(C) There is no normal force supporting them as they are in 'free fall'.
(D) The pull of the earth on them is cancelled by the pull of the moon.
(E) Inside of the spacecraft is made gravitation free.

4. Two particles each of mass 1 kg are placed 1 m apart. The force of attraction between them is
    (A) 1 N   (B) 9x10^9 N   (C) 6.67 x 10^{-11} N   (D) 18x10^9 N   (E)13.34x10^{-11} N

5. The gravitational force of attraction between two particles is F. If each of the masses and the distance between them is <u>halved</u> the new force between them would be
    (A) same   (B) ½ F   (C) 2F   (D) 1/8 F   (E) 1/16 F

6. The SI unit for the Universal Gravitational Constant G is
    (A) m/s^2 (B) N kg^2/m^2 (C) N m^2 /kg^2 (D) N m^2 kg^2 (E) N/m^2

7. The SI unit for the acceleration due to gravity g is
    (A) m/s^2 (B) N kg^2/m^2 (C) N m^2 /kg^2 (D) N m^2 kg^2 (E) N/m^2

Chapter 12
Work, Energy, and Power

WORK:

A force **F** is applied on a block of mass m. The displacement of the block is **x**. The angle between the force and the displacement is θ. The force is said to have done work on the block and this work is defined as follows:

Work (W) done by a force (F) is defined as
$$W = F \, x \, \cos\theta$$

where x is the displacement and θ is the angle between the force and the displacement.
- Work is a scalar quantity.
- Its SI unit is joule (J).
- Work depends on three quantities: F, x, and θ.
- Work can be positive, negative or zero.
- Work is zero if angle θ = 90°
- W = Fx if θ = 0°
- W = -Fx if θ = 180°

Energy: It is defined as *Capacity of doing work* or *Ability to do work.*

Law of Conservation of Energy: *Energy cannot be created nor can it be destroyed; it can change from one form to another.*

Power:

Power (P) is defined as 'rate of doing work' or as 'rate of emission of energy'.
If work done on a body is W in time t, the powered delivered in this process is

$$\text{Power} = \text{work/time} \rightarrow P = W / t$$

Also, $\qquad\qquad$ Power = force x velocity $\rightarrow$ P = F v
[Because P = W/t = Fx/t = F(x/t) = Fv]

where W is the amount of work done, t is the time in which the work is done, F is the applied force, and v is the velocity of the body on which the work is being done.

- Power is a scalar quantity.
- Its SI unit is watt (W); 1 W = 1 J/s.
- Another common unit of power in everyday use is horsepower (hp)
- **1 hp = 746 W**

Work done in gravitational field

If a force F lifts a body of weight mg through a vertical height of h, the <u>work done by the applied force</u> is, and

$$W \text{ (force)} = mgh$$

the <u>work done by gravity</u> is

$$W \text{ (gravity)} = -mgh.$$

Work done as area under F vs. x graph

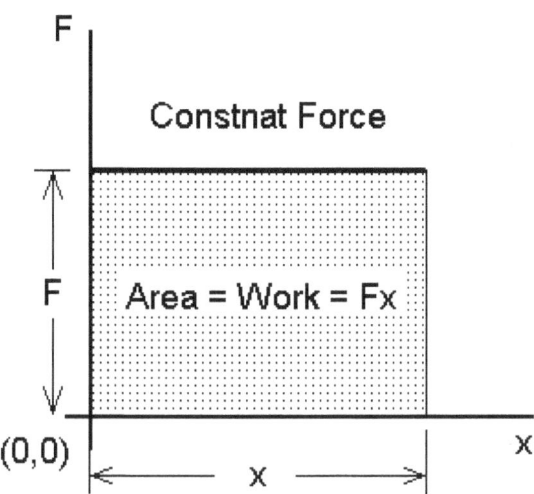

Constnat Force

Area = Work = Fx

(0,0)

Variable Force

Area = Work

Work done by elastic force of a spring

For a spring, $F = kx$. Thence F is variable. The work done in stretching or compressing a spring can be easily found from the area under the F vs. x graph.

In the case of force applied to stretch or compress a spring of elastic constant k, the force is not constant but proportional to the displacement x, and the work done by the applied force is

$$W = \tfrac{1}{2} \, k \, x^2$$

This equation assumes that initially the spring was relaxed.
If the extension or compressions was changed from x_1 to x_2 the work done during this change is

$$W = \tfrac{1}{2} \, k \, (x_2{}^2 - x_1{}^2)$$

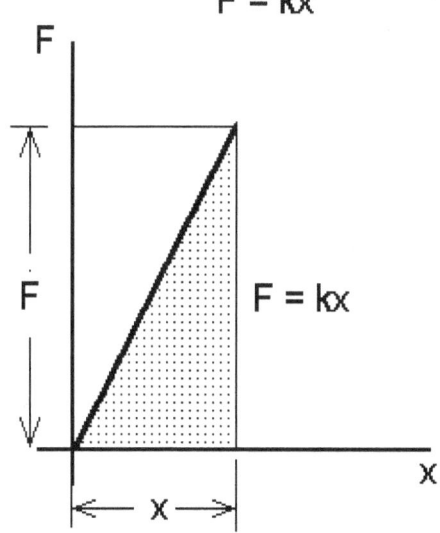

Spring
$F = kx$

$F = kx$

$$\text{area} = \frac{1}{2} F x = \frac{1}{2} (kx) \, x \; = \frac{1}{2} kx^2$$

$$\therefore W = PE \text{ (elastic)} = \frac{1}{2} kx^2$$

KINETIC ENERGY: Kinetic Energy is defined as the energy of motion. A body in motion is said to posses kinetic energy by virtue of its motion. The kinetic energy of a body of mass m traveling with a speed v is defined as $KE = \frac{1}{2} m v^2$

- KE is a scalar quantity
- The SI unit for KE energy is same as that for work, i.e., joule (J)

Some interesting facts & figures
• The most powerful laser in the world, the Nova laser at Lawrence Livermore National Laboratory, CA, USA, generates a pulse of energy equal to 100,000,000,000,000 watts of power for .000000001 second to a target the size of a grain of sand.
• Solar energy from100 miles by 100 miles in the Mojave Desert (USA) could replace all the fuel now burned to generate electricity in the entire U.S.

Problems on WORK:

1. A force of 20 N acts on a body and the displacement is 10 m. Determine the work done by the force in the following cases:

    a. $\theta = 0$ deg   *[200 J]*

    b. $\theta = 60$ deg   *[100 J]*

    c. $\theta = 90$ deg   *[0 J]*

    d. $\theta = 135$deg   *[-141.4 J]*

    e. $\theta = 180$ deg   *[-200 J]*

2. An object is moving with a KE of 20 J. Determine its KE if its velocity is changed from v to

    a. 2 v

    b. 0.5 v

    c. 10 v

    d. 0.1 v

Problems 3, 4, and 5 moved to the end of the chapter 13, as problems, 13, 14, and 15

6. A force of 2 N stretches a spring by 3 cm. Force is now increased slowly to stretch the spring by an additional 3 cm.
(a) Determine the new force acting on the spring.

(b) Determine the amount of work done in stretching this spring from 3 cm to 6 cm.

7. For a horizontal spring, an extension of 10 cm requires 20 J of work done on it. What will be additional stretch for an additional work of 20 J done on it?   *[0.0414 m]*

PROBLEMS ON POWER:

8. An engine is hauling a load of 20 N vertically up at constant speed of 2.0 m/s. Determine the power developed by the engine, in watts and in horsepower.

9. A motor can deliver a maximum of 10 hp. At what maximum speed can it lift a person of mass 60 kg?

   [12.7 m/s]

10. A car develops 150 hp to maintain a constant speed of 40 mph (17.8 m/s). Determine the net <u>resistive</u> force acting on the car.

Multiple Choice:

1.   For what value of angle θ between F and x is the work done zero?
(A) $0°$ (B) $30°$ (C) $90°$ (D) $180°$ (E) $45°$

2. Amount of work done to stretch a spring by x is W. If instead the spring were stretched by an amount 2x the amount of work done would be
(A) 2W   (B) 4W   (C) 8W   (D) 16W   (E) $\sqrt{2}$ W

3. The total work done by the centripetal force (F_c) on the particle undergoing UCM of radius r is
(A) $F_c\, 2\pi r$ (B) $F_c\, r$ (C) $F_c \pi r$ (D) $F_c \pi r^2$ (E) zero

4. Work done by kinetic friction on an object is
(A) always zero
(B) always positive
(C) always negative
(D) positive or negative
(E) either zero or positive

5. Work done by static friction on an object is
(A) always zero
(B) always positive
(C) always negative
(D) positive or negative
(E) either zero or positive

6-7 A spring of constant k is initially relaxed. It is stretched by an amount 2x.
   6. The final force acting on the spring is
(A) kx   (B) 2kx   (C) 4 kx   (D) 8 kx   (E) ½ kx

   7.   The work done from the extension x to 2x by the external force is
(A) ½ kx^2 (B) kx^2 (C) 1½ kx^2 (D) 2 kx^2 (E) 4 kx^2

Chapter 12
Work, Energy, and Power

WORK:

A force **F** is applied on a block of mass m. The displacement of the block is **x**. The angle between the force and the displacement is **θ**. The force is said to have done work on the block and this work is defined as follows:

Work (W) done by a force (F) is defined as
$$W = F \, x \cos \theta$$

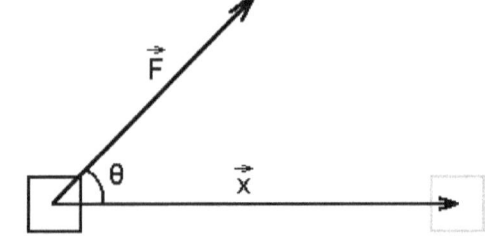

where x is the displacement and θ is the angle between the force and the displacement.

- Work is a scalar quantity.
- Its SI unit is joule (J).
- Work depends on three quantities: F, x, and θ.
- Work can be positive, negative or zero.
- Work is zero if angle $\theta = 90°$
- $W = Fx$ if $\theta = 0°$
- $W = -Fx$ if $\theta = 180°$

Energy: It is defined as *Capacity of doing work* or *Ability to do work*.

Law of Conservation of Energy: *Energy cannot be created nor can it be destroyed; it can change from one form to another.*

Power:

Power (P) is defined as 'rate of doing work' or as 'rate of emission of energy'.
If work done on a body is W in time t, the powered delivered in this process is

$$\text{Power} = \text{work/time} \rightarrow P = W / t$$

Also, $$\text{Power} = \text{force x velocity} \rightarrow P = F \, v$$
[Because P = W/t = Fx/t = F(x/t) = Fv]

where W is the amount of work done, t is the time in which the work is done, F is the applied force, and v is the velocity of the body on which the work is being done.

- Power is a scalar quantity.
- Its SI unit is watt (W); 1 W = 1 J/s.
- Another common unit of power in everyday use is horsepower (hp)
- **1 hp = 746 W**

Work done in gravitational field

If a force F lifts a body of weight mg through a vertical height of h, the <u>work done by the applied force</u> is, and

$$W \text{ (force)} = mgh$$

the <u>work done by gravity</u> is

$$W \text{ (gravity)} = -mgh.$$

Work done as area under F vs. x graph

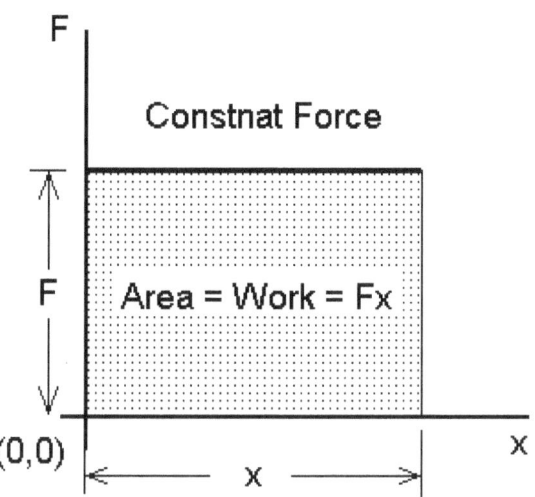

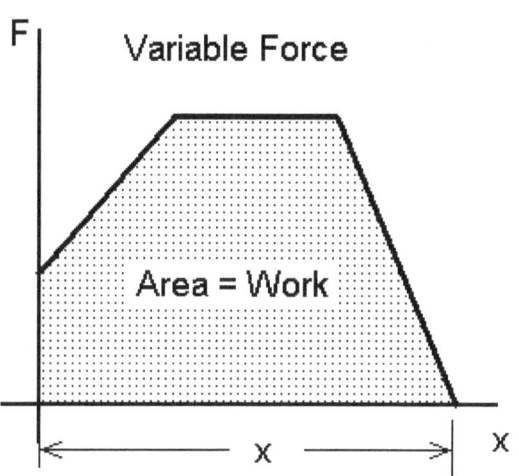

Work done by elastic force of a spring

For a spring, $F = kx$. Thence F is variable. The work done in stretching or compressing a spring can be easily found from the area under the F vs. x graph.

In the case of force applied to stretch or compress a spring of elastic constant k, the force is not constant but proportional to the displacement x, and the work done by the applied force is

$$W = \tfrac{1}{2} \, k \, x^2$$

This equation assumes that initially the spring was relaxed.

If the extension or compressions was changed from x_1 to x_2 the work done during this change is

$$W = \tfrac{1}{2} \, k \, (x_2^2 - x_1^2)$$

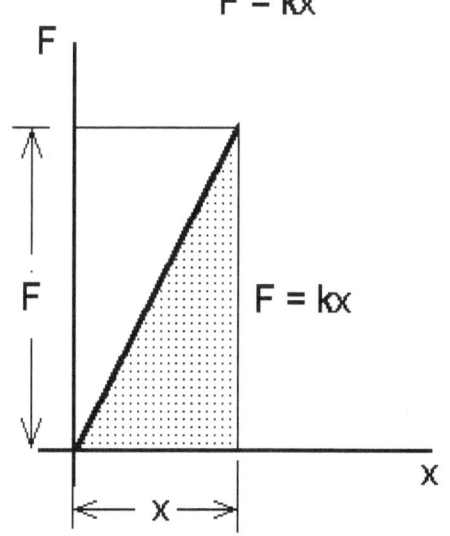

$$\text{area} = \frac{1}{2} F x = \frac{1}{2} (kx) \, x = \frac{1}{2} kx^2$$

$$\therefore W = PE \text{ (elastic)} = \frac{1}{2} kx^2$$

KINETIC ENERGY: Kinetic Energy is defined as the energy of motion. A body in motion is said to posses kinetic energy by virtue of its motion. The kinetic energy of a body of mass m traveling with a speed v is defined as $KE = \frac{1}{2} m v^2$

- KE is a scalar quantity
- The SI unit for KE energy is same as that for work, i.e., joule (J)

Some interesting facts & figures
• The most powerful laser in the world, the Nova laser at Lawrence Livermore National Laboratory, CA, USA, generates a pulse of energy equal to 100,000,000,000,000 watts of power for .000000001 second to a target the size of a grain of sand.
• Solar energy from100 miles by 100 miles in the Mojave Desert (USA) could replace all the fuel now burned to generate electricity in the entire U.S.

Problems on WORK:

1. A force of 20 N acts on a body and the displacement is 10 m. Determine the work done by the force in the following cases:

    a. $\theta = 0$ deg   *[200 J]*

    b. $\theta = 60$ deg   *[100 J]*

    c. $\theta = 90$ deg   *[0 J]*

    d. $\theta = 135$deg   *[-141.4 J]*

    e. $\theta = 180$ deg   *[-200 J]*

2. An object is moving with a KE of 20 J. Determine its KE if its velocity is changed from v to

    a. 2 v

    b. 0.5 v

    c. 10 v

    d. 0.1 v

Problems 3, 4, and 5 moved to the end of the chapter 13, as problems, 13, 14, and 15

6. A force of 2 N stretches a spring by 3 cm. Force is now increased slowly to stretch the spring by an additional 3 cm.
(a) Determine the new force acting on the spring.

(b) Determine the amount of work done in stretching this spring from 3 cm to 6 cm.

7. For a horizontal spring, an extension of 10 cm requires 20 J of work done on it. What will be additional stretch for an additional work of 20 J done on it?   *[0.0414 m]*

PROBLEMS ON POWER:

8. An engine is hauling a load of 20 N vertically up at constant speed of 2.0 m/s. Determine the power developed by the engine, in watts and in horsepower.

9. A motor can deliver a maximum of 10 hp. At what maximum speed can it lift a person of mass 60 kg?

     [12.7 m/s]

10. A car develops 150 hp to maintain a constant speed of 40 mph (17.8 m/s). Determine the net <u>resistive</u> force acting on the car.

Multiple Choice:

1. For what value of angle θ between F and x is the work done zero?
(A) $0°$ (B) $30°$ (C) $90°$ (D) $180°$ (E) $45°$

2. Amount of work done to stretch a spring by x is W. If instead the spring were stretched by an amount 2x the amount of work done would be
(A) 2W   (B) 4W   (C) 8W   (D) 16W   (E) $\sqrt{2}$ W

3. The total work done by the centripetal force (F_c) on the particle undergoing UCM of radius r is
(A) $F_c\, 2\pi r$ (B) $F_c\, r$ (C) $F_c \pi r$ (D) $F_c \pi r^2$ (E) zero

4. Work done by kinetic friction on an object is
(A) always zero
(B) always positive
(C) always negative
(D) positive or negative
(E) either zero or positive

5. Work done by static friction on an object is
(A) always zero
(B) always positive
(C) always negative
(D) positive or negative
(E) either zero or positive

6-7 A spring of constant k is initially relaxed. It is stretched by an amount 2x.
   6. The final force acting on the spring is
(A) kx   (B) 2kx   (C) 4 kx   (D) 8 kx   (E) ½ kx

   7. The work done from the extension x to 2x by the external force is
(A) $\frac{1}{2} kx^2$ (B) kx^2 (C) $1\frac{1}{2}\ kx^2$ (D) $2\ kx^2$ (E) $4\ kx^2$

Chapter 13
Work-Energy Theorem
Conservation of Mechanical Energy

Kinetic Energy: For a particle of mass m moving with a velocity v, its kinetic energy, KE, is given by

$$KE = \frac{1}{2}mv^2$$

Work-Energy Theorem: The total work done on a body equals the change in its kinetic energy. Thus,

$$W_{total} = \Delta KE = KE_2 - KE_1$$

[Derivation: W = F.x = ma.x = ½ m(2ax)
= ½ m ($v_2^2 - v_1^2$) = ½ mv_2^2 – ½ mv_1^2 = $KE_2 - KE_1$]

The term W_{total} includes work done by *conservative forces* (such as elastic, gravitational, electric and magnetic) and *non-conservative* forces (such as friction, viscous drag, and air resistance etc).

Potential Energy (PE): The potential energy of two or more particles is due to the positions or configuration of particles. The potential energy exists only in the fields of conservative forces, for example, gravitational, elastic, and electric forces.

The PE for a particle at a point is defined as the negative of the work done by the conservative field in bringing the particle from infinity to that point.

The potential energy change ΔPE when a particle moves from one point to another due to the action of a conservative force is thus equal to the negative of the work done by the conservative force (W_{cons}). Thus,

$$\Delta PE = - W_{cons}$$

The equation for the work-energy theorem can thus be modified to include potential energy as below:

Thus,
$W_{total} = KE_2 - KE_1$
$W_{cons. forces} + W_{frictional forces} = KE_2 - KE_1$
$-(PE_2 - PE_1) + W_{frictional forces} = KE_2 - KE_1$
$KE_1 + PE_1 + W_{frictional forces} = KE_2 + PE_2$

This equation can be written in a more usable form as below:

$$KE_1 + PE_1 = KE_2 + PE_2 + \left| W_{frictional forces} \right|$$

This equation states that the sum KE and PE of a particle at one point or at some instant equals the sum of KE and PE at another point or at a later instant PLUS energy lost to friction. The absolute value for frictional work prevents the mistake of putting a negative sign for the frictional work on the right hand side.

Gravitational and Elastic Potential Energies:

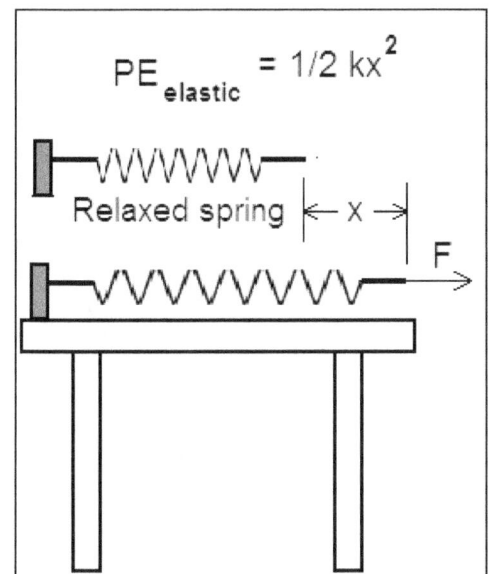

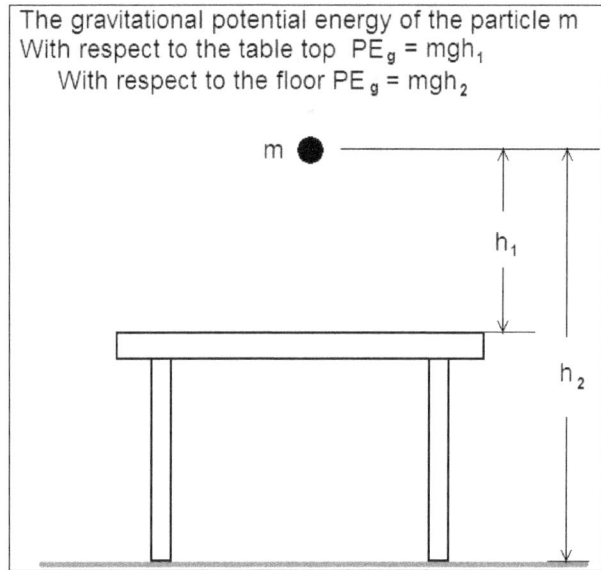

The gravitational potential energy of the particle m
With respect to the table top $PE_g = mgh_1$
With respect to the floor $PE_g = mgh_2$

Gravitational potential energy for <u>small heights</u> h <<R_E (earth radius): Gravity is approximately uniform for small heights (even several kilometers). PE_g can be considered as 0 for any horizontal level (say floor, ground, or table top). For a particle m at a height h above this surface PE_g is given by

$$PE_g = mgh$$

Elastic potential energy: If a spring of elastic constant, k, is stretched (or compressed) by an amount x from over its relaxed length, the elastic potential energy stoed in the spring is given by

$$PE_{elastic} = \tfrac{1}{2} kx^2$$

Gravitational potential energy over <u>large distances</u>: If a body of mass m is at a distance r from the center of the earth (outside of the earth) its PE_g is given by

$$PE_g = -\frac{GmM_E}{r}$$

Escape Velocity V_e: If an object is launched away from the surface of the earth with a kinetic energy $\tfrac{1}{2} mv_e^2 = GmM_E/r$ or greater, it will escape from the earth. The escape velocity from the earth surface is about 11,000m/s or 11 km/s or 6.7 miles/s or, 25,000 miles/hr.

Mechanical Energy (ME):
For a system of particles, the sum of its **kinetic energy** and **potential energy** is called its mechanical energy.

Thus, **ME = KE + PE**.

Principle of Conservation of Mechanical Energy:

If frictional forces are absent or negligible (i.e. $W_{frictional\ forces} = 0$), the ME of a system is conserved. Thus

$$KE_1 + PE_1 = KE_2 + PE_2$$

Thus if frictional losses are negligible or zero, any <u>gain/loss</u> of KE will result in the equal amount of <u>loss/gain</u> of PE, leaving the total ME constant.

Examples for conservation of Mechanical Energy

Roller-coaster (Ignoring friction):

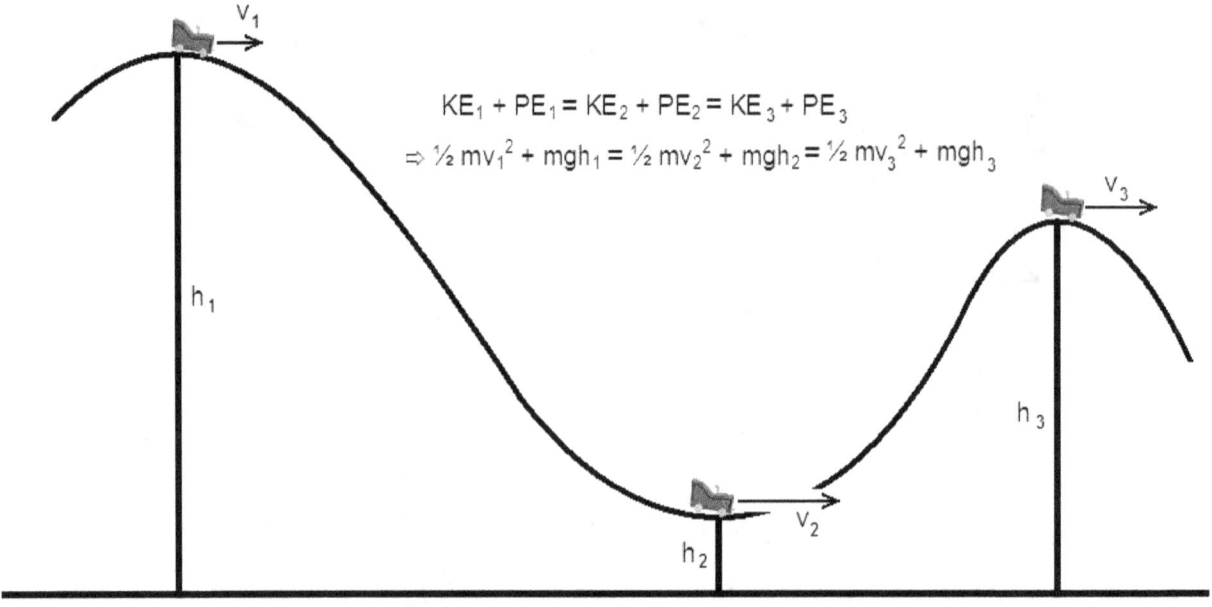

$$KE_1 + PE_1 = KE_2 + PE_2 = KE_3 + PE_3$$
$$\Rightarrow \tfrac{1}{2} mv_1^2 + mgh_1 = \tfrac{1}{2} mv_2^2 + mgh_2 = \tfrac{1}{2} mv_3^2 + mgh_3$$

Maximum compression of the spring after colliding with a block:

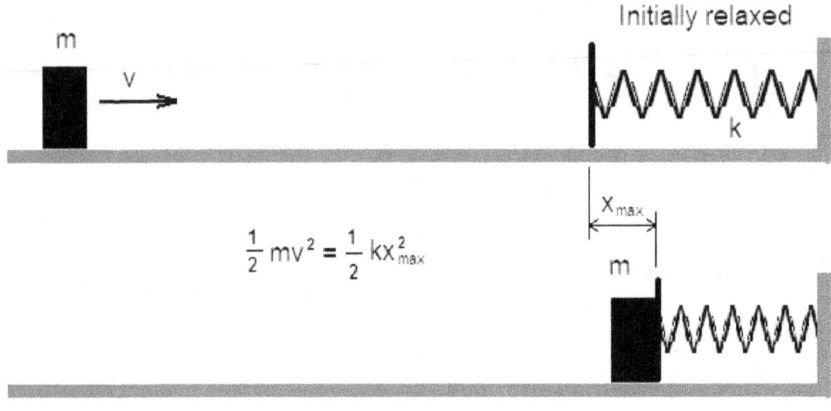

$$\frac{1}{2} mv^2 = \frac{1}{2} kx_{max}^2$$

Simple Pendulum:

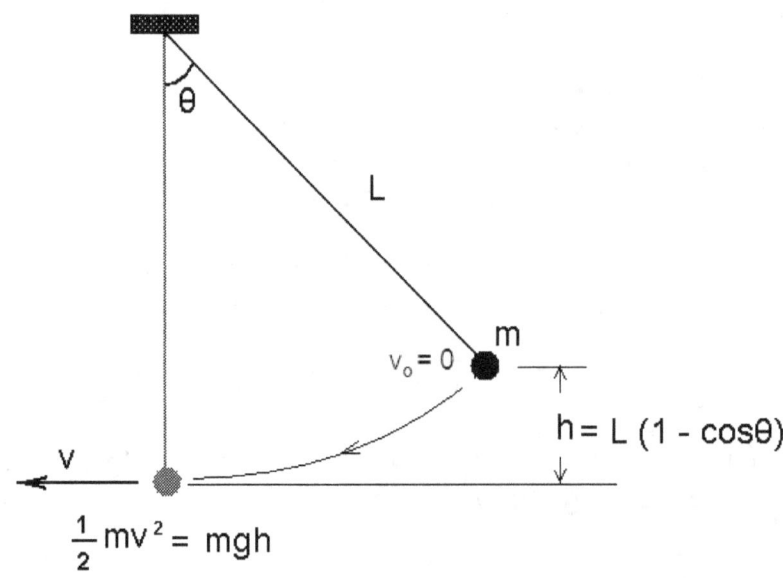

$$h = L\,(1 - \cos\theta)$$

$$\frac{1}{2} mv^2 = mgh$$

Determination of work from **F vs. x Graph:**

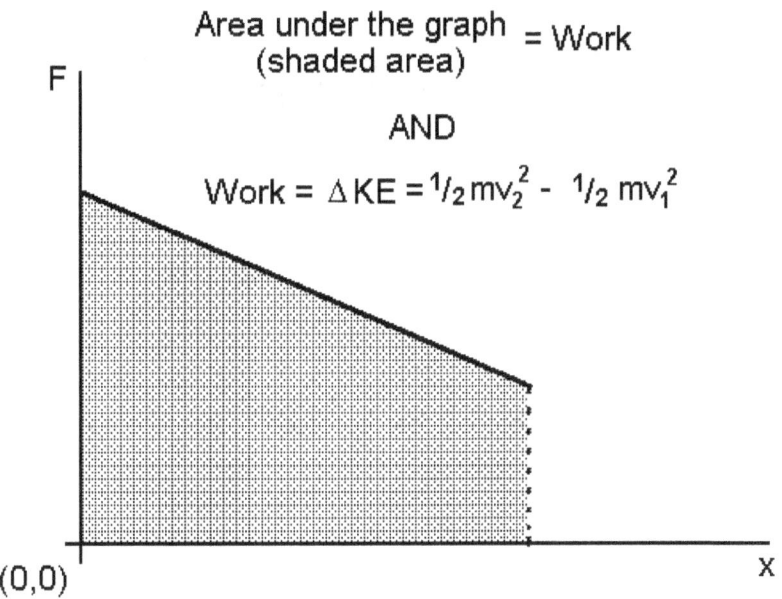

Area under the graph = Work
(shaded area)

AND

$$\text{Work} = \Delta KE = \tfrac{1}{2}mv_2^2 - \tfrac{1}{2}mv_1^2$$

F vs x graph for force F acting on a body of mass m
and causing displacement of x

Some interesting facts & figures
• **The United States consumes 25% of all the world's energy.**
• **There is enough fuel in a full tank of a Jumbo Jet to drive an average car four times around the world.**
• **Ten minutes of one hurricane contains enough energy to match the nuclear stockpiles of the world.**
• **A car traveling at 80 km/h (50mph) uses half its fuel to overcome wind resistance.**

Problems:

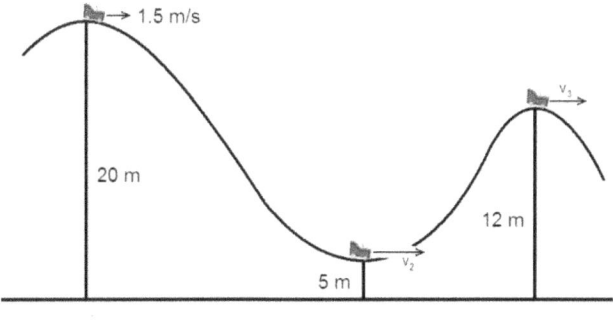

1. A roller coaster has a speed of 1.5 m/s at the highest point. It moves down to the lowest point 2 and rises to another peak at point 3. The points 1, 2, and 3 are at heights of 20 m, 5m, and 12 m respectively. Determine the speeds of the roller coaster at points 2 and 3 if frictional losses are ignored. [v_2=17.2 m/s, v_3=12.6 m/s]

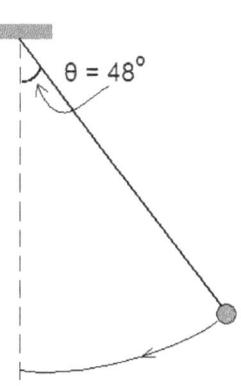

2. A simple pendulum has a length of 1.8 m. The bob of mass 0.25 kg is moved from the mean position such that the string makes an angle of 48° with the vertical. If now the bob is released from rest, determine

   (a). the speed of the bob as it passes the lowest point

   (b). the tension in the string as the bob passes the lowest point.

3. A block of mass 0.8 kg slides along a horizontal frictionless surface with a speed of 4.0 m/s. It then collides with a horizontal spring of elastic constant k = 50 N/m.

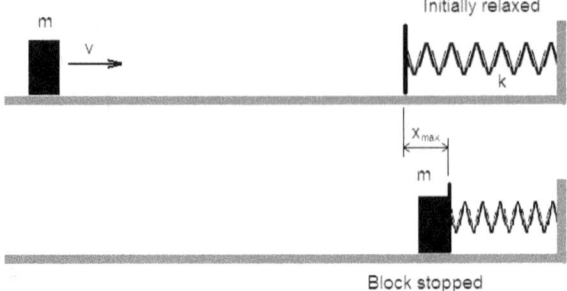

Initially relaxed

Block stopped

(a). Determine the maximum compression of the spring.
[0.506 m]

(b). Determine the compression of the spring when it slowed down the block to a speed of 2.0 m/s.
[0.438 m]

4. A block of mass 0.2 kg, starting from rest, slides down a frictionless track from a height of 2.5 m. The track, at the bottom, curves and becomes horizontal. The block travels along the horizontal part of the track and collides with a horizontal spring of elastic constant 60N/m.
(a). Determine the speed of the block as it reaches the horizontal surface

(b). Determine the maximum compression of the spring.

5. In the problem #4 if the spring is removed and the horizontal part of the track is rough with μ_k = 0.3, determine the distance traveled by the block along the horizontal surface before it comes to rest.
[8.33 m]

**6.

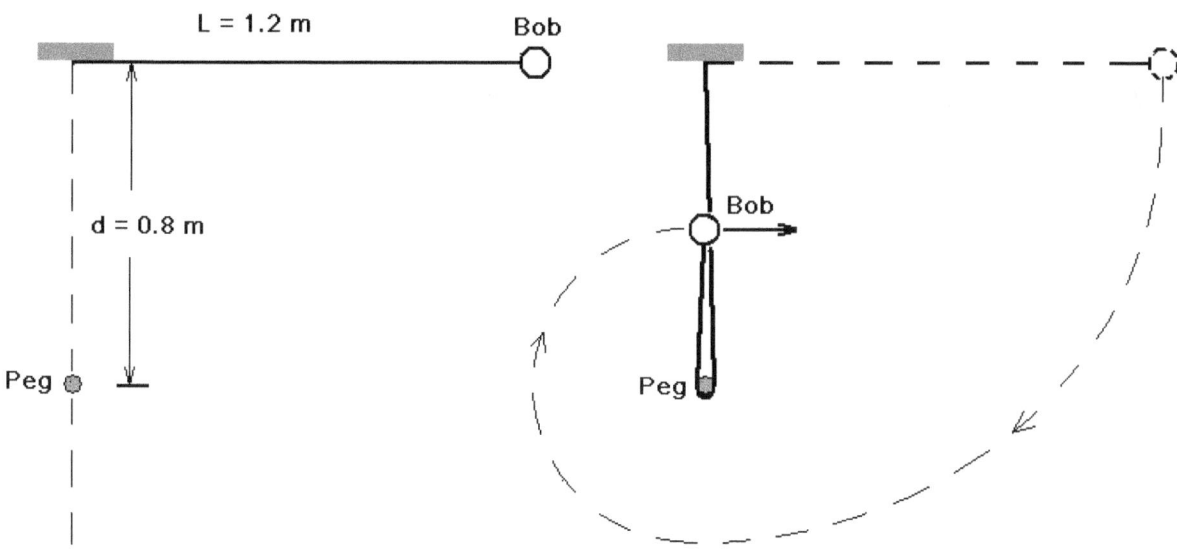

A simple pendulum of length L = 1.2 m is released from a horizontal position. A horizontal peg is placed at d = 0.8 m vertically below the point of suspension
    (a). Determine the speed of the bob at its lowest position.

    (b). Determine the speed of the bob at the highest point after the string catches on the peg.

    (c). Show that if the ball is to swing completely around the peg, then d > 3L/5.

**7.   A spring is cut in half across its length. Determine the elastic constant (k') of each half if that of the whole spring was k.   *[k' = 2k]*

**8. Determine the elastic constant of two springs of equal lengths and constants k_1 and k_2
    (a). if they are connected in parallel

    (b). if they are connected in series

9. A block of mass 2 kg, starting from rest, slides down from the top of a 30-cm radius semicircular track (concave side up, symmetric about vertical axis).
    (a). Assume the track to be frictionless. Determine the speed of the block after it passes the lowest position and reaches a position, which makes an angle of 60 degrees with the vertical.   *[1.71 m]*

    (b). If the surface is instead rough so that the block comes to rest after losing 15% of its mechanical energy to friction, determine the angular position at which the block comes to rest.   *[81.4°]*

10. A simple pendulum of length L has a bob of mass m. It is initially held in the horizontal position and then released.

(a) Show that the velocity of the bob as it passes through the lowest position is

$$v = \sqrt{2gL}$$

(b) Show that the tension in the string as the bob passes through the lowest position is $T = 3mg$

11. LOOPING THE LOOP: A track starting from certain height curves into a circle of radius r at the end. The whole track lies in a vertical plane and is frictionless. A small block slides down the track and follows the circular loop at the end.

(a) What is the minimum velocity v_3 of the block at the point 3 if the block is not to loose contact with the track?   [√(gr)]

(b) What is the minimum height from which the block must be released so that it does not leave contact with the track at the top of the loop?
[2.5r]

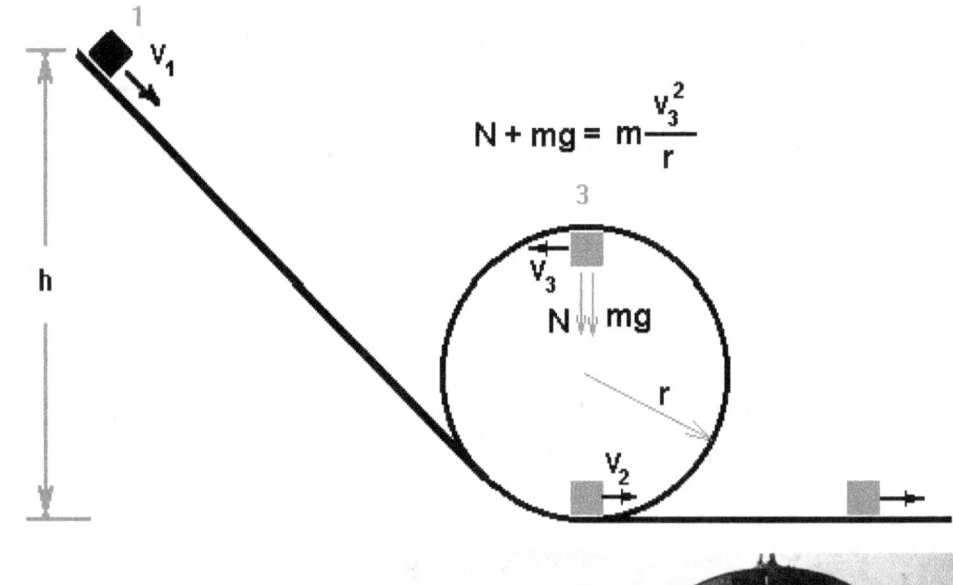

**12. As shown in the figure above, two small particles m_1 and m_2 ach of mass m are connected by thin rigid straight rods L_1 and L_2 each of length L. The system is free to rotate about a pivot P in a vertical plane. The system is held horizontal and released. At the lowest position, m_2 has twice the speed of m_1. Give your answers in terms of m, g, and L.

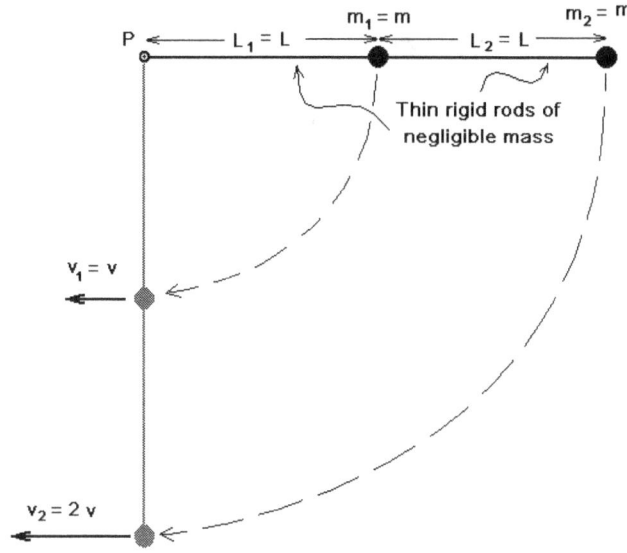

(a) Calculate the speed of each ball at the lowest position.

(b) Calculate the tension in each rod at the lowest position.

13. An object is released from a height of 20 m and falls vertically down to the ground. Determine the velocity of the object just before it hits the ground
      a. using the equations of motion for uniform acceleration.   *[19.8 m/s]*

      b. using the relationship between work and kinetic energy.   *[19.8 m/s]*

14. A particle of mass 3 kg is initially at rest at the origin (0,0). It is acted upon by a Force along x-axis, as shown in the diagram below. Use work-energy theorem to determine the velocity of the particle at x = 5 m.

15. Do problem 14 for two more cases for which the F vs. x graphs are shown below   *[4.83 m/s, 3.65 m/s]*

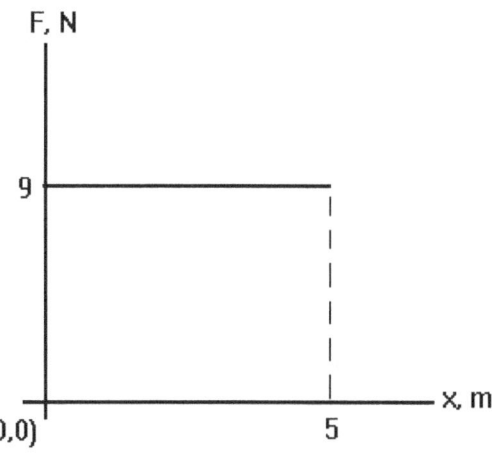

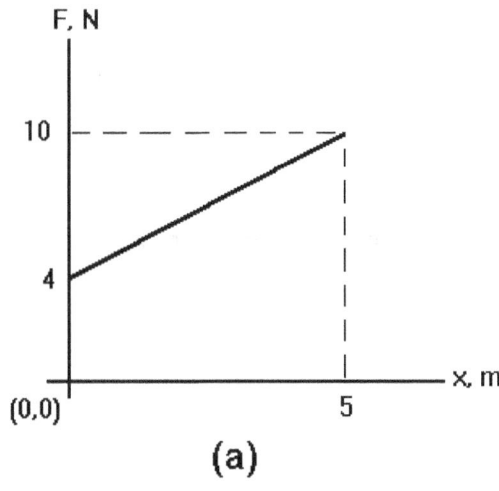

(a)

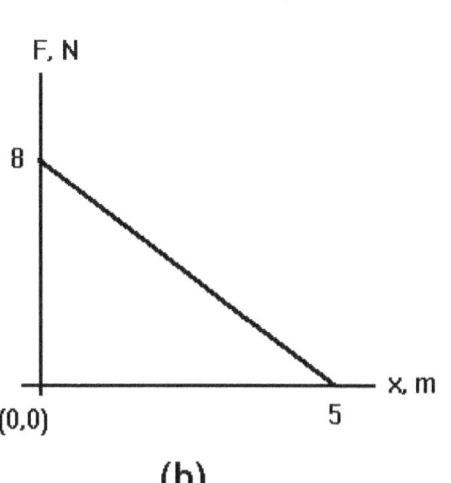

(b)

Multiple Choice:

1. Consider a football kicked from a level ground. The ball reaches the maximum height at B and returns to the ground at C.

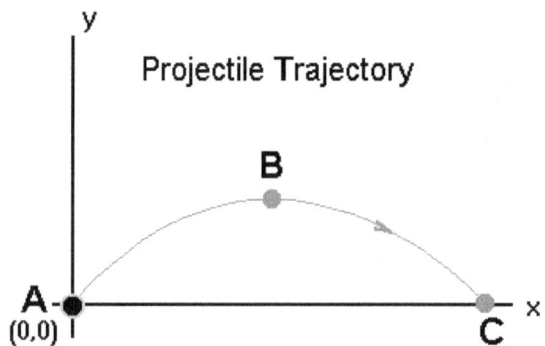

(i) At which point(s) does the ball have maximum gravitational potential energy?
(A) at A and C   (B) at A only   (C) at C only   (D) at B only
(E) The gravitational potential energy is constant over the entire trajectory.

(ii) At which point(s) does the ball have maximum kinetic energy?
(A) at A and C   (B) at A only   (C) at C only   (D) at B only
(E) The kinetic energy is constant over the entire trajectory.

(iii) At which point(s) does the ball have minimum gravitational potential energy?
(A) at A and C   (B) at A only   (C) at C only   (D) at B only
(E) The gravitational potential energy is constant over the entire trajectory.

(iv) At which point(s) does the ball have minimum kinetic energy?
(A) at A and C   (B) at A only   (C) at C only   (D) at B only
(E) The kinetic energy is constant over the entire trajectory.

(v) At which point(s) does the ball have zero kinetic energy?
(A) at A and C   (B) at A only   (C) at C only   (D) at B only
(E) at none of these

(vi) At which point(s) does the ball have maximum total mechanical energy?
(A) at A and C   (B) at A only   (C) at C only   (D) at B only
(E) The mechanical energy is constant over the entire trajectory.

2. A particle has twice the mass and half the speed of a second particle. What is the ratio of the KE of the first particle to KE of the second particle?
(A) 1   (B) 2   (C) ½   (D) ¼   (E) 4

3. A force stretches a spring from its relaxed length by an amount x. If the amount of elastic potential energy stored in the spring is U the elastic constant of the spring must be
(A) ½ Ux^2 (B) ½ x^2/U   (C) 2U/x^2 (D) ½ U/x^2 (E) 2x^2/U

4. Which of the following is a non-conservative force?
(A) Elastic force   (B) Air resistance   (C) Gravity   (D) Electric Force   (E) None of these

Chapter 14
Impulse (I), Linear Momentum (P), Conservation of Linear Momentum

Linear Momentum (p): It is the product of mass and velocity of a body. It is a vector quantity.

$$\mathbf{p} = m\mathbf{v} \quad \text{(SI unit: kg.m/s)}$$

Impulse: If a force F acts on a body for a time interval Δt an impulse (I) is said to have been applied on the body. This impulse is given by

$$I = F\,\Delta t \quad \text{(SI unit: Ns)}$$

Impulse from the F vs. t graph: If a graph of Force (variable or constant) vs. time is given the impulse is the area under the curve.

Impulse = Area under the F vs. t graph

<u>Impulse-Momentum theorem:</u> An impulse (I) applied on a body of mass **m** brings about a change in momentum (ΔP) of the body and is equal to it, i.e.

$$I = \Delta P$$
$$I = mv_2 - mv_1$$

Law of Conservation of Linear Momentum: In the absence of an external force the total linear momentum of a particle or a system of particles is conserved, i.e.

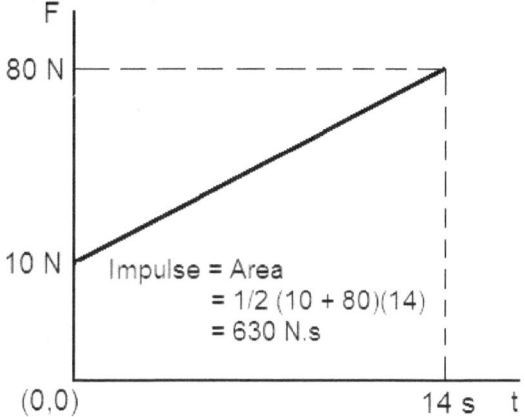

F

80 N

10 N — Impulse = Area
= 1/2 (10 + 80)(14)
= 630 N.s

(0,0)   14 s   t

Total initial momentum = Total final momentum
$$\Sigma p_i = \Sigma p_f$$

Collision between two bodies: For all types of collision (elastic, inelastic, and perfectly inelastic collisions) between two bodies, the total linear momentum of the two bodies before and after collision is the same.

In one dimension:

$$m_1 v_1 + m_2 v_2 = m_1 v_1' + m_2 v_2'$$

where, v_1, and v_2 are the velocities before collision and v_1' and v_2' are the velocities after collision between the two particles with masses m_1 and m_2 respectively.

In two dimension (xy-plane):
$\Sigma P_{ix} = \Sigma P_{fx}$ and $\Sigma P_{iy} = \Sigma P_{fy}$
i.e.,

$$m_1 v_{1x} + m_2 v_{2x} = m_1 v_{1x}' + m_2 v_{2x}'$$
$$m_1 v_{1y} + m_2 v_{2y} = m_1 v_{1y}' + m_2 v_{2y}'$$

Example: Below, m_1 is traveling in x direction with velocity v_1 (with components $v_{1x} = v_1$ and $v_{1y} = 0$) and m_2 is stationary ($v_2 = v_{2x} = v_{2y} = 0$). After the collision, their final velocities are v_1' (v_{1x}' and v_{1y}' as components) and v_2' (v_{2x}' and v_{2y}' as components)

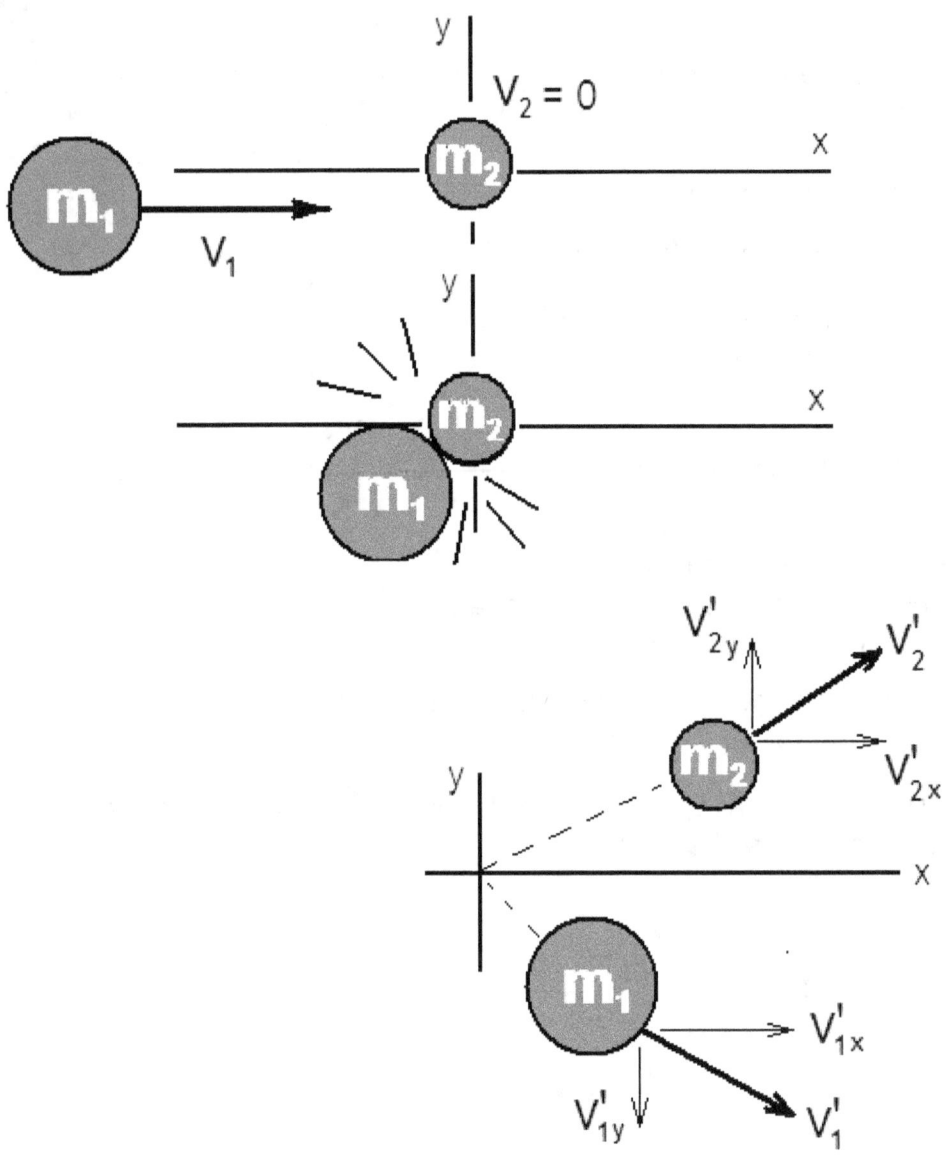

Here' $m_1 v_{1x} + m_2 v_{2x} = m_1 v_{1x}' + m_2 v_{2x}'$ and $m_1 v_{1y} + m_2 v_{2y} = -m_1 v_{1y}' + m_2 v_{2y}'$
$\Rightarrow m_1 v_1 + 0 = m_2 v_{1x}' + m_2 v_{2x}'$ and $0 + 0 = - m_1 v_{1y}' + m_2 v_{2y}'$

Three Types of Collisions: The collisions between two particles can be divided into three categories:

Elastic Collision: Here KE is conserved. $\Sigma KE_i = \Sigma KE_f$

Inelastic Collision: Here the KE is not conserved: $\Sigma KE_i \neq \Sigma KE_f$

Completely Inelastic Collision: Here the bodies stick together after the collision. Hence, $m_1 v_1 + m_2 v_2 = (m_1 + m_2)V$
where $V = v_1' = v_2'$

Loss or gain of KE in inelastic and perfectly inelastic collision:
The change in the KE due to the collision is given by $\Delta KE = KE_f - KE_i$
Loss if ΔKE is negative and gain if ΔKE is positive.

At Large Hadron Collider (LHC), above right, CERN, Switzerland, scientists hope to detect the particle Higgs Boson. This particle explains why something has mass. The diagram above left is simulation of proton-proton completely inelastic collision resulting in the production and then decay of Higgs Boson.

Some interesting facts & figures
• The 21-year-old Andy Roddick playing in 2004 Davis Cup, broke his own serve speed record of 152 mph by a 155-miles-per-hour serve.
• On March 30, 2010, Large Hadron Collider achieved a landmark by colliding two protons with an energy output 3 times greater than the largest collision energy before.

PROBLEMS:

1. A ball of mass 350 grams moving at 20 m/s is struck by a bat to move in the opposite direction at 20 m/s.
(a) Determine the change in the linear momentum of the ball.   *[14 kg m/s]*

(b) Determine the impulse applied by the bat on the ball.   *[14 kg m/s]*

(c) Determine the impulse applied by the ball on the bat.   *[-14 kg m/s]*

(d) If the collision lasted for 0.15 s, determine the average force applied by the bat on the ball during the collision.   *[93.3 N]*

2. A bullet of mass 35 g travelling at 250 m/s is fired horizontally into a 5.0 kg wooden block at rest on a frictionless horizontal surface. The bullet stays lodged in the block.
a. Determine the speed of the bullet+block system after the collision

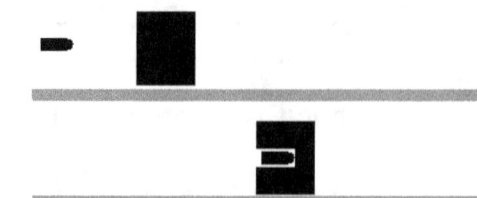

b. Determine the change in the KE of the system due to collision.

3. The block+bullet system of the problem 2 then strikes a horizontal spring of spring constant k = 200 N/m. The spring is initially relaxed. Determine the maximum compression of the spring.
[0.28 m]

4. If in problem 2 the surface is rough and the combined system comes to rest after travelling a distance of 8.0 m, determine the coefficient of kinetic friction for the surface.

5. A block of mass 4.0 kg is suspended by a string of length 2.2 m. A bullet of mass 50 g is fired horizontally into the block. The bullet traveling at 400 m/s enters the block and comes to rest inside it in a short time.
a. Determine the velocity of the block+bullet system immediately after the collision. [4.94 m/s]

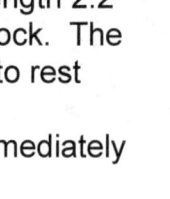

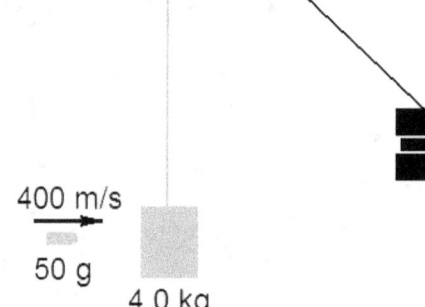

b. Determine the angle, θ, made by the string with the vertical as the block swings up and reaches the highest position. [64.4°]

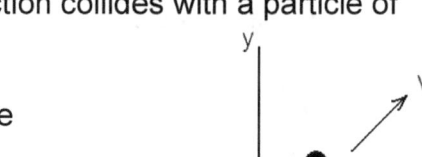

6. A particle of mass m_1 = 8 kg travelling at 10 m/s in the +y direction collides with a particle of mass m_2 = 6 kg travelling at 20 m/s in the +x direction. After the collision the particles stick together and move together.
a. Determine the magnitude and the direction of the velocity of the combined particles after the collision.

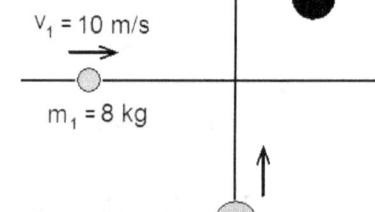

b. Determine the loss in the KE due to the collision.

7. A bullet of mass 25 g is fired from a rifle of mass 4.0 kg, with a velocity of 150 m/s. Determine the recoil velocity of the rifle.   *[-0.94 m/s]*

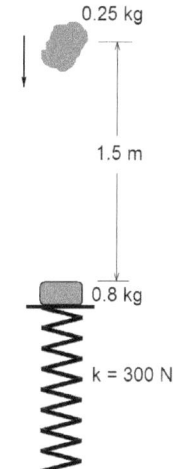

**8. A putty of mass 0.25 kg is dropped from a height of 1.5 m on to another mass of 0.8 kg at the top of a vertical spring (negligible mass) of elastic constant 300 N/m. Determine approximately, the maximum compression of the spring (i.e. neglect the small height through which the spring is compressed). *Note: The collision is perfectly inelastic*

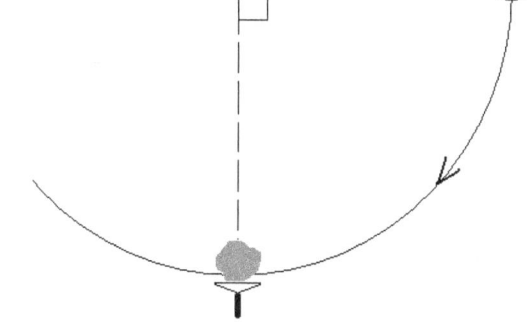

9. The bob of a 1.5-m long simple pendulum has a mass of 0.2 kg. The pendulum is moved to a horizontal position and released. The bob swings down and collides with a stationary lump of putty of mass 1.2 kg when passing the mean position. The collision is perfectly inelastic. The bob swings to the other side carrying with it the putty. Determine the maximum angle to which the pendulum swings after the collision.
          [11.7°]

10. Jim of mass 70 kg is roller-skating at a speed of 1.8 m/s along a straight sidewalk. He sees his little brother Tom of mass 25 kg standing in his way. Jim while passing near his brother picks him up. Determine their speed as they move on.

11. A 7-kg block is resting on a frictionless table. A 20-g bullet of velocity 1200 m/s is fired horizontally into the block. The bullet enters the block and leaves horizontally at a reduced speed of 900 m/s.
(a) Calculate the speed of the block after the collision.
          [0.857 m/s]

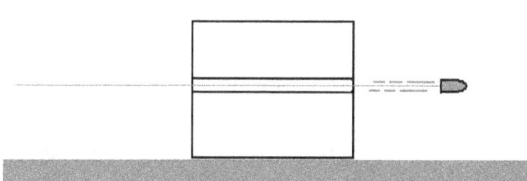

(b) Calculate the gain or loss of the KE of the bullet+block system due to the collision.
          [- 6300 J (loss)]

12. Two blocks $m_1 = 3$ kg and $m_2 = 5$ kg are pushed against each other with a light spring in between on a frictionless table, as shown in the figure below. The spring is compressed by 0.09 m. The blocks are then released. The block m_1 recoils with a speed of $v_1' = 10$ m/s.
(a) Calculate the magnitude and direction of the velocity v_2' of m_2.

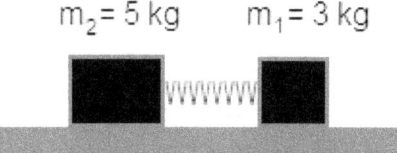

(b) Calculate the total kinetic energy of the two blocks after they are released.

(c) Calculate the elastic constant k of the spring.

13. Below is the F vs. t graph for a variable force F acting on an 8-kg block for time 30 seconds. The block is initially at rest.

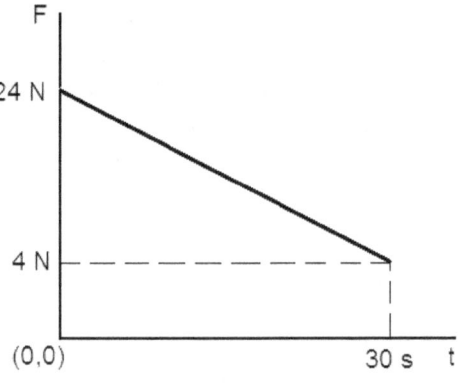

(a) Calculate the initial and final acceleration of the block.
[3 m/s² ; 0.5 m/s²]

(b) Calculate the impulse imparted to the block over 30 seconds.
[420 N.s]

(c) Calculate the speed of the block at t = 30 s.
[52.5 m/s]

Multiple Choice:

1. A particle of mass 5 kg moving at 7 m/s collides with another particle of mass 7 kg moving at 5 m/s in opposite direction. If the collision is perfectly inelastic the velocity of the combined mass after collision is
(A) 0m/s   (B) 1 m/s   (C) 2 m/s   (D) 6 m/s   (E) 12 m/s

2-4: For a collision between a number of particles, in the absence a net external force:

2. The linear momentum is conserved in

(A) Elastic collision only   (B) Inelastic collision only
(C) Perfectly inelastic collision only (D) In all the collisions
(E) In none of the collisions

3. The kinetic energy is conserved in

(A) Elastic collision only   (B) Inelastic collision only
(C) Perfectly inelastic collision only (D) In all the collisions
(E) In none of the collisions

4. Energy is conserved in

(A) Elastic collision only   (B) Inelastic collision only
(C) Perfectly inelastic collision only (D) In all the collisions
(E) In none of the collisions

Chapter 15
Torque and Mechanical Equilibrium

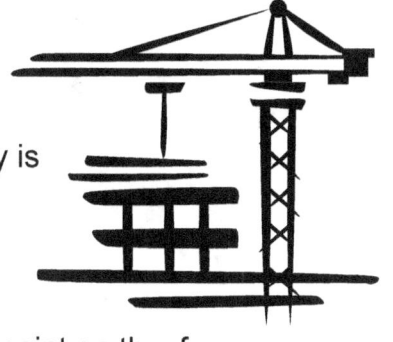

Torque is a physical quantity responsible to create angular acceleration of a body.

Torque produced by a force F applied on a body is given by

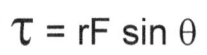

$$\tau = rF \sin \theta$$

F is the force acting on the body
r is the length of the position vector from O to any point on the force

θ is the angle between **r** and **F**

Torque plays the same role in rotational motion as force plays in translational motion.

Here, the treatment for torque will be confined to cases where the vector **r** and all the forces acting on the body lie in one plane and the body is free to rotate in the same plane. Thus the rotation of the body can be either **clockwise** or **counterclockwise** in the plane of the diagram.

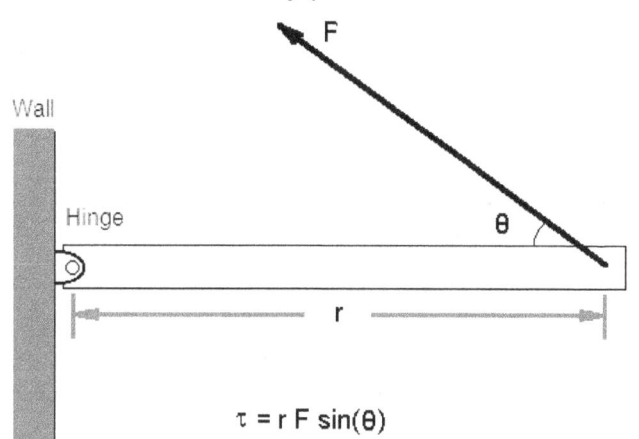

$\tau = r\,F\,\sin(\theta)$

Each force applied on a body produces its own torque on the body. The net torque acting on the body is the sum of all the torques. The sum is carried out taking into account the sign convention stated below.

Sign Convention for the Torques:

The torque, which rotates or <u>tends</u> to rotate the body clockwise, is
negative and the torque, which rotates or <u>tends</u> to rotate the body counterclockwise, is positive.

Torque is zero if
- F = 0
- r = 0, i.e. the force is passing through the point about which torque is taken
- θ = 0° or 180°

Gravitational torque, that is, the torque due to the weight of the object

The torque due to the weight of an object can be found in the following way:

Assume the weight of the object to be concentrated at its center of gravity. For a symmetric and uniform object, the center of gravity is its geometric center. Regard the weight as the force and calculate the torque caused by it by the equation

$$\tau = r\,(mg)\,\sin \theta$$

Example – Determining gravitational torque:

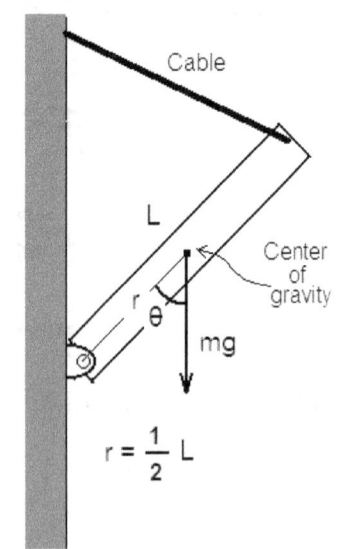

Cable

L

Center of gravity

mg

$r = \frac{1}{2} L$

$\tau = r F \sin\theta$

Gravitational Torque $= \tau = -(\frac{1}{2} L)(mg)\sin\theta$

Mechanical Equilibrium (No rotation No translation!)

Any motion of a rigid body is a combination of **rotational** and **translational** motion.
A rigid body is said to be in mechanical equilibrium if it is in translational and rotational equilibrium

The conditions for translational equilibrium are: $\Sigma F_x = 0$ and $\Sigma F_y = 0,$

The condition for the rotational equilibrium is: About any point

$$\Sigma\tau = 0$$

or

$$\left| \Sigma\tau_{(clockwise)} \right| = \left| \Sigma\tau_{(counterclockwise)} \right|$$

Reaction Force (R) on beam from the hinge:

Since the magnitude and direction of the reaction force R applied by the hinge is unknown <u>assume</u> it has positive components R_x and R_y. Use these as unknowns in the equations below and solve
$$\Sigma F_x = 0 \text{ and } \Sigma F_y = 0.$$
If the directions of one or both of R_x and R_y is wrong the answer will b negative but the magnitude will be correct. The magnitude of R can be found from the equation

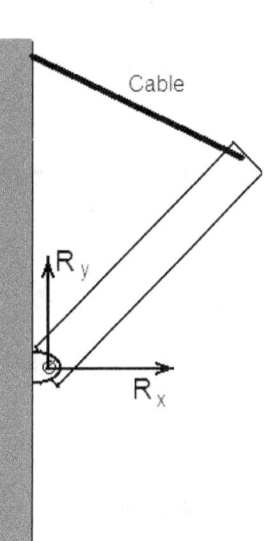

Cable

R_y

R_x

$$R = \sqrt{R_x^2 + R_y^2}$$

and the direction can be found from the correct signs of R_x and R_y and using the equation

$$\theta = \tan^{-1}\left(\frac{R_y}{R_x}\right)$$

PROBLEMS:

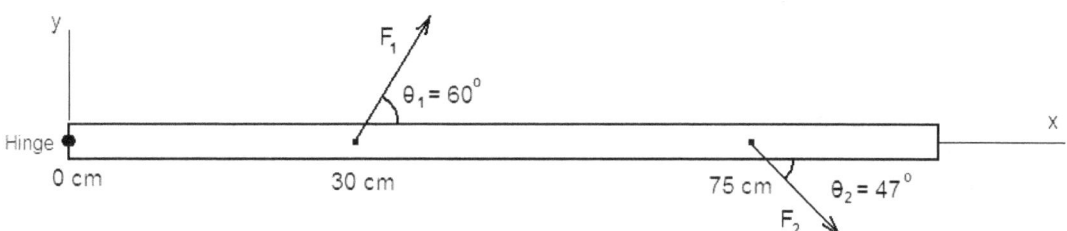

1. A light meter stick is hinged at the 0-cm mark at the origin and is lying along the positive x-axis. It is free to rotate in the xy- plane. Two forces one of 20 N and the other of 16 N are applied at the 30-cm and 75-cm marks respectively. The 20-N force is at an angle of 60 degrees and lies in the first quadrant and the 16-N force is applied at angle of 47 degrees and is in the fourth quadrant. Determine the net torque acting on the meter stick.
   [-3.6N m]

2. Determine the torque about hinge H acting on the body in each of the following cases. State whether the torque is CW or CCW.

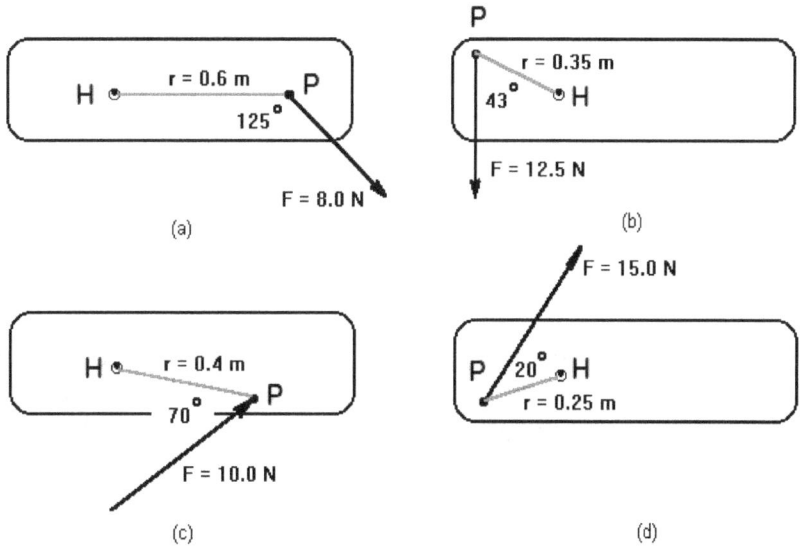

3. A 0.15 kg meter stick is supported at the 50 cm mark. A mass of 0.5 kg is attached at the 80 cm mark.

a. How much mass should be attached to the 40 cm mark to keep the meter stick horizontal?   *[1.5 kg]*

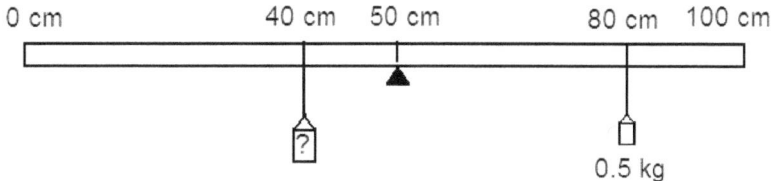

b. Determine the supporting force from the fulcrum on the meter stick.   *[21.07 N]*

77

4. A meter stick of mass 45 g is to be placed on a fulcrum at 20 cm mark. Determine the position on the meter stick from which a mass of 90 g be suspended so that the meter stick stays horizontal. Also determine the magnitude and the direction of the reaction force from the fulcrum.

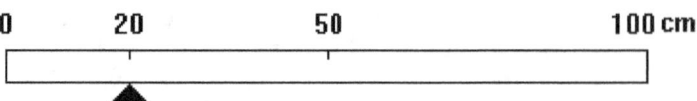

5. A uniform rod of length 2 m and weight 40 N is hinged at one end to a vertical wall.
It supports a weight of 25 N suspended 0.5 m from the hinged end. The other end is supported by a cable, which makes an angle of 35 degrees with the rod, which is held horizontal. Determine the tension in the cable and the reaction force from the hinge on the rod. *[45.8 N, 53.9 N @ 45.9°]*

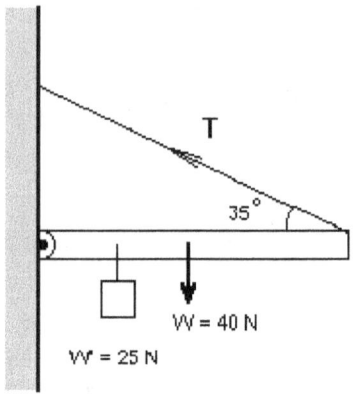

6. A 2-m long uniform rod AB is suspended horizontal by two vertical strings attached to the ends A and B. The rod has a mass of 0.6 kg. An owl of mass of 2 kg is sitting on the rod 0.8 m from the end A. Determine the tension in each string.

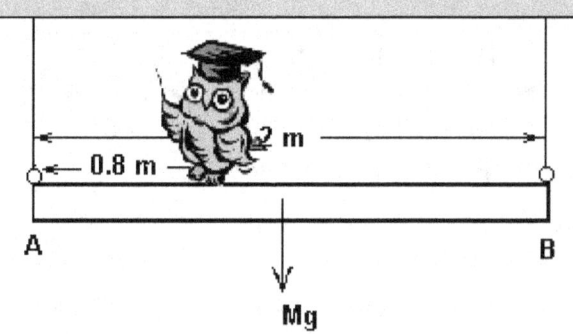

7. A mobile is formed using light rods, light strings and three fish. The mass of the baby fish is 2 kg. Calculate the masses of the Mama fish and Papa fish.
[Mama = 6.0 kg ; Papa = 5.33 kg]

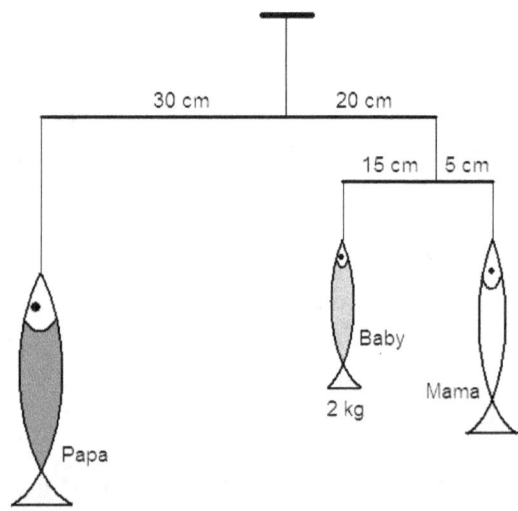

Multiple Choice:

1. Two equal and opposite forces are applied on a rod. Which of the following <u>must be</u> true about the net force F_{net} and net torque τ_{net} acting on the rod?
(A) $F_{net} = 0$ and $\tau_{net} = 0$ (B) $F_{net} = 0$ (C) $\tau_{net} = 0$ (D) $F_{net} \neq 0$ (E) $\tau_{net} \neq 0$

2-3. A uniform beam is hinged at one end and kept horizontal by a vertical cable attached at the other end. The tension in the cable is F.

    2. The weight of the beam must be
    (A) ¼ F   (B) ½ F   (C) F   (D) 2F   (E) 4F

    3. The vertical cable is now attached at the center of mass of the beam the tension in the cable must be
    (A) 0   (B) ½ F   (C) F   (D) 2F   (E) 4F

4. The SI unit of torque is
    (A) N.s   (B) N/s   (C) N.m   (D) N/m   (E) τ

5. A uniform beam of weight W is hinged at one end to a wall and is held horizontal by a vertical cable attached to the other end. The reaction force from the hinge on the beam must be

(A) ½ W upward
(B) ½ W downward
(C) W upward
(D) W downward
(E) zero

Chapter 17
Simple Harmonic Motion (SHM)

Simple Harmonic Motion can be defined using a reference circle as follows:

'If a particle is undergoing uniform circular motion then its projection on any diameter of its circular path performs Simple Harmonic Motion.'

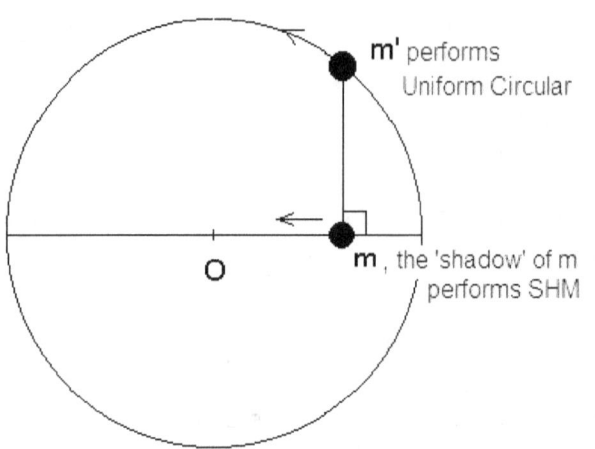

m' performs Uniform Circular

m, the 'shadow' of m performs SHM

Reference Circle and SHM

Some Characteristics of SHM: A particle performing SHM is confined to a line segment between two extreme points. The center of the line segment is called the **mean position** of the SHM.

Amplitude (X_o): The maximum displacement of the particle from the mean position

Oscillation: When the particle along the reference circle completes one revolution its projection performing SHM completes one oscillation. At the end of one oscillation the particle returns to its original state of motion at the original position.

Time Period (T): The time required for one oscillation

Frequency (f): The number of oscillations per second.
$f = 1/T$

Maximum and minimum speeds: The particle has maximum speed at the mean position and minimum (zero) speed at the extreme positions.

Maximum and minimum accelerations: The particle has minimum (zero) acceleration at the mean position and maximum acceleration (in magnitude) at the extreme positions

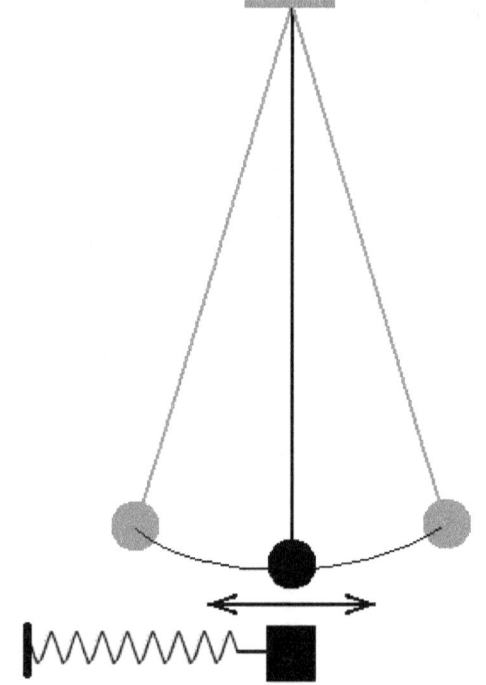

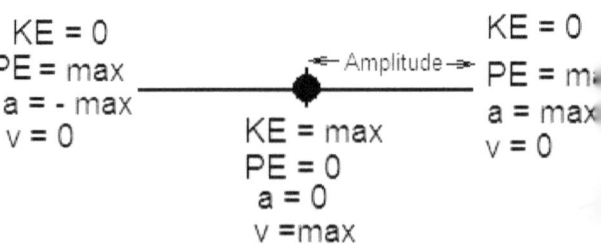

KE = 0
PE = max
a = - max
v = 0

KE = max
PE = 0
a = 0
v = max

Amplitude

KE = 0
PE = m
a = max
v = 0

Distance and displacement in SHM: Each oscillation corresponds to a distance traveled of d = 4 x amplitude. Displacement depends on the initial and final positions of the particle.

Note:

- *Displacement in one oscillation is zero; in fact it is zero in any number of complete oscillations*
- *Each oscillation occurs in one period T.*
- *Displacement in any integral number of periods is zero.*

Energy of SHM (Pendulum and Spring-mass system): The KE and PE of a particle in SHM is changing continuously but the total mechanical energy (i.e. ME = KE + PE) stays constant (see the graphs below) in the absence of any frictional losses.

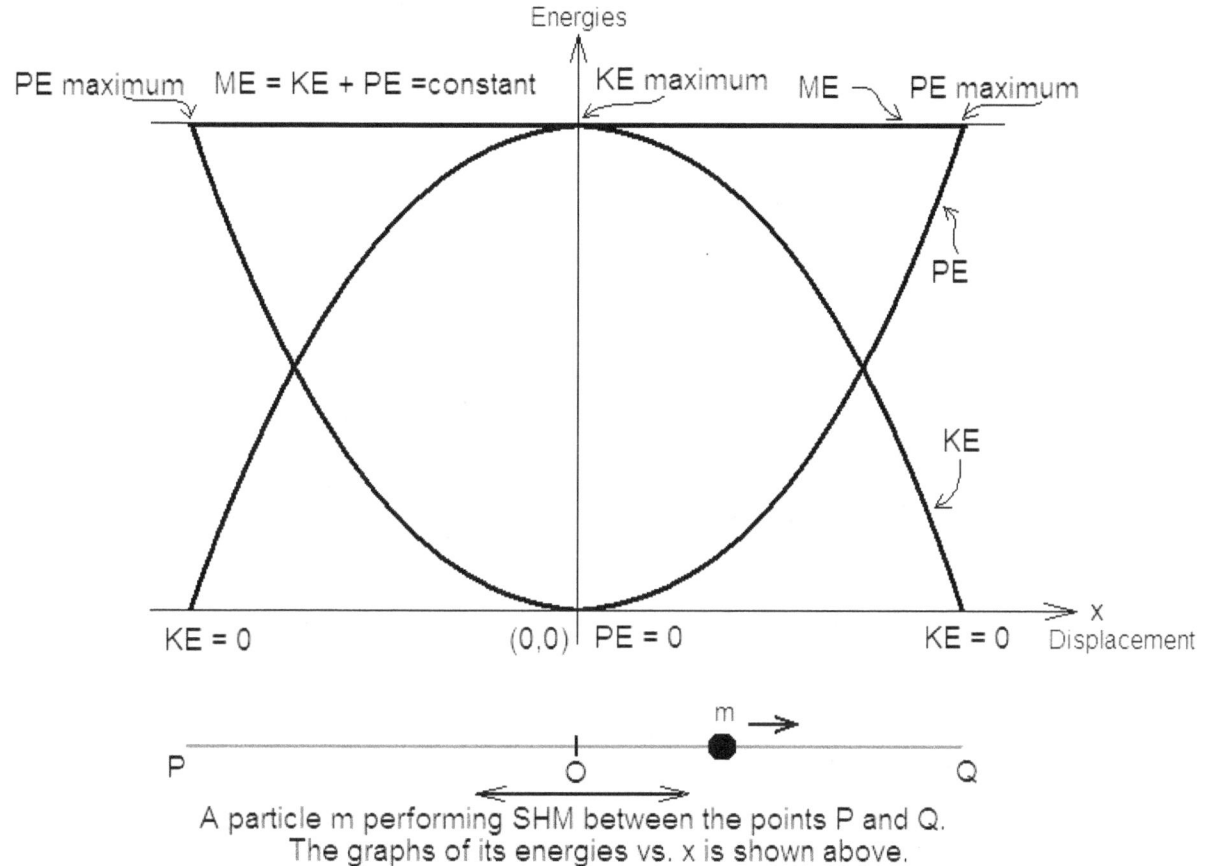

A particle m performing SHM between the points P and Q.
The graphs of its energies vs. x is shown above.

Hence, at any point x and at moment t, $\frac{1}{2} mv^2 + PE = ME$ constant

Kinetic Energy: The KE of the particle is maximum at the mean position and minimum (zero) at the extreme position.

Potential Energy: The PE of the particle is maximum at the extreme positions and minimum (zero) at the mean position.

Total Mechanical Energy: The total mechanical energy of the particle, which is the sum of its KE and PE, stays constant during the SHM.

Some interesting facts & figures
• **If a floating object is slightly depressed further into the liquid and released it performs up and down SHM**
• **A liquid poured into a U-tube will settle with the levels in the two arms at the same height. If this level is slightly disturbed it performs vertical SHM**

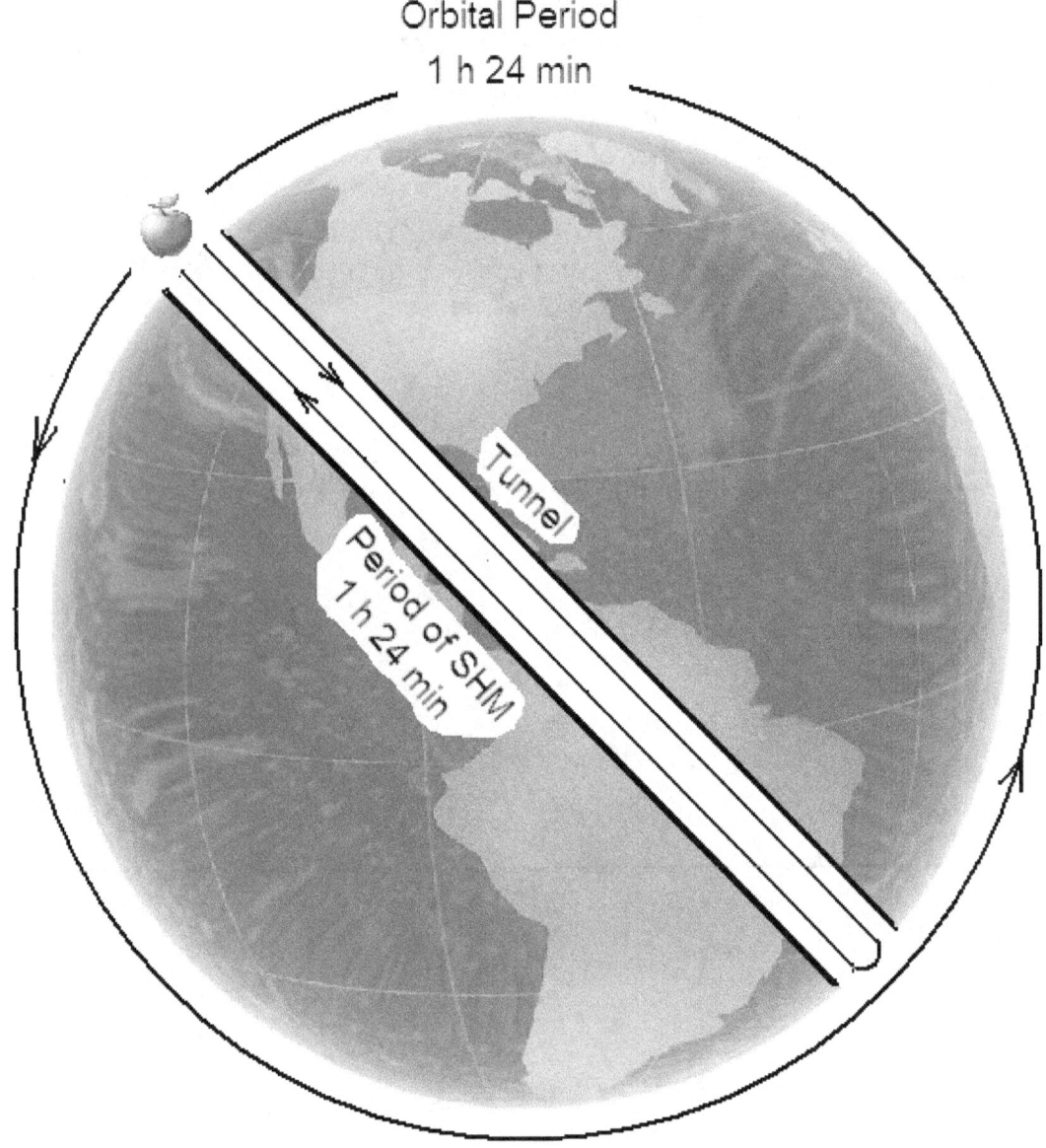

Orbital Period
1 h 24 min

Tunnel

Period of SHM
1 h 24 min

Multiple Choice:

1. A particle is performing SHM with amplitude of 8 cm. The distance traveled by the particle in one oscillation is
(A) 8 cm   (B) 16 cm   (C) 32 cm   (D) 4 cm   (E) 2 cm

2. For a particle performing SHM, _____ is maximum at the mean position (center).
(A) KE   (B) PE   (C) Total mechanical energy
(D) KE – PE   (E) acceleration

3. For a particle performing SHM, _____ is maximum at the extreme positions.
(A) KE   (B) PE   (C) Total mechanical energy
(D) KE – PE   (E) velocity

4. For a particle performing SHM, _____ is constant at all the positions.
(A) KE   (B) PE   (C) Total mechanical energy
(D) KE – PE   (E) velocity

5. The period of SHM of a particle is 4 seconds and its amplitude is 0.3 m. At t = 0 the particle is at the center O and moving in the positive x-direction. At t = 6 seconds the distance traveled d and the displacement **x** for the particle are
(A) d = 0; **x** = 0``   (B) d = 0.6 m; **x** = +0.6 m   (C) d = 1.8 m; **x** = 0   (D) d = 1.8 m; **x** = +1.8
(E) d = 1.8 m; **x** = - 1.8 m

Chapter 18
Spring-mass System and SHM

A particle of mass m attached to a light spring of elastic constant k performs simple harmonic motion when set in linear vibrations.

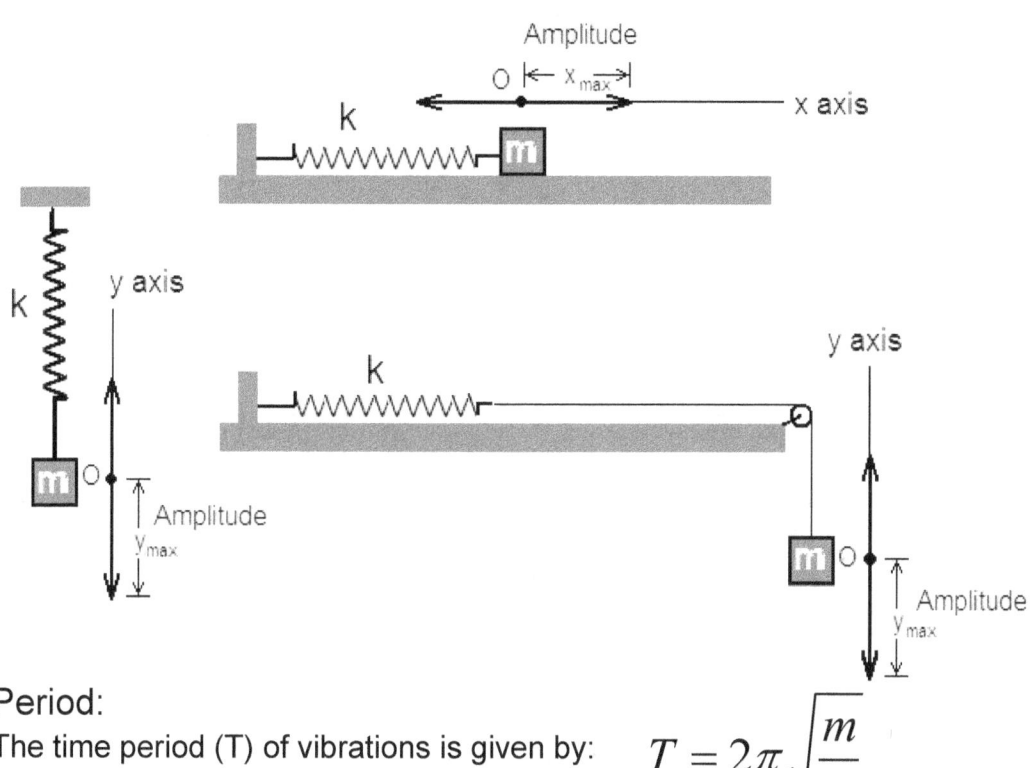

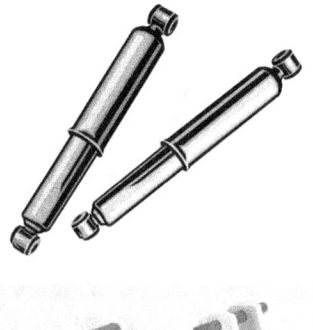

Period:
The time period (T) of vibrations is given by:

$$T = 2\pi\sqrt{\frac{m}{k}}$$

Frequency
The frequency (f) of vibration is equal to 1/T. Hence,

$$f = \frac{1}{2\pi}\sqrt{\frac{k}{m}}$$

Note:
- *The period of spring-mass system depends on the mass of the particle, (T α √m)*
- *The period of spring-mass system depends on the spring constant k, (T α 1/√k)*
- *The period of the spring-mass system <u>does not</u> depend on the acceleration due to gravity.*
- *Hence the period of a spring-mass system is the same on earth as it would be on moon.*
- *In fact gravity is not essential for the oscillations of the spring mass system.*
- *A spring-mass system will oscillate vertically as well as horizontally with the same period.*

Energy of SHM in Spring-mass oscillations

$$ME = KE + PE = \frac{1}{2} mv^2 + \frac{1}{2} kx^2$$

Maximum Speed: When speed is maximum, the PE is zero and all the energy is kinetic. Hence,

$$ME = \frac{1}{2} mv_{max}^2$$

v_{max} can be calculated from the equation above

Amplitude (maximum displacement x_{max} for the center of SHM): When displacement of the block is maximum, the velocity is zero and all the energy is potential. Hence,

$$ME = \frac{1}{2} kx_{max}^2$$

x_o can be calculated from the equation above

Some interesting facts and figures
• **Astronauts in the state of weightlessness (microgravity) cannot weight themselves in the usual manner of standing on a bath room scale.**
• **Oscillation period of a spring depends on the mass attached to it and does not depend on gravity. Astronaut can thus find their mass using a system based on spring-mass oscillation. This machine is called 'Body Mass Measuring Device' or BMMD.**

Problems:

1. A mass of 500 g is attached to a spring and the system is made to oscillate. The frequency of oscillations is measured to be 3.2 HZ. Determine the elastic constant of the spring.   *[202 N/m]*

2. A spring stretches by 3.0 cm when a mass m_1 = 200 g is suspended from it. Determine the additional stretch if another mass m_2 = 350 g is added. Determine the period of oscillations of the spring with both m_1 and m_2 attached to it.

3. A spring oscillates with frequencies of 5 Hz and 8 Hz for two unknown masses m_1 and m_2 respectively. Determine the ratio m_1 / m_2.   *[2.56]*

4. A spring extends by 1.5 cm when a mass is suspended from it. Determine the frequency of oscillations of the spring with the same mass attached to it.

5. A spring (k = 400 N/m) is suspended vertically. Now two masses m_1 = 2 kg and m_2 = 6 kg are attached to it as shown in the figure.

   (a) Calculate the extension of the spring.   *[0.196 m]*

   (b) Calculate the tension in the string.   *[58.8 N]*

   The string is now cut..

   (c) Calculate the period of oscillations of the spring.   *[0.444 s]*

    (d) Calculate the velocity with which the mass m_2 is going to hit the floor.   *[7.41 m/s]*

6. A spring-mass system is oscillating with amplitude of 3.0m. Mass m = 4.0 kg and k = 500 N/m.

(a) Calculate the mechanical energy of the system.   *[2250 J]*

(b) Calculate the maximum speed of the mass.   *[33.5 m/s]*

(c) Calculate the speed of the mass when it is 2 m from the center of the motion.   *[25 m/s]*

(d) Calculate the force the spring applies on the mass when the displacement is 2 m.   *[1000 N]*

(e) Calculate the maximum force the spring applies on the mass.   *[1500 N]*

Multiple Choice:

1. A mass m attached to a spring oscillates with a period of 3 s. If the mass m is replaced by a mass 4m the new time period of oscillations will be
(A) 3 s   (B) 12 s   (C) 6 s   (D) 48 s   (E) 1.5 s

2. The period of a given spring-mass system on earth is T. If the same system oscillates on the moon where the acceleration due to gravity is $1/6^{th}$ that on the earth, the new period T' will be.

(A) T (B) $6T$ (C) $\sqrt{6}T$ (D) $\frac{1}{6}T$ (E) $\frac{1}{\sqrt{6}}T$

3. In a spring-mass system, if the mass and spring constant are both doubled the time period will be _____ times the original time period.
(A) 1   (B) 2   (C) 4   (D) ½   (E) √2

4. A spring, initially relaxed, is stretched by an amount x. The force needed is F and the elastic potential energy stored in it is U. If the spring is stretched by an additional amount 2x which of the following is true about the new force F' and the new stored potential energy U'?
(A) F' = 2F; U' = 2U
(B) F' = 9F; U' = 3U
(C) F' = 4F; U' = 2U
(D) F' = 3F; U' = 6U
(E) F' = 3F; U' = 9U

5. A 0.08-kg particle is performing SHM. If the energy of the SHM is 0.36 J the maximum speed of the particle must be
(A) 4.5 m/s   (B) 3.0 m/s   (C) 2.4 m/s   (D) 1.5 m/s   (E) 0.5 m/s

6. A 5 kg particle is performing SHM and has a maximum speed at 4 m/s. If its PE at the center is zero, its maximum potential energy must be

(A) 40 J   (B) 20 J   (C) 10 J   (D) 5 J   (E) 2.5 J

Chapter 19
Simple Pendulum and SHM

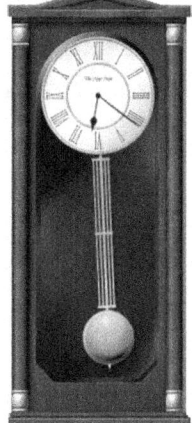

A simple pendulum consists of a small bob suspended by a light inextensible spring from a fixed support.

For a simple pendulum of length L, oscillating with small angular amplitude θ (usually less than 5°) the motion of the bob is SHM.

Simple Pendulum

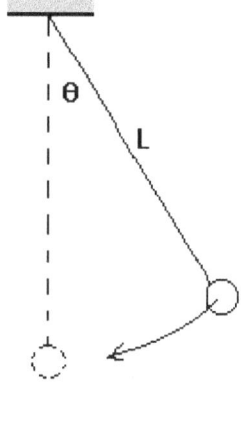

Period (T):
The period of oscillation = $T = 2\pi\sqrt{\dfrac{L}{g}}$

Frequency (f = 1/T)
The frequency of oscillation =.
Hence, $f = \dfrac{1}{2\pi}\sqrt{\dfrac{g}{L}}$

Note:
- The period of simple pendulum depends on the length of the pendulum, (T α √L)
- The period of simple pendulum depends on the gravity, (T α 1/√g)
- In fact gravity is essential for the oscillations of the simple pendulum.
- The period of simple pendulum <u>does not</u> depend on the mass of the bob..
- A pendulum will oscillate slower (greater T) on moon as compared to the earth.

Maximum Speed (v_{max}) : When speed is maximum (at the center of the motion), the gravitational PE of the bob is zero and all its energy is kinetic.
Hence,   ME = ½ mv_{max}^{2}

v_{max} can be calculated from the equation above

Some interesting facts & figures
• **A pendulum clock runs slower in summer and faster in winter.**
• **A Foucault pendulum reveals the rotation of the earth.**
• **The World's longest pendulum is 4440 feet (1353 m). It was constructed in 1901 in the #4 Tamarack Mine shaft. What is the period of this pendulum?**

Problems:

1. Determine the time period and frequency of vibration of a simple pendulum of length 75 cm.
   [1.74 s, 0.58 Hz]

2. Determine the length of a second's pendulum on (a) earth and (b) moon.

3. Determine the length of a simple pendulum if it completes 25 oscillations in 30 s.

[0.36 m]

**4. Determine the percentage change in the time period of a simple pendulum if its length is increased by 50%.

**5. The period for a simple pendulum at a place is 2.001s where the value of g is 9.80921 m/s^2. When the pendulum is taken to a different place the time period changes to 2.004 s. Determine the value of g at the second place.   *[9.77986 s]*

6. A simple pendulum of length 2.4 m and bob of mass 0.44 kg is oscillating with total mechanical energy of 22 J.
(a) Calculate the speed of the bob at the center of the motion.

(b) Calculate its period.

MULTIPLE CHOICE:

1. The period of a simple pendulum on earth is T. If the same pendulum oscillates on the moon where the acceleration due to gravity is 1/6th that on the earth, the new period T' will be.

(A) T   (B) 6T   (C) ($\sqrt{6}$) T   (D) (1/6) T   (E) (1/$\sqrt{6}$) T

2. Which of the following affect(s) the period of a simple pendulum?
    I.   mass of its bob
    II.   its length
    III.   gravity

(A) all of them   (B) I only   (C) II only   (D) III only   (E) II and III only

3. For a simple pendulum, if the mass and length are both doubled the time period will be _____ times the original time period.
(A) 1   (B) 2   (C) 4   (D) ½   (E) $\sqrt{2}$

4. A simple pendulum makes 18 oscillations in 25 seconds. If the length of the pendulum is made 9 times it will make _____ oscillations in 25 second.
(A) 2   (B) 3   (C) 6   (D) 54   (E) 162

Chapter 20
Temperature

Temperature may be defined as the degree of 'hotness' or 'coldness' of a body. It will be defined concretely in Thermodynamics and Kinetic Theory of Gases.

There are three scales for temperature measurement commonly in use, namely, Fahrenheit, Celsius, and Kelvin scales as shown below:

TEMPERATURE SCALES

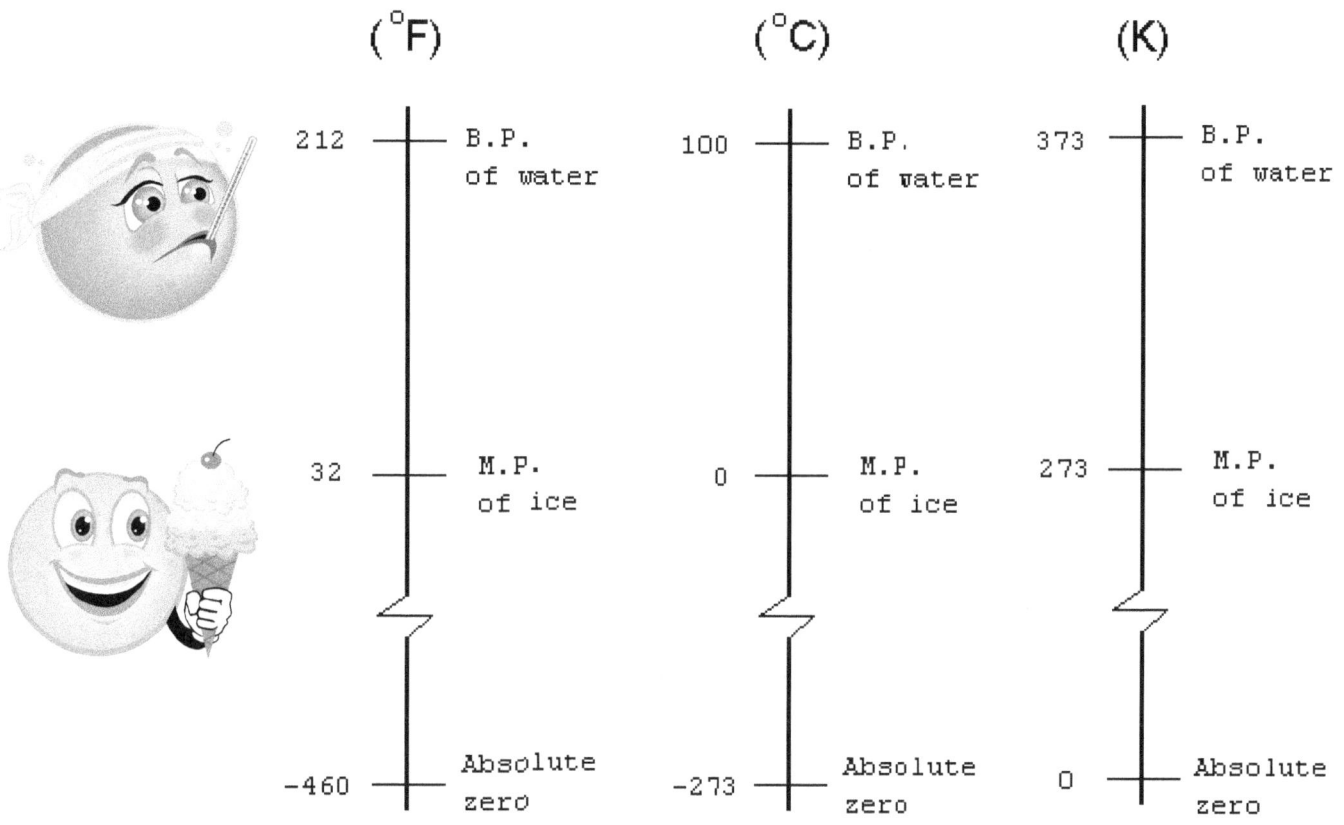

Of these three, only the Kelvin scale is the absolute scale for measurement of temperature because the zero on this scale is the lowest possible temperature. Kelvin is SI unit for measurement of temperature.

Equations for conversion of temperature between different scales:

$$\frac{T_F - 32}{9} = \frac{T_C}{5} \qquad ; \qquad T_K = T_C + 273$$

Equations for temperature <u>change</u>:

$$\frac{\Delta T_F}{9} = \frac{\Delta T_C}{5} \qquad ; \qquad \Delta T_K = \Delta T_C$$

Some interesting information facts & figures

- The center of the Sun is about 15 million °C.
- The temperature of space is 2.725 K. It is due to the microwave background radiation in the present state of expansion of the universe.
- To determine the temperature from cricket chirps in °C, use $T_C = 50 + (N - 40) / 4$, where N is number of chirps in one minute.
- The highest temperature produced in a laboratory was 920,000,000 F (511,000,000 C) at the Tokamak Fusion Test Reactor in Princeton, NJ, USA.
- The hottest planet in the solar system is Venus, with an estimated surface temperature of 462 oC.
- Certain frogs that can survive the experience of being frozen.
- A bolt of lightning is about 54,000°F (30,000°C); six times hotter than the surface of the Sun.
- Dry ice is solid CO_2 at or below -78.5°C. It sublimes (converts directly from solid to gas phase) at temperature of -78.5 °C at normal atmospheric pressure. The opposite of 'sublimation' is 'deposition'.

PROBLEMS:

1. Convert the following temperatures from °F to °C:

   (a) 0   (b) -200   (c) 32   (d) 98.6   (e) 212   (f) –40

   [-17.8 °C, -129 °C, 0 °C, +37 °C, 100 °C, -40 °C]

2. Convert the following temperatures from °C to °F:

   (a) 0   (b) -200   (c) -18   (d) 50   (e) 100   (f) -40   (g) -273

3. Convert the following temperatures to K.

   (a) 10 °C   (b) 100 °C   (c) -70°C (d) 32 °F   (e) 0 °F (f) -100 °F

   [283 K, 373 K, 203 K, 273 K, 255.2 K, 199.7 K]

4. The boiling points and melting points of some substances are given below in K. Express these on Celsius scale.

Substance	Boiling/Melting point	Temperature (K)	Temperature (°C)
Nitrogen	Boiling point	77	
Oxygen	Boiling point	90	
Hydrogen	Boiling point	20	
Water	Melting point	273	
Zinc	Melting point	693	
Silver	Melting point	1235	
Gold	Melting point	1338	
Copper	Melting point	1356	

5. The temperature of water in a beaker is increased by 32 °F. Determine this rise of temperature on Celsius scale.   *[17.8 °C]*

6. The highest temperature ever recorded on the earth was 58° C in 1922 in Libya. How much is this temperature in °F ?

7. The lowest temperature ever recorded on the earth was -129°F at Vostok, Antarctica on July 21, 1983. How much is this temperature in °C?   *[-89.2 °C]*

Multiple Choice:

1. Which of the following temperatures is meaningless?
(A) 1 million °C   (B) 1 million °F   (C) – 273 °F   (D) -273 K   (E) 0.0001 K

2. Which of the following is the lowest temperature?
(A) 0 °F   (B) 0 °C   (C) 0 K   (D) – 10 °C   (E) – 40 °C

3. The temperature of a piece of metal is raised by 25° C. What is this rise of temperature on Kelvin scale?
(A) 25 K   (B) (273 + 25) K   (C) (273 – 25) K   (D) (25 – 273) K   (E) – 25 K

Chapter 21
Thermal Expansion

Most substances expand on heating and contract on cooling. When a substance expands/contracts its linear dimensions as well as volume increases/decreases.

Linear Expansion:

The change in linear dimension of an object on heating/cooling depends on:

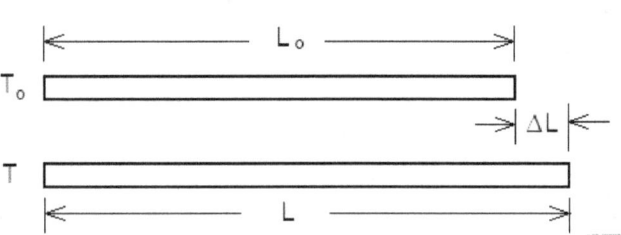

- original length (L_o)
- he temperature rise (ΔT)
- nature of material (represented by a constant α)
- α is called coefficient of linear expansion

Change in length $\quad \Delta L = \alpha L_o \Delta T$
New length $\quad\quad\quad L = L_o(1 + \alpha \Delta T)$

Volume Expansion:

The change in volume of an object on heating/cooling depends on:
- original length (V_o)
- the temperature rise (ΔT)
- nature of material (represented by a constant β)
- β is called coefficient of volume expansion

Change in volume $\quad \Delta V = \beta V_o \Delta T$
New volume $\quad\quad\quad V = V_o(1 + \beta \Delta T)$

Relationship between β and α: $\quad \beta \approx 3\alpha$

Material	α (°C⁻¹)	$\beta \approx 3\alpha$ (°C⁻¹)
Invar	1.2×10^{-6}	3.6×10^{-6}
Steel	12×10^{-6}	36×10^{-6}
Aluminum	23×10^{-6}	69×10^{-6}
Brass	19×10^{-6}	57×10^{-6}
Glass	9×10^{-6}	25.5×10^{-6}
Pyrex	3.2×10^{-6}	9.6×10^{-6}
Mercury	-	182×10^{-6}
Water	-	207×10^{-6}
Gasoline	-	950×10^{-6}

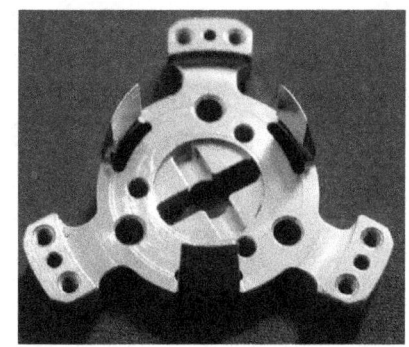

An aerospace part made out of invar alloy

Some interesting facts & figures
• **Anomalous expansion of water:** Cooling water from 100°C results in decreasing its volume until 4°C below which water begins to expand. Expansion is considerably greater when water turns into ice.   • The density of water is 999.9720 kg/m³ at 4°C, 999.8395 kg/m³ at 0°C, and 958.4 kg/m³ at 100°C.   • A shape memory alloy remembers its shape when it was cold-forged. If it is deformed due to heating it will return the original shape when cooled. This alloy is known by various names: SMA, smart metal, memory metal, memory alloy, muscle wire, and smart alloy.   • The 1.3 km length of the Golden Gate bridge made of steel expands by more than ¼ meter when temperature changes by 20 °C. Why doesn't it warp?

PROBLEMS:

1. A brass sphere has a radius (r_o) of 8.00 cm at a temperature (T_o) 20°C. If the sphere is heated to a temperature of (T) 100°C, determine Δr and ΔV for the sphere.

[Δr = 0.012 cm, ΔV = 9.78 cm³]

2. Steel rail segments each of length 10 m are laid when the temperature is -15°C. How much minimum space should be left between the rails to allow for the expansion during the summer when the rails might get heated to a temperature of 50°C?

3. A Pyrex® beaker is marked 30 ml at 20 °C.
(a) Calculate the capacity of the beaker at 150 °.

[30.037 ml]

(b) This beaker is filled with water at 20 °C. The whole system is gradually heated to 100 °C. Calculate the volume of the overflowing water at 100 °C.

[0.474 ml]

4. A steel rod is 1 mm longer than an aluminum rod at 20 °C. Their lengths become equal when both are heated to a temperature of 190 °C. Calculate their lengths at 20 °C.
α_{steel} = 12 X 10^{-6} °C^{-1} ; $\alpha_{aluminum}$ = 23x10^{-6} °C^{-1}.

5. A steel hoop is to be fitted as a rim on a wooden wheel. The radius of each of them is 30 cm at room temperature. To slip the steel ring over the wooden wheel the ring needs to be heated so that it can expand to a radius 1.5 mm greater than the rim of the wooden wheel. To what temperature should the steel ring be heated?   *[417 °C above the room temp]*

6. A steel-nickel alloy known as 'Invar' was invented in 1896 by Swiss scientist Charles Eduard Guillaume. This alloy has remarkably low coefficient of linear expansion, 1.2x10^{-6} /°C. Calculate the increase in length of one-meter invar bar when it is heated from 20°C to 100°C.   *[9.6 x 10^{-5} m]*

Multiple Choice:

1. A brass rod PQ is bent into the form of a part-rectangle as in the figure below. Now if the bent rod is heated uniformly the gap between its ends P and Q will

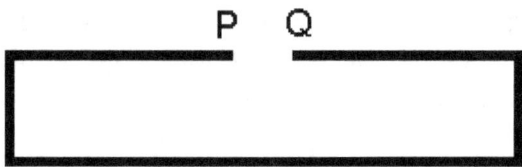

(A) increase   (B) decrease   (C) stay the same
(D) decrease until the gap is filled then increase   (E) information insufficient

2. A hole is drilled in a metal plate. If the plate is now cooled the size of the hole will
(A) increase   (B) decrease   (C) stay the same
(D) decrease until the hole is filled then increase   (E) information insufficient

3. The pendulum of a clock is made of brass. The clock runs at the correct speed in winter. In summer, the clock will
(A) run faster   (B) run slower   (C) run at the same speed
(D) run in the reverse   (E) stop

Chapter 22
Mechanical Equivalent of Heat

Heat is a form of energy. Among the commonly used units for heat are **calorie** (cal) and **joule** (J).

One **calorie** is the amount of heat required to raise the temperature of 1 gram of water through 1 degree Celsius.

James Joule (1818-1889) established a relationship between energy in the form of work and energy in the form of heat. His landmark experiment established a conversion factor between 'work' and 'heat' and in terms of the modern units it is

1 cal = 4.186 joules

For example, 4.186 joules of work can be converted by friction into 1 calorie of heat.

Food Calorie (Cal): Energy supplied by food is also measured in Calories. However it is 1000 times the calorie defined above. The food calorie is written with capital 'C'.

1 food calorie = 1 Cal = 1,000 cal

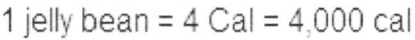

1 jelly bean = 4 Cal = 4,000 cal

Some interesting facts & figures
• One joule is about the energy required to lift a small apple one meter straight up.
• One μjoule (7 TeV) is about the energy per collision in the Large Hadron Collider (LHC) experiments.

PROBLEMS

1. A piece of candy supplies 27.5 Calories. If this energy is converted into work for lifting a load, how much mass can be lifted to a height of 5 ft (~1.52 m) with it?   *[7717 kg]*

2. Butter supplies 6.0 Cal/g. Calculate the maximum height a 65-kg person can climb with the energy available from 5 g of butter?

3. A 5-kg block is sliding over the rough surface of table. The friction brings the block to a stop from its initial velocity of 4.4 m/s. Calculate the amount of heat generated, in J and calories.   *[11 J, 2.63 J]*

Multiple Choice
1. 1 cal equals
(A) 1 J   (B) 4.186 J   (C) 1,000 J   (D)   0.239 J   (E) 0.53 J

Chapter 23
Heat Transfer
Conduction, Convection, and Radiation

Conduction: It is the mode of heat transfer that takes place in material media – solids, liquids, and gases. The energy is transferred from one particle to the next without overall movement of the medium.

In fluids, especially gases, conduction may be overwhelmed by convection, which is another mode of transfer of heat.

Convection: In this mode, heat is transferred between two points by actual movement of the medium. Hence this mode occurs only in fluid mediums.
Examples:
- Heating water in a beaker over Bunsen burner
- Heating of the earth's atmosphere
- Warming of air in a room
- Fanning

Radiation: Radiation does not require material medium; it can take place through vacuum. The radiation can also pass through some material media such as water, glass, and air etc. The solar heat comes to the earth through millions of kilometers of vacuum.

Heat conduction through rod or disc
Here we consider heat conduction only through solid bodies such as along the length of a rod or across the thickness of a plate.

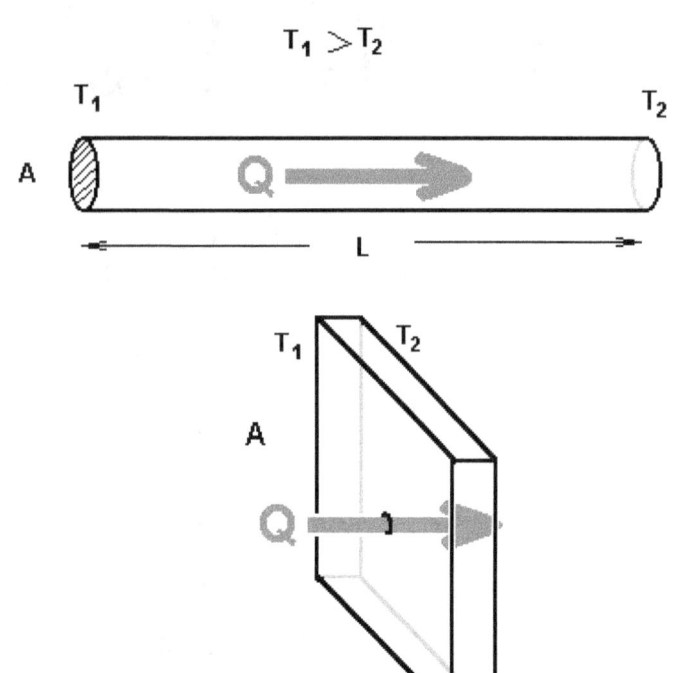

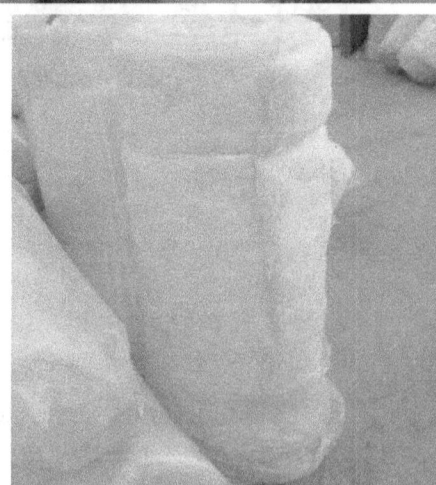

Glass wool: k = 0.024 W/m/K

Rate of Conduction of Heat (Q/t): The rate of conduction of heat, **Q/t**, along a rod or across the thickness of a plate is

- directly proportional to the temperature difference, $T_1 - T_2 = \Delta T$
- directly proportional to the area, **A**
- inversely proportional to length or thickness, **L**
- dependent on the nature of the material defined by the constant **k** called thermal conductivity of the material.

Thermal conductivity of material, k, is measured in J/s/m/K or W/m/K

$$\frac{Q}{t} = \frac{kA\Delta T}{L}$$

Thus,

Q → Amount of heat
t → time for which the heat flows
k → thermal conductivity
ΔT → change in temperature
T → absolute temperature
L → length along which the heat is conducted

Materials	Conductivity ($J.s^{-1}m^{-1}.K^{-1}$)
steel	14
aluminum	235
Brass	452
copper	401
fiber glass	0.024
air	0.026
helium	0.15
Styrofoam	0.033
window glass	1

Thermal image of a house. The brighter parts show a greater rate of loss of heat

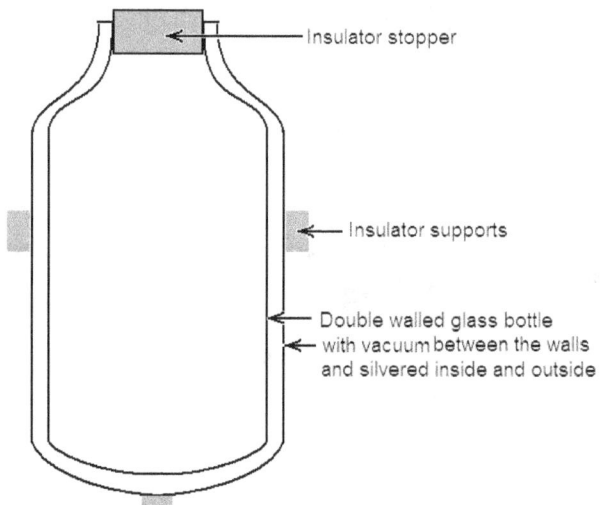

Insulator stopper

Insulator supports

Double walled glass bottle with vacuum between the walls and silvered inside and outside

The inside glass bottle of a vacuum flask. The three modes of transfer of heat are minimized

Vacuum Flask

Temperature (°F)

Wind (mph)	Calm	40	35	30	25	20	15	10	5	0	-5	-10	-15	-20	-25	-30	-35	-40	-45
	5	36	31	25	19	13	7	1	-5	-11	-16	-22	-28	-34	-40	-46	-52	-57	-63
	10	34	27	21	15	9	3	-4	-10	-16	-22	-28	-35	-41	-47	-53	-59	-66	-72
	15	32	25	19	13	6	0	-7	-13	-19	-26	-32	-39	-45	-51	-58	-64	-71	-77
	20	30	24	17	11	4	-2	-9	-15	-22	-29	-35	-42	-48	-55	-61	-68	-74	-81
	25	29	23	16	9	3	-4	-11	-17	-24	-31	-37	-44	-51	-58	-64	-71	-78	-84
	30	28	22	15	8	1	-5	-12	-19	-26	-33	-39	-46	-53	-60	-67	-73	-80	-87
	35	28	21	14	7	0	-7	-14	-21	-27	-34	-41	-48	-55	-62	-69	-76	-82	-89
	40	27	20	13	6	-1	-8	-15	-22	-29	-36	-43	-50	-57	-64	-71	-78	-84	-91
	45	26	19	12	5	-2	-9	-16	-23	-30	-37	-44	-51	-58	-65	-72	-79	-86	-93
	50	26	19	12	4	-3	-10	-17	-24	-31	-38	-45	-52	-60	-67	-74	-81	-88	-95
	55	25	18	11	4	-3	-11	-18	-25	-32	-39	-46	-54	-61	-68	-75	-82	-89	-97
	60	25	17	10	3	-4	-11	-19	-26	-33	-40	-48	-55	-62	-69	-76	-84	-91	-98

Frostbite Times [ ] 30 minutes [ ] 10 minutes [ ] 5 minutes

$$\text{Wind Chill (°F)} = 35.74 + 0.6215T - 35.75(V^{0.16}) + 0.4275T(V^{0.16})$$

Where, T = Air Temperature (°F) V = Wind Speed (mph)   *Effective 11/01/01*

Wind Chill Index

Some interesting facts & figures
•   40 to 50 percent of body heat can be lost through the head (no hat) as a result of its extensive circulatory network.
•   NASA Ames Research Center has developed Ultra High Temperature Ceramics (UHTC). Ultra High Temperature Ceramics are a family of ceramic materials with extremely high melting temperatures,

PROBLEMS:

1. One end of a copper rod of length 0.6 m and area of cross-section $5 \times 10^{-4} m^2$ is kept at a temperature of 300°C and the other end is kept in the ice at 0°C. The loss of heat from the lateral side is prevented by insulation.

a. Determine the rate of flow of heat along the length of the rod. *[100 J/s]*

b. Determine the rate at which the ice melts (in grams/second) [OMIT] *[1.26 g/s]*

c. Determine the temperature of the rod 40 cm from the hot end. *[100 °C]*

**2. A steel and a lead rod of identical

100°C [Steel] [Lead] 0°C

geometrical dimensions (L = 0.4m, and A = 0.001 m²) are placed in contact end to end. The lateral sides are insulated. The free end of the steel rod is kept at a temperature of 100°C and the free end of the lead rod at 0°C.

a. Determine the temperature of the junction of the two rods.

b. Determine the rate of flow of heat through the composite rod.

3. Solve problem 2 if the temperatures of the free ends of the two rods are switched around.
[71.4 °C, 2.5 J/s]

4. Solve problem 2 (b) if the two rods are kept in contact along their lateral sides rather than end to end and for this new composite rod one end is at 100°C and the other end is at 0°C.

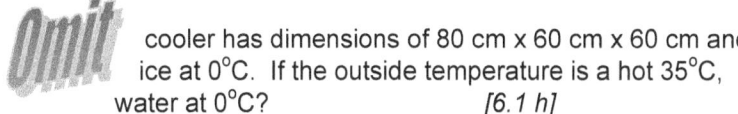

100°C   Steel / Lead   0°C

5. A Styrofoam cooler is taken for a picnic. The ~~Omit~~ cooler has dimensions of 80 cm x 60 cm x 60 cm and an average thickness of 2 cm. It carries 10 kg of ice at 0°C. If the outside temperature is a hot 35°C, how long will it be before all the ice melts into water at 0°C? *[6.1 h]*

**6. A composite rod consists of aluminum and steel rods of identical area of cross section placed in thermal contact end-to-end. The free ends of the steel and aluminum rods kept at 100 °C and 0 °C respectively. Calculate the ratio of the lengths L_{steel} / $L_{aluminum}$ so that the temperature of the junction is at 50 °C.

7. A window pan in a room is of 1.5 m x 0.8 m rectangular shape with thickness of 4 mm as shown in the diagram here. Its thermal conductivity, k, is 0.75 W/m/K. For 12 hours of a wintry night, the average temperatures inside and outside the room were 20°C and -10° C. Calculate the heat lost through this window in the 12 hours duration. (Ignore the insulation effect of air film in contact with the window pan inside and outside which is normally quite significant). *[2.916x10⁸ J]*

0.8m

1.5 m

4 mm

Multiple Choice:

1. The two ends of a rod are maintained at 100°C and 0°C. The rate of flow of heat through the rod is found to be 4 J/s. If the temperatures of the two ends are increased to 200°C and 100°C respectively, the rate of flow of heat through the rod
(A) will increase by a factor of 2   (B) will decrease   (C) will be the same
(D) will increases by a factor of 100   (E) will increase by a factor of 200

2. Which of the following statements are true regarding the rate of flow of heat along a rod
I. It is directly proportional to the temperature difference between its ends
II. It is inversely proportional to the length of the rod.
III. It is directly proportional to the are of cross-section of the rod.
(A) I only   (B) II only   (C) III only   (D) I and III only   (E) All of them

3. Two rods X and Y are made of same metal and have same temperature difference between their ends. The rod X is twice as long and twice as thick (in area) as the rod Y. The ratio of the heat flow rate of the rod X to that of the rod Y must be
(A) ¼   (B) ½   (C) 1   (D) 2   (E) 4

4. Conduction cannot take place through

(A) Solids only   (B) Liquids only (C) Gases only (D) only Vacuum   (E) Both liquids and gases

Chapter 24
Ideal Gas Laws

Avogadro's number (N_A): It is exactly equal to the number of atoms in 12 grams of carbon-12 isotope.

$$N_A = 6.02 \times 10^{23} \text{ molecules}$$

Molecular mass (M): It is defined as the mass of one Avogadro number of molecules (N_A) of a substance

Mole (n): The number of moles for a given quantity of gas is defined as the ratio of its mass (m) to its molecular mass and ALSO the ratio of the number of molecules in the substance to Avogadro's number. Thus,

$$m = nM \qquad N = nN_A$$

Boyle's Law:

For a given quantity of gas at constant temperature

$$PV = Const. \quad \rightarrow \quad P_1V_1 = P_2V_2$$

Charles' Law:
For a given quantity of gas at constant pressure

$$\frac{V}{T} = const. \quad \rightarrow \quad \frac{V_1}{T_1} = \frac{V_2}{T_2}$$

T, T_1, and T_2 are absolute temperature (in Kelvin).

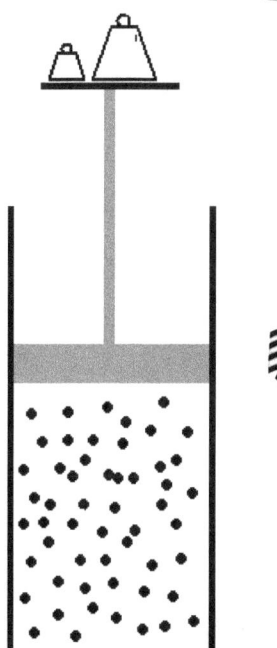

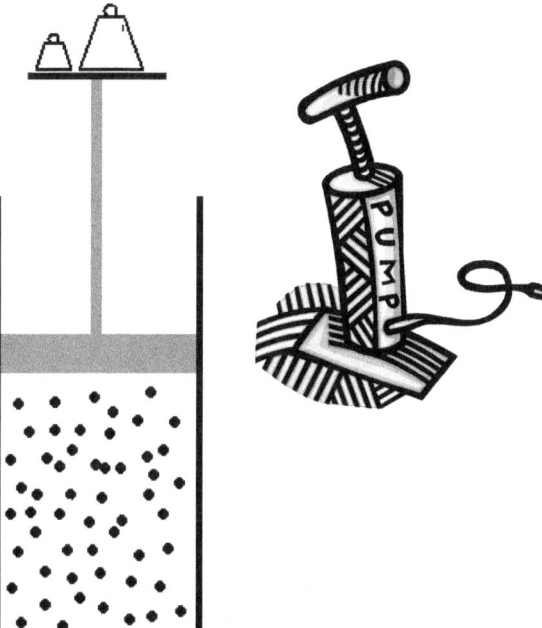

Gay-Lussac's Law:
For a given quantity of gas

$$\frac{P}{T} = const. \quad \rightarrow \quad \frac{P_1}{T_1} = \frac{P_2}{T_2}$$

The Universal Ideal Gas Law:

$$\frac{PV}{T} = const. \quad \rightarrow \quad \frac{P_1V_1}{T_1} = \frac{P_2V_2}{T_2}$$

The above quantity, PV/T, depends on the amount and nature of the gas. However, if the amount of gas is taken in moles (n), then it becomes independent of the nature of the gas. The equation can then be written as

$$PV = nRT$$

P = pressure of the gas
V = volume of the gas
T = Absolute temperature in Kelvin
R = Universal gas constant = **8.314 J/(mol.K)**

Substance	Molecular mass, M (approx.)
H	1 g/mol
H_2	2 g/mol
O	16 g/mol
O_2	32 g/mol
C	12 g/mol
N	14 g/mol
N_2	28 g/mol
H_2O	18 g/mol
CO_2	44 g/mol

Some interesting facts & figures
<ul><li>In second or third century China, hot air balloon Kongming lantern was developed for military communications.</li><li>The first recorded manned flight in a hot air balloon occurred onNovember 21, 1783 in Paris. It was built by the Montgolfier brothers. The flight rose to 500 feet and traveled 5.5 miles in 25 minutes. The pilots were Jean-François Pilâtre de Rozier and François Laurent d'Arlandes.</li><li>Mole Day: Celebrated on October 23, at 6:02 AM to 6:02 PM to commemorate 6.02×10^{23}.</li><li>Absolute zero is OK.</li></ul>

PROBLEMS:

1. The pressure on an ideal gas in a cylinder is reduced from p to 1/3 p at a constant temperature. If the final volume is 3.5 m³ what was the initial volume?   *[1.2 m³]*

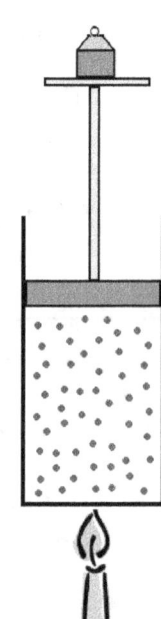

2. A piston applies a constant pressure on 1.78 m³ of ideal gas at 20°C in a container. If the temperature of the gas is increased to 80°C determine the new volume of the gas.

3. A cylinder contains 0.300 m³ of O_2 at a pressure of 4×10^6 Pa and a temperature of 20°C. For this gas determine,

   a. the number of moles   *[493 moles]*

   b. the number of molecules   *[2.97x10²⁶ moles]*

   c. the amount in grams   *[1.58x10⁴ g]*

                                                                                                          0.5 m

4. A tube closed at one end contains a 0.50 m column of air trapped under a small pellet

of mercury at a temperature of 25°C. If the tube is taken outdoors where the temperature is -25°C, determine the new length of the air column.

MULTIPLE CHOICE:

1. The pressure of a given sample of an ideal gas is halved and the volume is quadrupled. The new absolute temperature of the gas must be how many times the original temperature?
(A) 1   (B) ½   (C) 2   (D) 8   (E) 1/8

2. The pressure of an ideal gas is doubled at constant volume. If the initial temperature of the gas was 4 °C the final temperature will be
(A) 1 °C   (B) 2 °C   (C) 4 °C   (D) 8 °C   (E) 281 °C

3. A quantity of ideal gas at -60 °C expands to 3 times its initial volume at a constant pressure. The final temperature of the gas is

(A) 639° C   (B) 366° C   (C) -20° C   (D) -180° C   (E) -202° C

4. The number of molecules in 18 g of H_2O contains x number of molecules and 44 g of CO_2 contains y number of molecules. Which of the following is true of the relationship between x and y?
(A) x = y only if both are gases
(B) x = y only if both are in the same state
(C) x = y irrespective of their physical state
(D) x = (44/18)y irrespective of their physical state
(E) x = (18/44)y irrespective of their physical state

5. 11 g of CO_2 equals
(A) 11 mol   (B) 4 mol   (C) 0.5 mol   (D) 44 mol   (E) 0.25 mol

6. A graphite rod contains $5N_A$ number of atoms. Its mass must be
(A) 3.01×10^{24} g   (B) 220 g   (C) 60 g   (D) 5 g   (E) 0.12 g

7. 27 g of ice is heated to convert it into steam. The number of moles of steam thus formed must be
(A) 27   (B) $27N_A$ (C) 1.5   (D) $1.5N_A$ (E) zero

Chapter 25
Kinetic Theory of Gases

According to the kinetic theory, gas consists of tiny particles called molecules. The molecules move around randomly with all possible speeds. Their collisions with each other and with the walls of the container are elastic in nature. The **pressure** of the gas is caused by the bombardment of the container walls by the molecules. The **temperature** of the gas is due to the kinetic energy of the molecules.

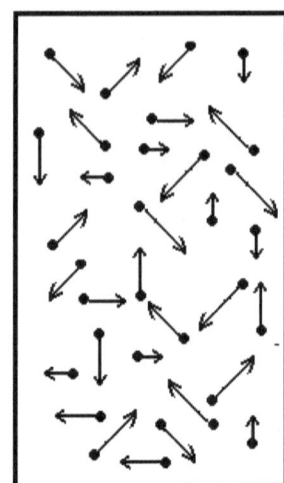

The Root-Mean-Square speed (v_{rms}) : The average speed of the molecule that can relate to the KE_{av} is not the arithmetic mean but the root-mean-square speed (v_{rms}). The root-means-square speed is obtained by finding the square root of the average of the squares of the speeds of all the molecules.

For example if there are 5 molecules with different speeds v_1, v_2, v_3, v_4, and v_5 then v_{rms} is given by

$$v_{rms} = \sqrt{\frac{v_1^2 + v_2^2 + v_3^2 + v_4^2 + v_5^2}{5}}$$

The speed v_{rms} depends on the absolute temperature and the molecular mass of the gas. It is given by,

$$v_{rms} = \sqrt{\frac{3RT}{M(kg/mol)}} = \sqrt{\frac{3kT}{m}}$$

Average KE per Molecule: The average kinetic energy per molecule of an ideal gas depends only on its absolute temperature and is given by

$$\frac{1}{2}mv_{rms}^2 = KE_{av} = \frac{3}{2}kT$$

Internal Energy: The total kinetic energy of all the molecules is called its internal energy and is given by

$$U = \frac{3}{2}nRT$$

Change in the Internal Energy: The change in the internal energy **ΔU** of a gas is independent of its pressure and volume. It depends only on the temperature change **ΔT**. Thus

$$\Delta U = \frac{3}{2}nR\Delta T$$

In the above equations:
KE_{av}= average KE for one molecule of the gas
U = internal energy of the given gas = total KE of all the molecules
M = molecular mass of the gas
m = mass of one molecule of the gas
v_{rms}= root mean square speed of the molecules of the gas
R = universal gas constant = 8.32 J/(mole.K)
T = absolute temperature of the gas
N = number of molecules of the given amount of gas
k (Boltzmann's constant) = R/N_A= 1.3807 x 10^{-23} J/K
N_A (Avogadro's number) = 6.02 x 10^{23} molecules/mole

Some interesting facts & figures
• **The space at the surface of the moon is practically a vacuum. The density of atmosphere at the lunar surface is 1×10^{-14} or a hundred trillionth of the density of atmosphere at the earth surface.** • **Any gas formed at the lunar surface quickly escapes the moon's relatively weak gravity because v_{rms} of these gases is much greater than the escape velocity from the lunar surface.**

Problems:

1. Determine the rms speeds for O_2 and N_2 molecules in the earth's atmosphere at 20 °C.

[478 m/s, 511 m/s]

2. A cylinder contains a mixture of He and Ar gases. Determine the ratio of rms-speed of He atoms to that of Ar atoms

3. At what temperature is the rms-speed of the O_2 molecules?
   (a) 20 mph (8.9 m/s) (b) 60 mph (26.8 m/s) (c) 750 mph (335 m/s) *[0.1 K, 0.92 K, 144 K]*

4. A cylinder contains 1.5 m^3 of Ar gas (M = 40 g/mole) at a pressure of
   3×10^5 Pa and a temperature of 30 °C. Determine,
   (a) the rms-speed of the Ar atoms

   (b) the total internal energy (U) for the gas

Multiple Choice:

1. The rms-speed of the molecules of a gas in a container is 600 m/s. If the absolute temperature of the gas is increased to 3 times the original value the new rms-speed for the molecule will be about
A. 1800 m/s   B. 200 m/s   C. 5400 m/s   D. 67 m/s   E. 1040 m/s

2. The rms-speed for a gas in a container is 200 m/s. If its pressure is doubled and the volume is halved the rms-speed for the gas becomes
A. 400 m/s   B. 800 m/s   C. 50 m/s   D. 100 m/s   E. remains at 200 m/s

3. The rms-speed and the internal energy of an ideal gas are v_{rms} and U respectively. If the absolute temperature of the gas were decreased to ¼ th the original value the new values for rms-speed and internal energy would be
A. ¼ v_{rms} and ¼ U   B. ½ v_{rms} and ¼ U
C. ¼ v_{rms} and ½ U   D. ½ v_{rms} and ½ U
E. 1/8 v_{rms} and ¼ U

4. For three molecules having speeds of 1 m/s, 2 m/s, and 3 m/s, the v_{rms} must be
(A) 2 m/s   (B) $\sqrt{(14/3)}$ (C) 2/3   (D) $\sqrt{2}$ (E) $\sqrt{2/3}$

5. The rms-speed for a gas is 800 m/s at 819° C. Its rms-speed at 0° C will be
(A) 0 m/s
(B) Infinity
(C) 1386 m/s
(D) 462 m/s

(E) 400 m/s

Chapter 26
First Law of Thermodynamics

If a gas is allowed to expand under constant pressure (see the figure below), the gas does work on the environment by applying a force F = pressure x area of the piston on the piston displacing it by Δx. Hence the work done by the gas is

$$W = F \Delta x = PA\Delta x = P(A\Delta x) = P\Delta V.$$

However since the gas is doing the work at the expense of its internal energy this work done is taken as negative. Hence $W = - P\Delta V$

- The work done is **positive** when the work is done **on** the gas, that is, when ΔV is **negative** or during **compression**.

- The work done is **negative** when work is done **by** the gas, that is, when **ΔV** is positive or during **expansion**.

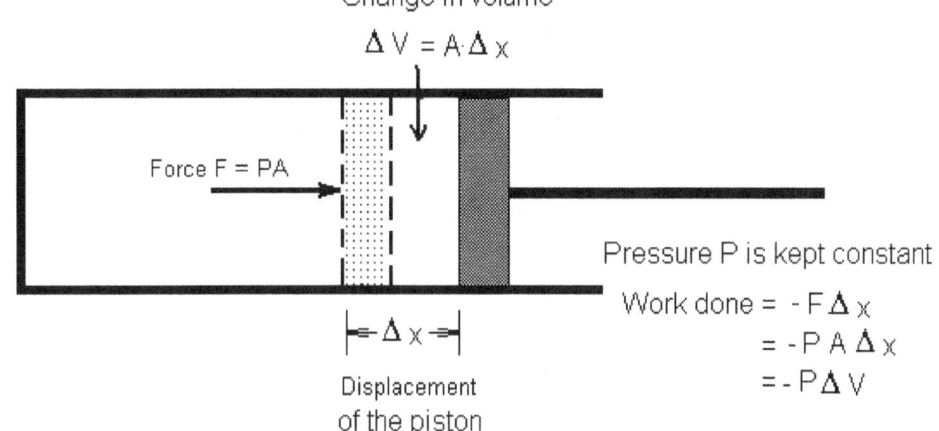

Change in volume
$\Delta V = A \cdot \Delta x$

Force F = PA

|← Δx →|

Displacement of the piston

Pressure P is kept constant

Work done = $- F \Delta x$
    $= - P A \Delta x$
    $= - P \Delta V$

U denotes the internal energy of the gas.
ΔU is the change in the internal energy of the gas. It depends only on the absolute temperature (T) of the gas
ΔU = 0, if the temperature is constant OR the process is one complete cycle.

Work done from P vs. V graph (PV-diagram)

If P vs. V graph is given for a gas, one way to determine the work done is to obtain the area under the curve which is same as the work done. This method is particularly useful if the pressure (P) of the gas is varying with the volume.

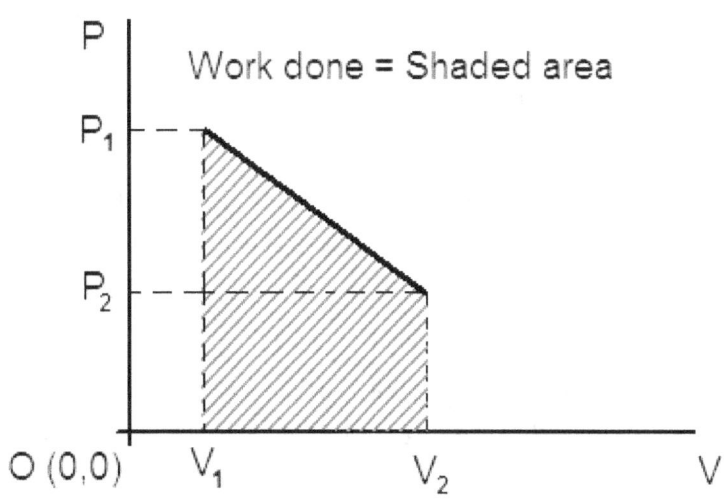

Work done = Shaded area

Example: The pressure, P, of a gas is varying with volume, V, as shown in the PV-diagram below. Calculate the amount of work done.

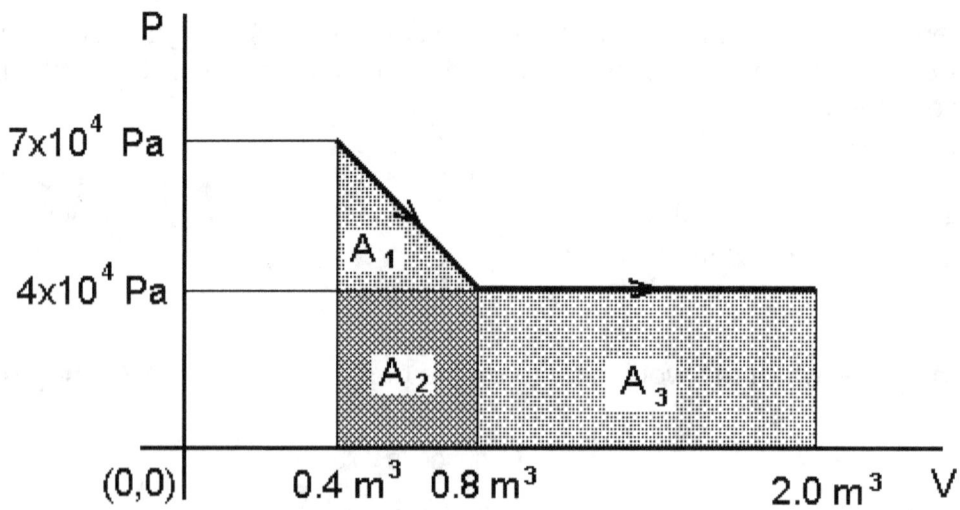

Solution: Work = area under the P vs. V graph.

Hence,
W = A
  = $A_1 + A_2 + A_3$
  = ½ (0.8 - 0.4)($7 \times 10^4 - 4 \times 10^4$) + (0.8 - 0.4)($4 \times 10^4 - 0$) + (2.0 - 0.8)($4 \times 10^4 - 0$)
  = 6,000 J + 16,000 J + 48,000 = 70,000 J
However, since the gas is expanding, the work is done BY the gas and it is negative. Hence

W = - 70,000 J

First law of Thermodynamics
(basically the Law of Conservation of Energy applied to thermodynamic processes)

$$\Delta U = Q + W$$

Q is <u>positive</u> if the heat flows <u>into</u> the system
     is <u>negative</u> if the heat flows <u>out</u> of the system

W is <u>positive</u> if the work is done <u>on</u> the system by the environment
         (i.e. when gas is compressed)
     is <u>negative</u> if the work is done <u>by</u> the system on the environment
         (i.e. when gas expands)

ΔU is <u>positive</u> if internal energy increases
     is <u>negative</u> if internal energy decreases

Some Reversible thermodynamic processes :

1. Isobaric process:
(Constant Pressure)

$$W = -P\Delta V = -P(V_2 - V_1)$$

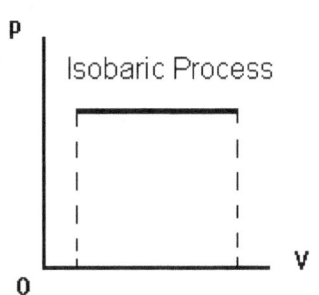

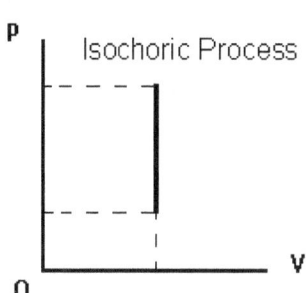

2. Isochoric Process:
(Constant Volume)

$$\Delta V = 0 \text{ hence } W = 0$$

Hence $Q = \Delta U$
(All the supplied heat goes in increasing the internal energy of the system.)

3. Isothermal Process:
(Constant Temperature)

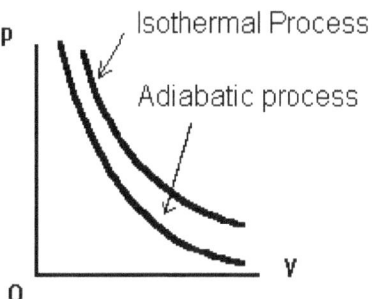

$$\Delta U = 0, \text{ Hence } Q = -W$$

4. Adiabatic Process:
(Thermally insulated system, Q = 0)

$$\Delta U = W$$

Specific Heat of Gases

Specific Heat at Constant Pressure (C_p): It is defined as the amount of heat required to raise the temperature of 1 mole of a gas through 1 K (or 1°C) <u>at a constant pressure</u>. Hence
$$Q = n\,C_p\Delta T$$

Specific Heat at Constant Volume (C_V): It is defined as the amount of heat required to raise the temperature of 1 mole of a gas through 1 K (or 1°C) <u>at a constant volume</u>. Hence
$$Q = n\,C_V\Delta T$$

For a monatomic gas (such as Ar, He, Ne, H, O etc.)

$$C_p = 20.78 \text{ J/(mol.K)}, \qquad C_V = 12.47 \text{ J/(mol.K)}$$

Internal Energy of Ideal Gas (U): The internal energy change (ΔU)of a given amount (n moles)of an ideal gas depends only on the temperature change (ΔT). It is independent of the path between its initial and final states.

$$\Delta U = (3/2)\,nR\Delta T$$

For an <u>isothermal process</u>, $\Delta U = 0$ because $\Delta T = 0$

Some interesting facts & figures
• As atmospheric air rises up it expands adiabatically. This is the main reason why air is colder at higher altitudes
• Water has the second highest specific heat capacity of all known substances, after ammonia.
• The Aioi Works of Japan's Diesel United, Ltd built the first The Wartsila-Sulzer RTA96-C turbocharged two-stroke diesel engine which is the most powerful and most efficient prime-mover in the world today.
• The Boeing 777 series of aircraft currently has the world's most powerful engines.

Problems:

1. An ideal gas is supplied with 1,000 J of heat at constant volume. Determine W and ΔU for the process. *[0, 1000 J]*

2. An ideal gas expands isothermally and does 300 J of work. Determine ΔU and Q for the process.

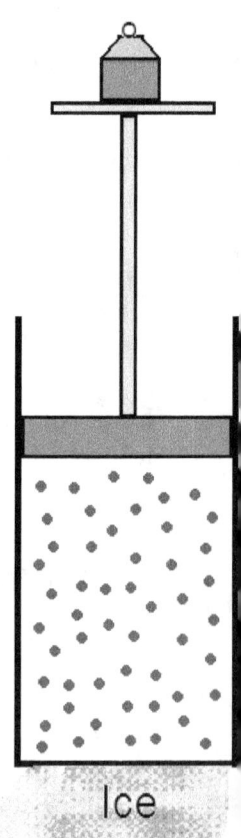

3. An ideal gas contracts isobarically at a pressure of 3.0×10^5 Pa from 5.5 m^3 to 3.4 m^3 when 600,000 J of heat is removed from the system. Determine W, Q and ΔU for the system. *[6.3x10^5 J, -6.0x10^5 J, 3x10^4 J]*

Ice

4. The temperature of 50 kg of Ne gas is raised from 20 °C to 100 °C at a constant pressure. [M(Ne) = 10 g/mole]

   (a) Determine the amount of heat required.

   (b) Determine the change in the internal energy of the gas.

   (c) Determine the work done by the gas on the environment.
5. Solve problem 4 if the volume and not the pressure stay constant.   *[5x10⁶ J, 5x10⁶ J, 0]*

6. A cylinder contains 3.0 moles of He gas at a temperature of 300 K. The state of the gas is changed through following two stages:

   I. The pressure is kept constant and the heat is added so that the Volume increases to 3 times the original volume.
   II. Then the volume is kept and the heat is removed so that the pressure drops to 1/2 the original value.

   (a) Determine the temperature of the gas at the end of the processes I and II.

   (b) Determine the amount of heat involved during the two processes.

   (c) Determine the change in the internal energy during the process I, the process II and the two processes together.

   (d) Determine the amount of work done by the gas on the environment in the two processes.

7. An ideal gas has a volume of 4.0 m³ at 5.0x10⁵ Pa pressure. It is taken through a reversible cycle through the 3 steps below: Draw a PV diagram for this cycle.

I. Its pressure increased to 10.0x10⁵ Pa at constant temperature.
II. Its volume is increased back to 4.0 m³ at constant pressure.
III. Its pressure is decreased back to 5.0x10⁵ Pa at constant volume

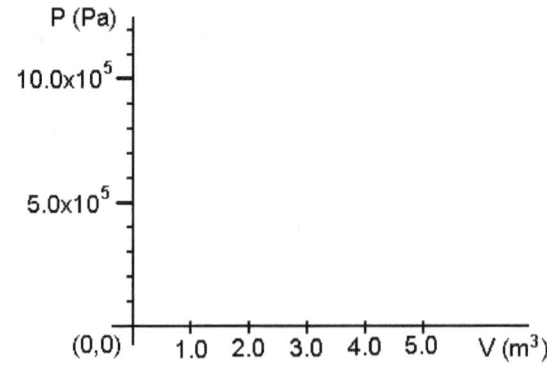

Multiple choice:

1. During a thermodynamic process the internal energy of the system remains unchanged. This kind of process is called
A. Isochoric Process   B. Isothermal Process   C. Isobaric process
D. Adiabatic Process   E. None of these

2. The PV diagram for a thermodynamic process is a vertical straight line. For this process,
A. $Q = 0$ B. $W = 0$ C. $\Delta T = 0$ D. $\Delta P = 0$ E. $\Delta U = 0$

3. The PV-diagram for a process is a horizontal straight line. For this process
A. $Q = 0$ B. $W = 0$ C. $\Delta T = 0$ D. $\Delta P = 0$ E. $\Delta U = 0$

4. In one complete cycle of a heat engine, the gas returns to its original state. Thus for a complete cycle,
(A) $W = 0$; $\Delta U = 0$
(B) $W < 0$; $\Delta U < 0$
(C) $W > 0$; $\Delta U > 0$
(D) $W > 0$; $\Delta U < 0$
(E) $W < 0$; $\Delta U = 0$

5. For an ideal gas, $W = 0$ for
(A) Isothermal Process
(B) Isobaric Process
(C) Isochoric Process
(D) Adiabatic process
(E) Cyclic Process

6. For an ideal gas, $Q = 0$ for
(A) Isothermal Process
(B) Isobaric Process
(C) Isochoric Process
(D) Adiabatic process
(E) Cyclic Process

7. For an ideal gas, T = constant for
(A) Isothermal Process
(B) Isobaric Process
(C) Isochoric Process
(D) Adiabatic process
(E) Cyclic Process

8. 1. For an ideal gas, $\Delta U = 0$ for
(A) Isothermal Process
(B) Isobaric Process
(C) Isochoric Process

(D) Adiabatic process
(E) Isotonic process

Chapter 27
Cyclical Thermodynamic Processes and Heat Engines & Refrigerators

Cyclical Thermodynamic Process: It is a process in which a gas starts from one state and returns to the same state after going through changes in P, V, and T. Since the system returns to the same temperature $\Delta T = 0 \rightarrow \Delta U = 0$. Thus applying the First Law of Thermodynamics, $\Delta U = Q + W$, we get $Q = -W$ for any thermodynamic cycle.

Both the heat engine and refrigerator work in cyclical thermodynamic processes.

- If W is **negative** and Q is **positive** the cycle represents a **heat engine** (Clockwise cycle on PV-diagram)

- If W is **positive** and Q is **negative** the cycle represents a **refrigerator** (Counterclockwise cycle on PV diagram)

Heat Engine:

A heat engine absorbs Q_H amount of heat from a hot reservoir at temperature T_H, converts an amount W into work and the rejects the rest Q_c to the cold reservoir at temperature T_c.

Hence, $Q_H = W + Q_C$ (<u>absolute values taken for all the quantities</u>)

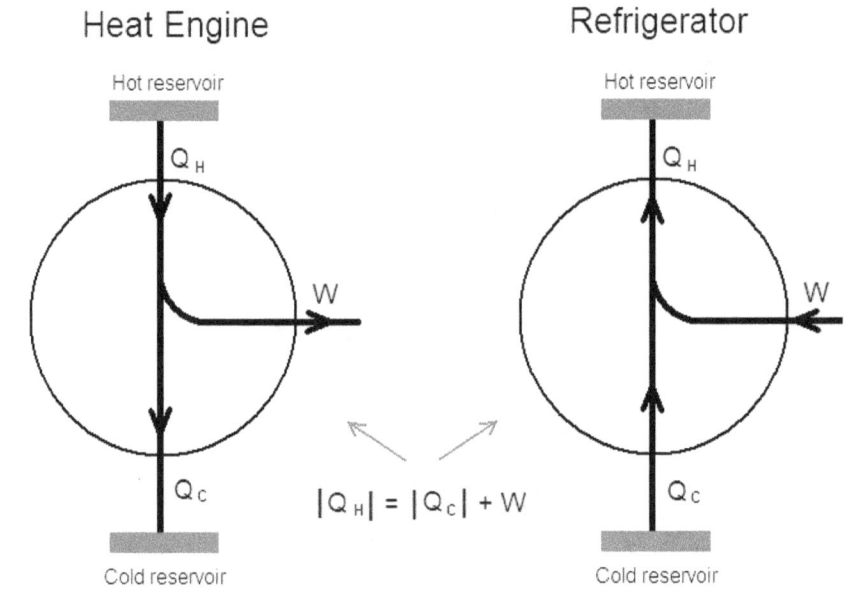

Efficiency (e) of a Heat Engine: The efficiency of any machine is given by

$$efficiency = \frac{UsefulOutput}{Input}$$

The efficiency of a heat engine is given by

$$e = \frac{W}{Q_H} \quad \text{or,} \qquad e = \frac{W}{Q_H} \times 100\% \qquad \text{or,} \qquad e = \frac{Q_H - Q_C}{Q_H} = 1 - \frac{Q_C}{Q_H}$$

Refrigerator: An amount of work, W, is put in it that helps extract heat Q_C from a cold reservoir. Both W and Q_C are rejected to a hot reservoir as heat Q_H

- For both these machines, $|Q_H| = |Q_C| + W$

For a cyclical process, at the end of one cycle the system is back to the same state and hence same temperature. Therefore, for one cycle

Carnot Engine: A Carnot engine is a theoretical ideal engine which works on a cycle (Carnot Cycle) that is bounded by two adiabatic and two isothermal processes. Carnot engine has the maximum possible efficiency for given temperatures, T_H and T_C, of the cold and hot reservoirs respectively. For the Carnot engine only,

$$\frac{Q_C}{Q_H} = \frac{T_C}{T_H}$$

The temperatures T_C and T_H are measured in Kelvin.

Hence the efficiency of the Carnot engine can be written as

$$e = 1 - \frac{T_C}{T_H}$$

Nicholas Leonard Sadi Carnot [1796-1932]

Upper Limit on the efficiency of a heat engine

A Carnot engine sets the upper limit on the efficiency of any real engine that works between the same two temperatures T_C and T_H as the Carnot engine.

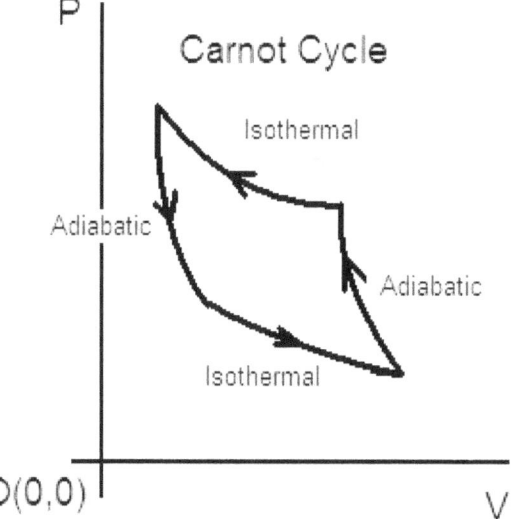

P

Carnot Cycle

Isothermal

Adiabatic

Adiabatic

Isothermal

O(0,0)

V

Problems:

1. In the pV diagram below, 28 g of He is taken through the cycle 12341.
The processes 1-2 and 3-4 are isobaric, and the processes 2-3 and 4-1 are isochoric.

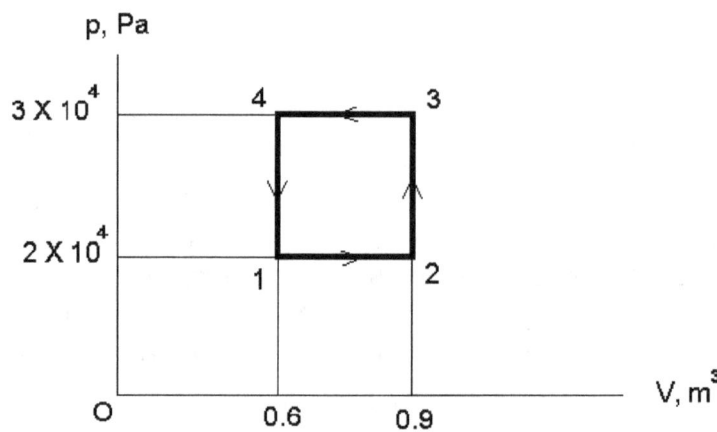

a. Determine the temperatures T_1, T_2, T_3, and T_4 *[206K, 309K, 464K, 309K]*

b. Determine the work done W_{12}, W_{23}, W_{34}, and W_{cycle}.
                              [-6000J, 0J, +9000J, 0J, +3000J]

c. Determine ΔU_{12} ΔU_{23}, ΔU_{34}, ΔU_4, and ΔU_{cycle}.   *[8992J, 13531J, -13531J, -8992J, 0J]*

d. Determine the amount of heat Q_{12}, Q_{23}, Q_{34}, Q_{41}, and Q_{cycle}. *[14992J, 13531J, -22531J, -8992J, -3000J]*

e. Is this a heat engine or a refrigerator? If heat engine, find its efficiency. *[Refrigerator]*

2. In the pV-diagram below, 400 moles of and ideal monatomic gas is taken through a thermodynamic reversible cycle 1-2-3-1.

The process 1-2 is isobaric.
The process 2-3 is isothermal
The process 3-1 is isochoric
During the isothermal process an amount of heat (Q_{23}) 3.30 x 10^6 J is <u>removed</u> from the system.

a. Determine the temperatures T_1, T_2, and T_3.

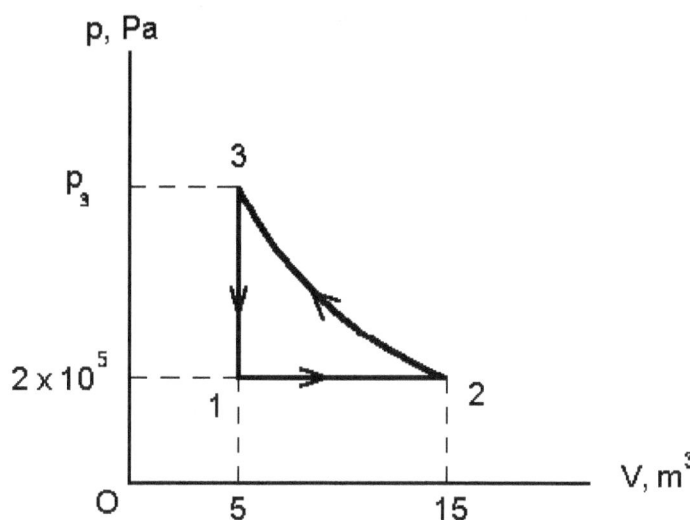

b. Determine the pressure p_3 for the state 3.

c. Determine the work done W_{12}, W_{23}, W_{31}, and W_{cycle}.

d. Determine ΔU_{12} ΔU_{23}, ΔU_{31}, and ΔU_{cycle}.

e. Determine the amount of heat Q_{12}, Q_{31}, and Q_{cycle}.

f. Is this a heat engine or a refrigerator? If heat engine, find its efficiency.

3. A heat engine has an efficiency of 44%. It is supplied with 5,000 J of heat. Determine the amounts of work output and the heat rejected by the engine.   [2200J, 2800J]

4. A Carnot engine has an efficiency of 60%. It absorbs 800 J of heat from a reservoir at 200 °C. Determine Q_C, T_C, and W for this engine.

5. A heat engine absorbs 3.5×10^5 J of heat to lift a 200-kg block through a height of 50 m. For this engine determine the amount of work done, the amount of heat rejected, and the efficiency.

[98000J, 2.52×10^5J, 28%]

6. A 900-MW power reactor has an efficiency of 23%. Determine the rates at which the energy is supplied to the reactor and the heat is rejected from the reactor to the atmosphere. [Hint: W=900 MW]

7. The PV-diagram below shows thermodynamic cycle for 21 moles of an ideal gas. The gas is initially at a pressure of 9×10^4 Pa and has a volume of 0.6 m³.

        The process 1 → 2 is isothermal
        The process 2 → 3 is isobaric
        The process 3 → 1 is isochoric
        Work done 1 → 2 = - 5.92×10^4 J

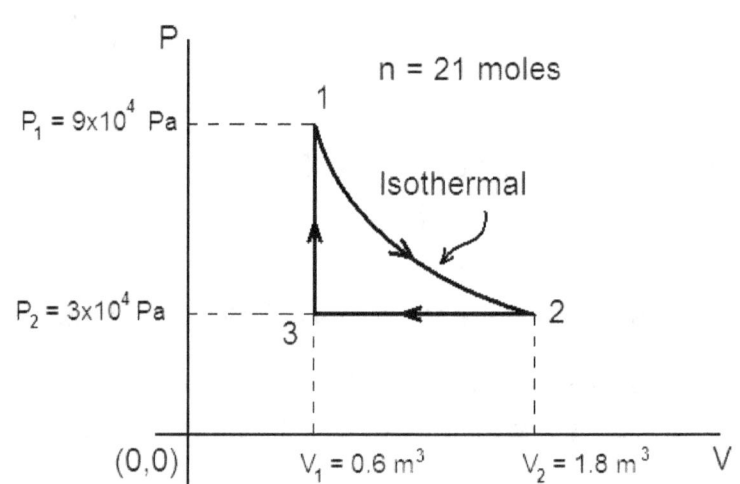

a. Determine the temperatures T_1, T_2, and T_3.

b. Determine the work done W_{23}, W_{31}, and W_{cycle}.

d. Determine ΔU_{12} ΔU_{23}, ΔU_{31}, and ΔU_{cycle}.

e. Determine the amount of heat Q_{12}, Q_{23}, Q_{31}, and Q_{cycle}.

f. Is this a heat engine or a refrigerator? If heat engine, find its efficiency.

Multiple Choice:

1. The efficiency of a heat engine is equal to
(A) Q_H/Q_C (B) Q_H/W (C) W/Q_H (D) W/Q_C (E) Q_C/W

2-3.

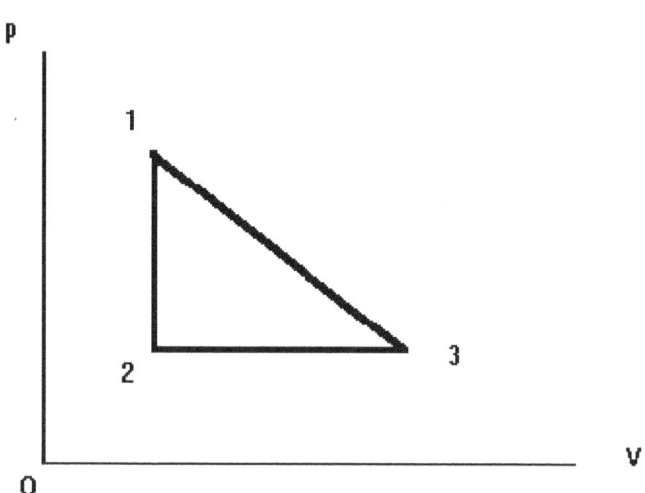

In the diagram above,

    2. Consider the process from 1 to 2. ΔU from 1-2 is <0. Hence,

    (A) W=0 and Q<0   (B) W=0 and Q>0   (C) W>0 and Q=0
    (D) W<0 and Q=0   (E) W=0 and Q=0

    3. Consider the process from 2-3. Q>0 for the process 2-3. Hence,

    (A) W>0 and ΔU<0   (B) W>0 and ΔU=0   (C) W<0 and ΔU<0
    (D) W<0 and ΔU=0   (E) W<0 and ΔU>0

    4. For a cyclical process, the area enclosed by the PV diagram has the units of

    (A) Kelvins   (B) Joules   (C) Pascals   (D) meter^3 (E) meter^2

5. A heat engine in a vehicle runs between the temperatures of 1200 K and 300 K. The upper limit on the efficiency of this engine is
(A) 4%   (B) 25%   (C) 36%   (D) 75%   (E) 100%

Chapter 28
Second Law of Thermodynamics
Entropy (S)

The Second Law of Thermodynamics can be stated in any of the following three different forms:

1. Clausius Statement: (No perfect refrigerator)

Heat, by itself, cannot flow from a body at lower temperature to a body at higher temperature.

2. Kelvin-Planck Statement: (No perfect heat engine)

No heat engine working in a cyclic process will absorb heat from a single reservoir and convert it completely into work

3. Entropy:

In any thermodynamic process that proceeds from one equilibrium state to another, the entropy of a (system + environment) either remains unchanged or increases.

Entropy is a thermodynamic quantity. Entropy is related to the randomness, chaos, or disorder of a thermodynamic system. An increase in the randomness, chaos, or disorder of a system implies an increase in the entropy of the system.

Entropy is a state variable i.e. its value depends on the state and not on the process by which the system is brought to that state.

Other state variables for a thermodynamic system include *temperature, pressure, volume,* and *internal energy.* Amount of *heat* and *work* are <u>not</u> state variables. Their values depend on the process connecting two states.

Entropy Change (ΔS):
For any thermodynamic process: ΔS (system + environment) ≥ 0

Reversible Processes: ΔS (system + environment) = 0
ΔS (system) = Q/T

Irreversible Process: ΔS (system + environment) > 0

Reversible Cycle: ΔS (cycle) = 0

Free Expansion: ΔS (system) > 0

Entropy is one of the arrows of *time*. The entropy of the Universe is always increasing. This may lead to what is called as 'heat death' of the Universe.

Energy of the Universe is constant; Entropy is not!
Energy obeys the conservation law; Entropy does not!

Thus in the closing on the topic of thermodynamics, we can say:

First Law of Thermodynamics : YOU CAN'T WIN!
Energy is conserved. You cant' get energy out of nothing.

Second Law of Thermodynamics: YOU CAN'T WIN; YOU CAN'T BREAK EVEN EITHER!
All the available energy cannot be converted into work.

Third Law of Thermodynamics:

It is not possible to lower the temperature of any system to absolute zero in a finite number of steps.

Some interesting facts & figures
• The 'chaos cloud' hoax originated in a *Weekly World News* article in September 2005. According to the article, the 'chaos cloud' is a massive object in outer space that "dissolves everything in its path, including comets, asteroids, planets and entire stars", and is due to reach Earth in 2014.
• Edward Lorentz laid the foundation for chaos theory; he showed that small changes in initial conditions can have large impacts. This idea is called the *butterfly affect*.

Multiple Choice:

1. Which law of thermodynamics states that even for an ideal heat engine the efficiency cannot be 100%?
(A) First Law of Thermodynamics   (B) Second Law of Thermodynamics
(C) Third Law of Thermodynamics   (D) Zeroth Law of Thermodynamics
(E) Law of Conservation of Energy

2. Which law of thermodynamics is based the Law of Conservation of Energy?
(A) First Law of Thermodynamics   (B) Second Law of Thermodynamics
(C) Third Law of Thermodynamics   (D) Zeroth Law of Thermodynamics
(E) Boyle's Law

3. Which quantity is a measure of disorder in a thermodynamic system?
(A) Temperature   (B) Pressure   (C) Volume   (D) Entropy   (E) Density

Chapter 29

FLUID STATICS

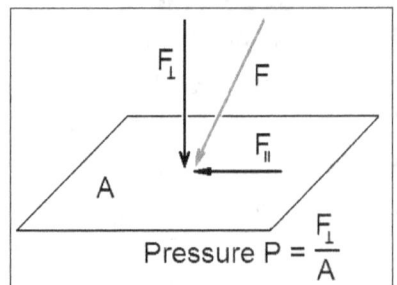

Pressure is defined as force per unit area on a surface. If the force F is perpendicular to the surface of area A then the pressure is given by

$$P = F/A$$

If the force **F** is not perpendicular then only its component $F_\perp$ which is perpendicular to the area **A** is effective in creating the pressure. Thus, $P = F_\perp/A$

Pressure $P = \dfrac{F_\perp}{A}$

The SI unit of pressure is N/m^2 and is called **Pascal (Pa)**

Gauge Pressure (P): Pressure due to liquid alone

Pressure **P** due to a liquid at a point depends on the <u>depth</u> (**h**) of the point below the free surface of the liquid, <u>even if the point is not directly below the open surface</u>. The pressure also depends on the density (**ρ**) of the liquid and the acceleration of gravity (**g**). It is given by

$$P = \rho g h$$

Area = A

Mass = m

F = W = mg

Presure due to the block on the bottom surface

$$P = \frac{F}{A} = \frac{mg}{A}$$

Absolute Pressure: However, if the open surface of the liquid is exposed to atmosphere of pressure P_{atm} then the <u>absolute</u> pressure, P_{abs} at the point **p** is given by

$$P_{abs} = P_{atm} + \rho g h$$

P_{abs} ➔ absolute pressure at the depth h
P_{atm} ➔ atmospheric pressure
$\rho g h$ ➔ gauge pressure

Pascal's principle:

Pressure applied to an enclosed fluid is transmitted undiminished to every portion of the fluid and to the walls of the container.

Hydraulic Jack (Press or Brakes): These devices work on the Pascal's Principle. They are used to amplify the applied force. There are two interconnected tubes; one tube is several times the diameter (area A_1) of the other tube (area A_2). A force (F_1) is applied on a piston in the small tube and the load (F_2) is placed on the piston of the large tube. By Pascal's Law the pressure applied at the small piston should be same as the pressure applied at the large piston. Thus, $F_1 / A_1 = F_2 / A_2$

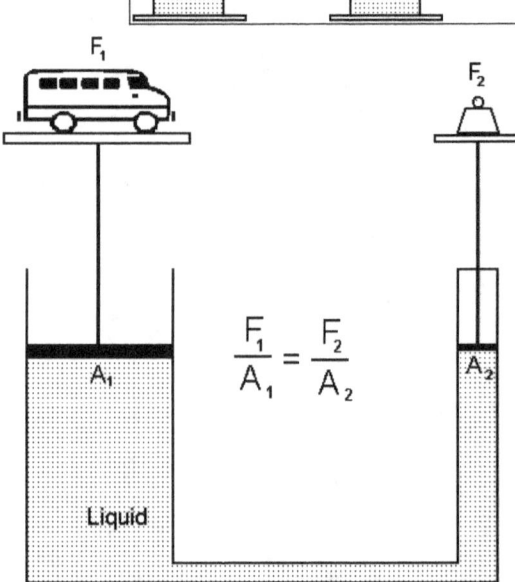

$$\frac{F_1}{A_1} = \frac{F_2}{A_2}$$

Archimedes Principle:

Eurika! Eurika!!

The **buoyant** force (**F$_B$**) due to a liquid on an object partly or completely immersed in it is given by

$$F_B = \rho_L V_{in}\, g$$

where,

ρ_L → density of the liquid
V_{in} → the volume of the object <u>inside</u> the liquid
g → the acceleration due to gravity

Buoyant force and Loss of Weight:

The upward buoyant force, **F$_B$**, acts against the downward weight **mg** of the solid immersed in a liquid. Hence, as shown in the figures below, if a solid suspended from a weight scale is lowered in a liquid, the scale registers a decrease (loss) in weight equal to **F$_B$** and reads **mg - F$_B$**. We will refer to mg as **W$_{air}$** and **mg − F$_B$ = W$_{air}$ − F$_B$** as **W$_{liquid}$**.

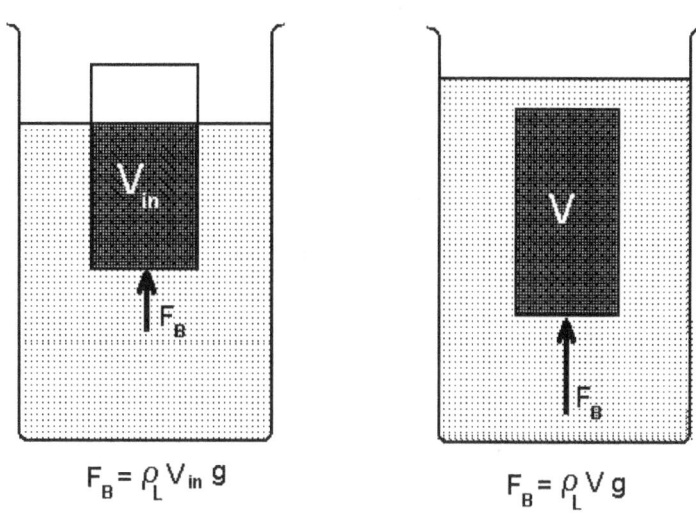

$$F_B = \rho_L V_{in}\, g \qquad\qquad F_B = \rho_L V\, g$$

Buoyant force on a partially and a completely immersed object in a liquid of density ρ_L

Determination of density using Archimedes Principle: Density of a solid or liquid can be found using the Archimedes' principle with any of the following equations:

$$\rho_s = \frac{W_{air}}{W_{air} - W_{liquid}}\, \rho_L$$

or

$$\rho_s = \frac{W_{air}}{F_{Buoyancy}}\, \rho_L$$

or

$$\rho_s = \frac{W_{air}}{Weight.of.liquid.displaced}\, \rho_L$$

W$_{air}$ of the body in air
W$_{liquid}$ = weight of the body when it is immersed <u>completely</u> in a liquid
ρ_L = density of the liquid is

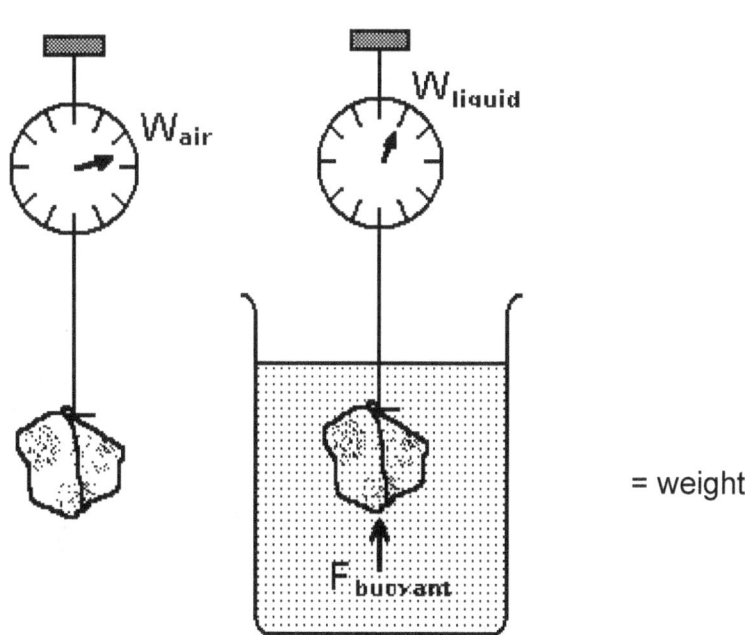

$$F_{buoyant} = W_{air} - W_{liquid}$$

= weight

Buoyancy on a Floating Body:

For a body floating in a liquid,

The weight of the floating body
= the weight of the liquid displaced
= the buoyant force on the floating body

$W_{floating\ body} = F_B$

Density of a Floating Body:

V_i = Volume of the floating body inside the liquid
V = Total volume of the floating body
ρ_L = Density of the liquid

$$\rho_{floating\ body} = \frac{V_i}{V}\rho_L$$

Free-body Diagram of a floating body

$$\rho_{floating\ body} = \frac{V_i}{V}\rho_{liquid}$$

$$F_{buoyant} = W_{floating\ body}$$

Cylindrical (or prismatic) body floating with axis vertical:

$$\rho_{floating\ body} = \frac{V_i}{V}\rho_L = \frac{h_i A}{h A}\rho_L$$

$$\rho_{Floating\ Body} = \frac{h_i}{h}\rho_L$$

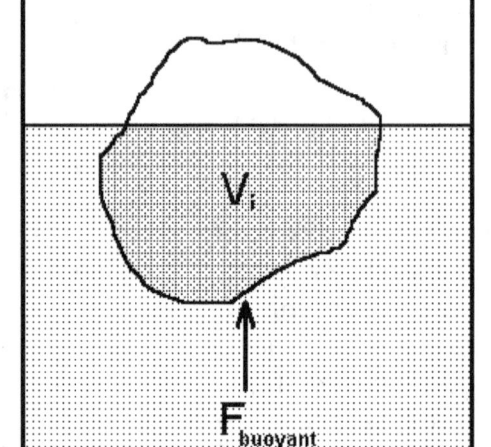
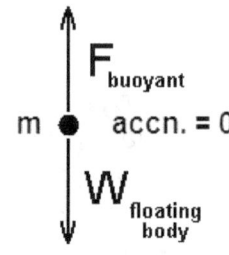

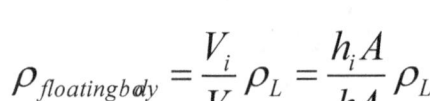

Dead sea water has a high concentration of salt making its water so dense that people cannot sink in it

Density of a liquid using a U-tube:

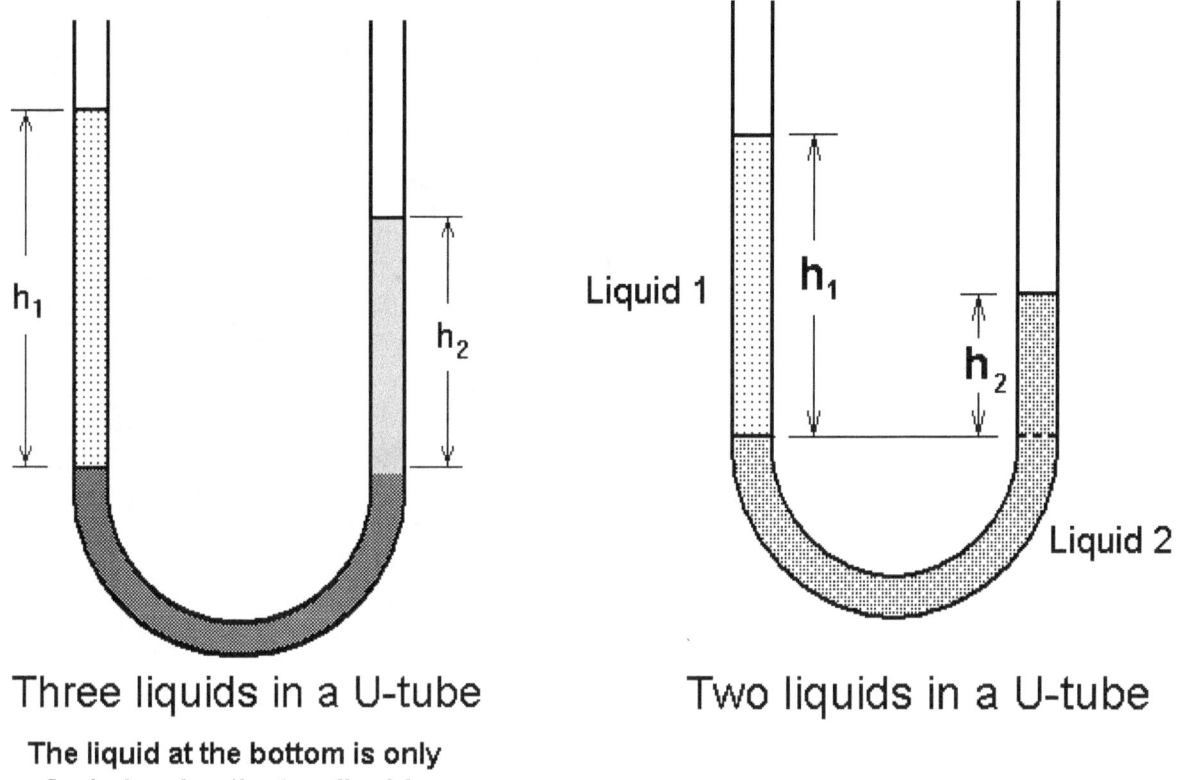

Three liquids in a U-tube

The liquid at the bottom is only
for balancing the two liquids

Two liquids in a U-tube

A U-tube can be used to compare densities of 2 liquids. One may use two or three different liquids with the U-tube method.

A column **h₁** of a liquid of density ρ_1 is balanced against a column **h₂** of another liquid of density ρ_2. The densities of the two liquids can be compared using the equation,

$$\rho_1 \, h_1 = \rho_2 \, h_2$$

Some interesting facts & figures
• The density of water is 999.8395 kg/m^3 at 0°C is and 999.9720 kg/m^3 at 4°C
• Porcupines float in water.
• The planet Saturn has a density lower than water. So, if placed in water it would float.
• The deepest part of the ocean is 35,813 feet (10,916 meters) deep and occurs in the Mariana Trench in the Pacific Ocean. At that depth the pressure is 18,000 pounds per square inch or about 1,000 atmospheres or 1.1x10^8 Pa.
• The largest man-made lake in the U.S. is Lake Mead, created by Hoover Dam.
• An inch (2.5 centimeters) of rain water is equivalent to 15 inches (38.1 centimeters) of dry, powdery snow.

PROBLEMS:

1. In a U-tube method form determination of density of an unknown liquid, a 10.5 –cm column of water is balanced by a 15.6-cm column of certain oil. Determine density of the oil.   *[673 kg/m³]*

2. Below is the diagrammatic representation of an experiment involving the Archimedes' Principle. A block of density ρ_S = 6,000 kg/m^3 and volume V = 0.005 m^3 is suspended from a spring scale. (Density of water ρ_W = 1,000 kg/m^3)

(a)

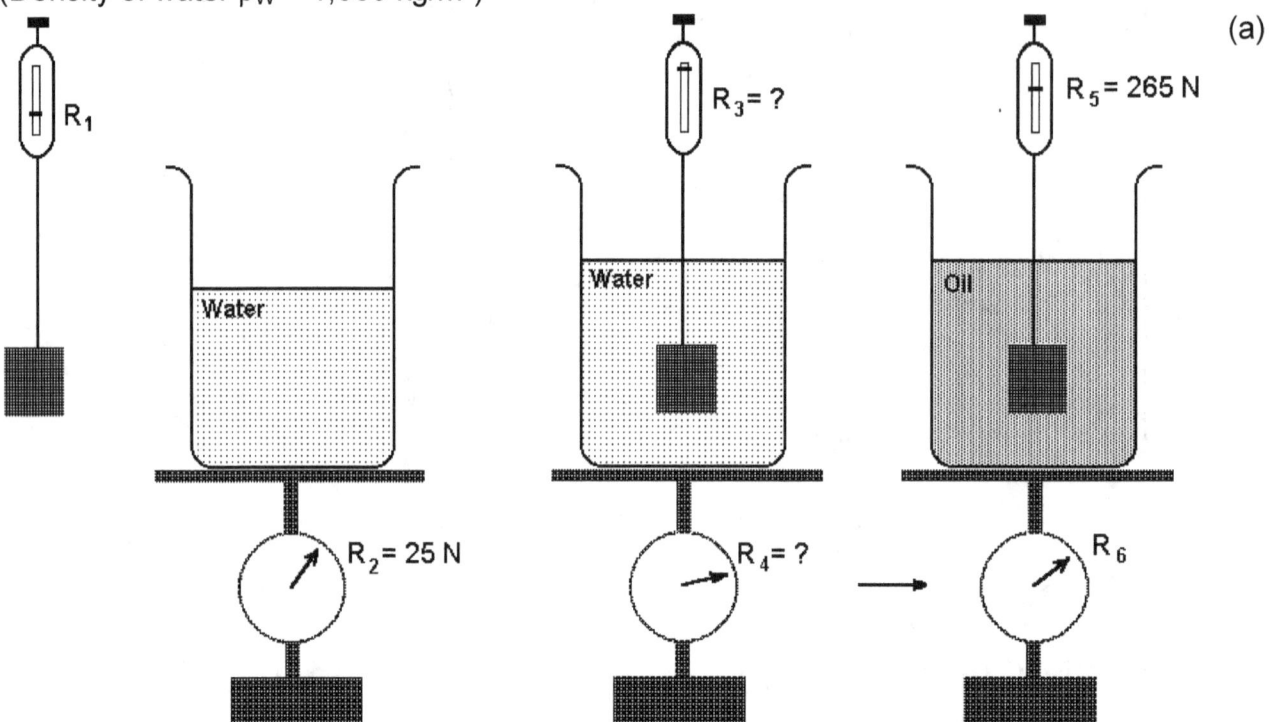

Determine the reading R_1 in newtons.

(b) A beaker with water is placed on another scale and the scale reads R_2 = 25 N. The solid is now immersed in water. Determine the readings R_3 and R_4.

(c) The solid is now immersed in unknown oil. The reading in the spring scale is 265 N. Determine the density (ρ_L) of the oil.

3. An experimental hydraulic lift is to be designed to lift a maximum of 2 metric tons (1 metric ton = 1,000 kg) load with a maximum applied force due to a mass of 50 kg. If the diameter of the larger cylinder is 1 meter calculate the diameter of the smaller cylinder.

4. A 0.4 m x 0.4 m x 0.4 m wooden cube has a density of 800 kg/m^3. It is floating in water (ρ_L) as shown in the figure.
(a) Calculate the height (h_o) of the block above the water surface.

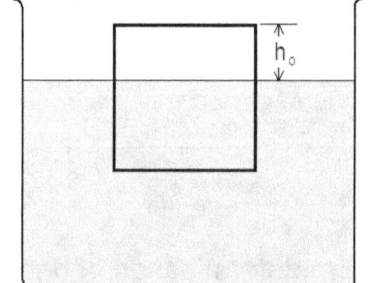
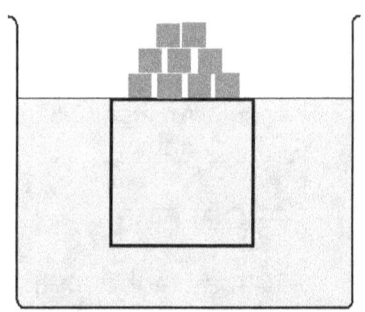

(b) Calculate how many iron cubes each of mass 0.25 kg can be placed on the cube before the cube sinks.

5. The diagram below shows a variation on Atwood's Machine. The pulley and the string are massless and frictionless. The two blocks are made of the same metal and have masses $m_1 = 8$ kg and $m_2 = 12$ kg. The block m_2 is completely immersed in water of density $\rho_L = 1,000$ kg/m^3. The whole system is stationary.

(a) Calculate the weights W_1 and W_2 of m_1 and m_2 respectively.   *[78.4 N, 117.6 N]*

(b) Draw free-body diagrams for m_1 and m_2.

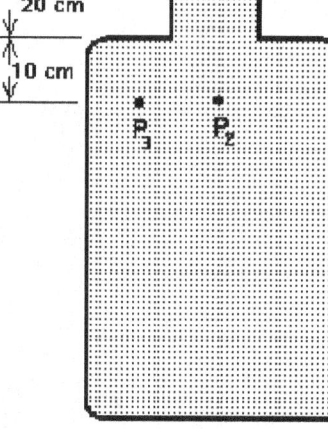

$m_1 = 8$ kg

$m_2 = 12$ kg

$\rho_L = 1,000$ kg/m^3

(c) Calculate the tension F_T in the string.   *[78.4 N]*

(d) Calculate buoyant force F_B acting on m_2.   *[39.2 N]*

(e) Calculate the volume of the block m_2.   *[4x10^{-3} m^3]*

(f) Calculate the density of the material of the two blocks.   *[3,000 kg/m^3]*

Multiple Choice:

1. A block of wood is floating in water with 25% of its volume above the water surface. If the density of water is 1000 kg/m^3 the density of the block is
(A) 0.25 kg/m^3 (B) 0.75 kg/m^3 (C) 250 kg/m^3 (D) 500 kg/m^3 (E) 750 kg/m^3

2. An open bottle is filled completely with water as shown in the diagram below. Which of the points shown have the same pressure?
(A) P_1 and P_2 (B) P_2 and P_3
(C) P_1 and P_3 (D) P_1, P_2, and P_3
(E) None of these

3. A block is suspended from a spring scale. The scale reads 25 N when the block is in the air. The scale reads 20 N when the block is completely immersed in water. The density of the block must be
(A) 5000 kg/m^3 (B) 4000 kg/m^3
(C) 1250 kg/m^3 (D) 800 kg/m^3 (E) 250 kg/m^3

4. A floating ball has 10% of its volume inside water of density 1,000 kg/m^3. The density of the ball must be
(A) 10,000 kg/m^3 (B) 1,000 kg/m^3 (C) 100 kg/m^3 (D) 10 kg/m^3
(E) 1 kg/m^3

5. An iceberg has a density of 950 kg/m^3 and a volume of 5,500 m^3. It is floating in water of density 1,100 kg/m^3. Which of the following is the best estimate for the iceberg's volume under water?
(A) 4,750 m^3 (B) 190 m^3 (C) 220 m^3 (D) 750 m^3 (E) 4,550 m^3

6. A helium balloon rises up and comes to rest in contact with the ceiling. If W is the weight of the balloon and F_N is the normal force on it from the ceiling, which of the following correctly gives the buoyant force F_B on the balloon?
(A) $F_B = W$ (B) $F_B = F_N$ (C) $F_B = W - F_N$ (D) $F_B = W + F_N$ (E) $F_B = F_N - W$

123

Chapter 30
FLUID DYNAMICS

It is assumed here that the fluid is incompressible, non-viscous, and that any flow of the liquid is streamlined (irrotational).

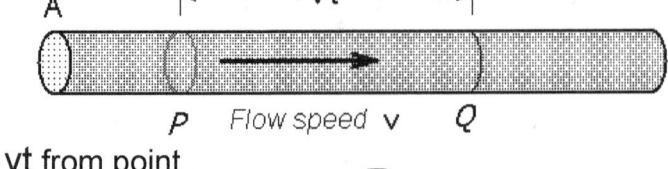

Volume flow rate of a liquid through a pipe

Let v be the flow speed of a liquid through a pipe of

cross sectional area A. The liquid travels a distance vt from point P to point Q in time t. The cylindrical volume between P and Q is the area of cross section, A, times the length, vt, of the cylinder, that is, Avt.

- The **volume flow rate** of the liquid through the pipe is
  hen
  volume/time = (Avt)/(t) = **Av**   in m^3/s
- The **mass flow rate** of the liquid through the pipe is then

  Density x volume /time = (ρAvt)/(t) = **ρAv**   in kg/s

Continuity equation: A liquid is flowing through a pipe of varying cross sections as shown in the diagram below. The rate at which liquid is flowing in the narrow section must be the same as the rate at which the liquid is flowing in the broad section. Hence,

$$A_1 v_1 = A_2 v_2$$

where,
A_1 ➔ area of cross section of narrow section
v_1 ➔ velocity of the liquid in the narrow section
A_2 ➔ area of cross section of the broad section v_2 ➔ velocity of the liquid in the broad section

Bernoulli's Equation: Let p_1 and p_2 be any two points in a streamlined flow of a liquid of density ρ. Then at these two points

$$P_1 + \tfrac{1}{2} \rho v_1^2 + \rho g y_1 = P_2 + \tfrac{1}{2} \rho v_2^2 + \rho g y_2$$

Here,

P_1 and P_2 are the pressures at the points p_1 and p_2
ρ is the density of the liquid
y_1 and y_2 are the heights of the points p_1 and p_2
above an arbitrary reference level.
v_1 and v_2 are the velocities of the fluid
at the points p_1 and p_2.

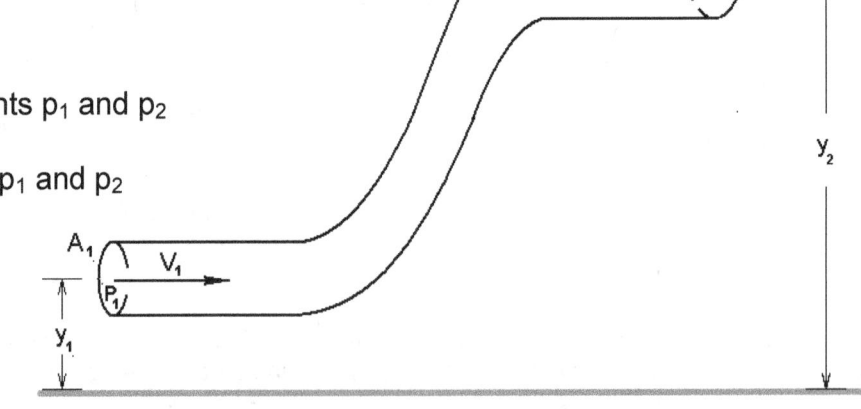

Bernoulli's equation for horizontal Flow:

If the points P_1 and P_2 are at the same height $y_1 = y_2$. Hence the Bernoulli's equation above reduces to

$$p_1 + \tfrac{1}{2} \rho v_1^2 = p_2 + \tfrac{1}{2} \rho v_2^2$$

Flow of liquid through a hole in a container:

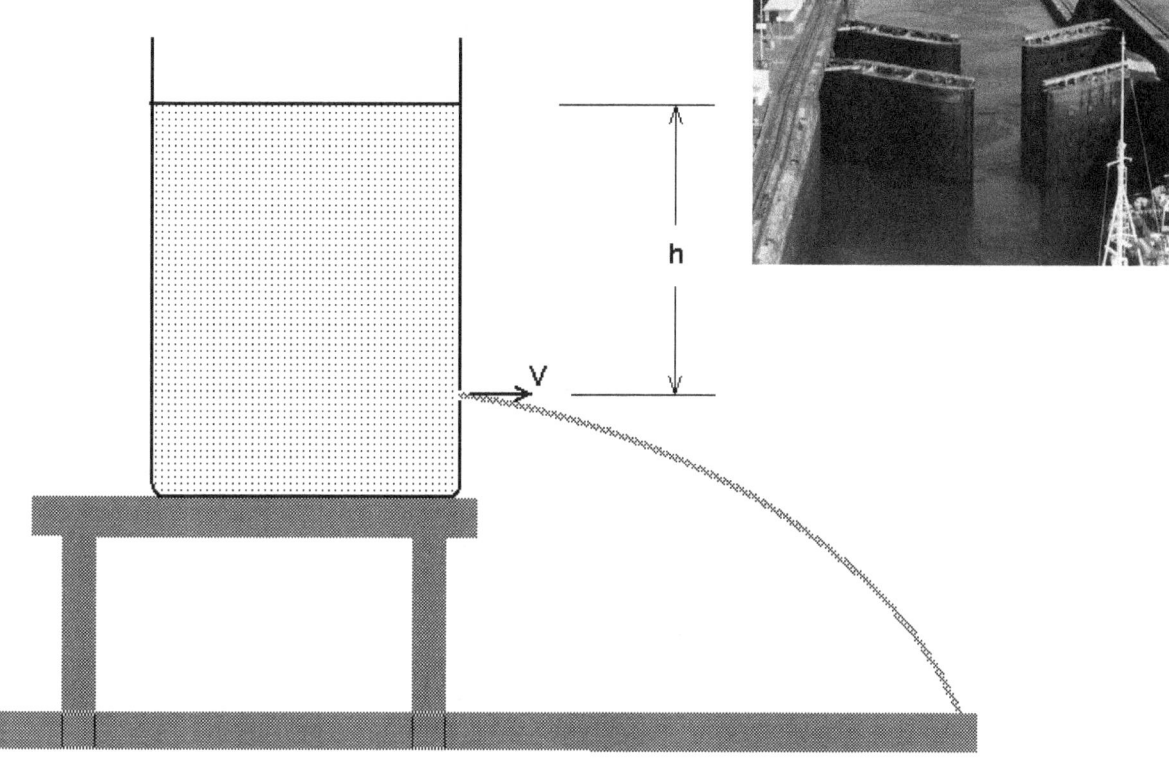

A container with a **large** area of cross section is filled with a liquid. There is a small hole a depth **h** below the liquid surface. As the water flows out, the depth, h, of the hole stays <u>approximately constant</u> because the area of cross section of the container is large. The pressures at the hole and over the surface of the liquid are equal to atmospheric pressure.

The velocity of water through the hole is given by

$$\boxed{v = \sqrt{2gh}}$$

Some interesting facts & figures
• Bernoulli was the first to use the word *integral* in solving Leibniz's problem of the isochronous curve.
• Bernoulli discovered the series of numbers that now known as Bernoulli numbers which are the coefficients of the exponential series expansion of $x/(1-e^{-x})$.

PROBLEMS:

1. A large tank with a large cross sectional area is filled to a height of 3 m. A faucet is connected to a small hole of 1 cm diameter at the bottom of the tank. The top of the tank is open to the atmosphere.

(a) Determine the speed with which the water flows out when the faucet is opened.   *[7.67 m/s]*

(b) Determine the volume flow rate of the water flowing out of the faucet.   *[0.000602 m³/s]*

(c) Determine the mass flow rate of the water flowing out of the faucet.   *[0.602 kg/s]*

(d) How long will it take to fill a 200-liter bucket of water?   *[332.2 s]*

2. A garden hose is held above ground so that the stream of water from it starts horizontally (see the figure below). The cross sectional area of the hose end is 2 cm² and the stream of water hits the ground 3 m away.

(a) If the end of the hose is 1.4 m above the ground determine the velocity at which the water stream leaves the end of the pipe.

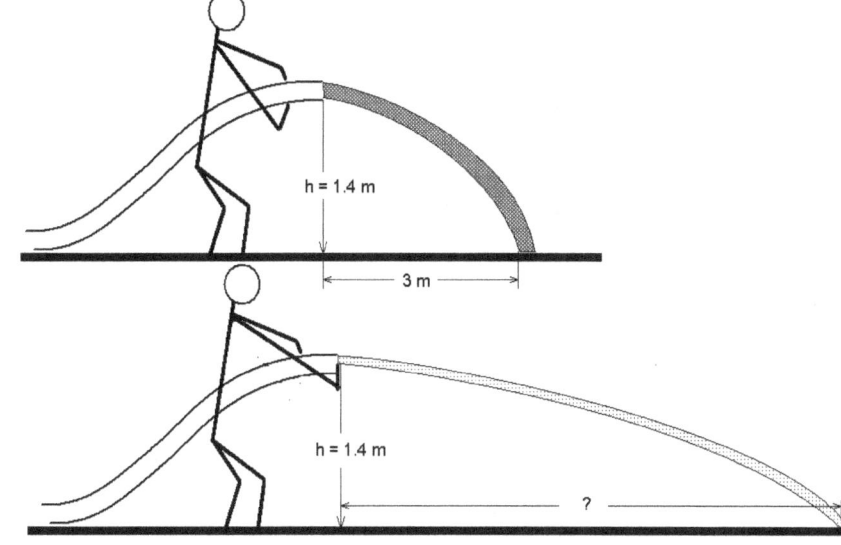

(b) The cross sectional area of the end of the hose is now reduced to 1 cm² by placing thumb over it. Assuming that the rate of flow is not affected, determine how far the stream of water will hit the ground.

3.

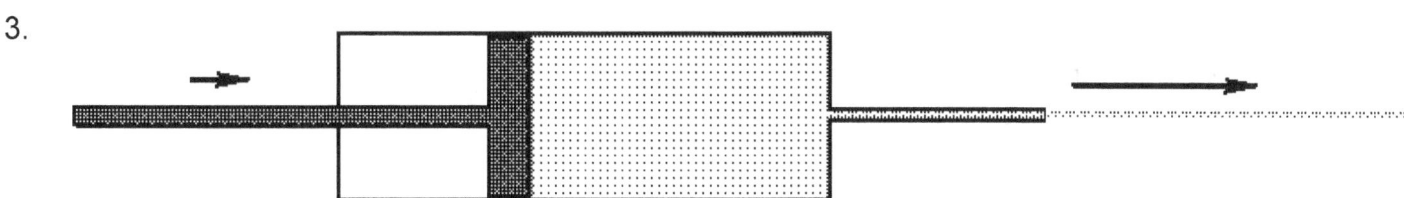

As shown above, a simple squirt gun consists of a cylinder of diameter D_1 = 3 cm and a nozzle of diameter 1mm. A kid holding the squirt gun pushes the piston into the cylinder at a speed of 2.5 cm/s.

(a) Calculate the speed of the water stream squirting from the nozzle.   *[225 cm/s]*

(b) If the cylinder contains 1 liter of water, how long will it take to empty the squirt gun at this speed?   *[56.6 s]*

Multiple Choice:

1. Water stream is flowing out at a speed of 6 m/s through a hole which is 0.8 m below the level of water in an open container. When the level of water has decreased to 0.2 m above the hole the speed of water stream will be
(A) the same   (B) 3 m/s   (C) 1.5 m/s   (D) 6√2 m/s   (E) 24 m/s

2. The diameter of a horizontal pipe changes along its length. As the water flows from the narrow section into the broader section of the pipe, which of the following is true about the rate of flow R and speed of water V in the broader section?
(A) R is same; v decreases
(B) R decreases; V decreases
(C) R increases; V increases
(D) R is same; V increases
(E) R increases; V decreases

Chapter 31
Static Electricity

Thales of Miletus

The science of electricity was known to the early Greeks - Thales of Miletus - around 600 BC. They knew that when a piece of resin called 'elektron' (Greek for amber) was rubbed with fur it acquired the property of attracting bits of straw.

In general anything, which acquires this property upon rubbing, is said to be 'electrified' or 'electrically charged'.

Benjamin Franklin later determined that there are two kinds of charges. He named them as positive (+) and negative (-).

Law of Electrostatic Interaction: *Like charges repel and the unlike charges attract.*

When a glass rod is rubbed with silk, the glass rod acquires a positive charge and silk acquires an equal amount of negative charge.

Benjamin Franklin

When an ebonite rod is rubbed with fur, the ebonite rod acquires a negative charge and fur acquires an equal amount of positive charge.

To sum up:
1. **There are two kinds of charges - positive and negative.**
2. **Like charges repel and unlike charges attract.**

The phenomenon of charging can be explained by atomic theory of matter. In an neutral atom there are as many electrons in the orbit as there are protons in the nucleus. A proton and an electron have equal an opposite charges; proton has a positive charge and the electron has a negative charge.

Electrons may be removed or added to an atom. When two objects are brought in contact or rubbed with each other, the object having greater *affinity* for electrons picks up electron from the other object and gets a net negative charge and the other object now deficient in electron is said to acquire a positive charge.

Conductors, Insulators, Semiconductors, and Superconductors:

Conductors are the materials, which allow the electric charges to flow through them. Examples: Most metals, carbon, electrolytes etc.
Insulators are materials, which offer high resistance to the flow of electric charges through them. Examples: plastics, rubber, Teflon, dry air, pure water, dry wood, glass etc.
Semiconductors: These are the materials whose resistance to electric current is between those of conductors and insulators. Such materials are called semiconductors. Examples: Germanium and Silicon
Superconductors: These materials offer no resistance to the flow of current.

Conservation of electric charges: In any interaction or process the total electric charge is conserved. Electric charges can be created or annihilated but always in equal and opposite amounts.

Gold-leaf Electroscope: It is used to detect charges on an object.

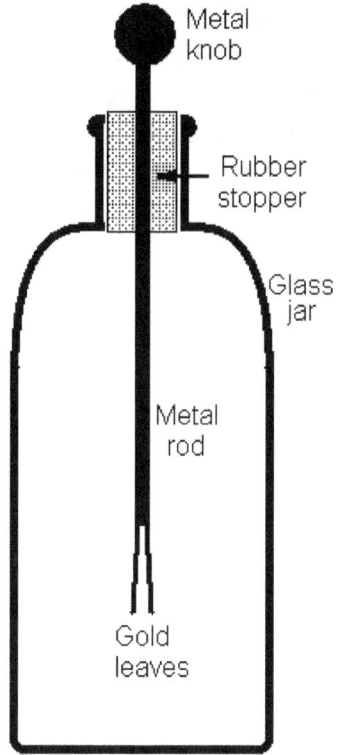

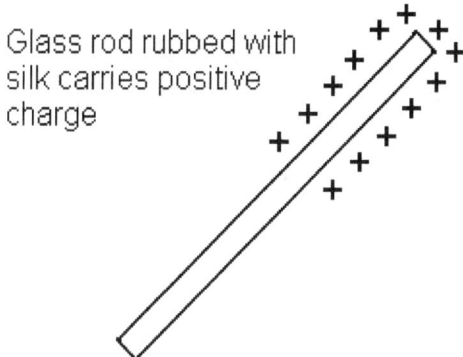

Glass rod rubbed with silk carries positive charge

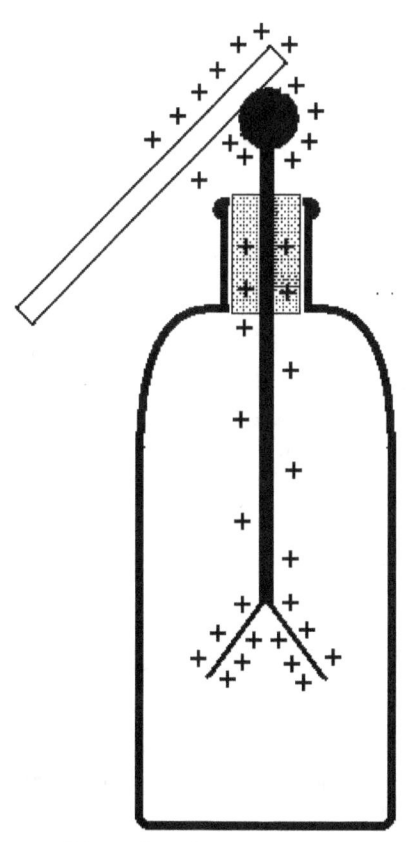

Charging by Conduction

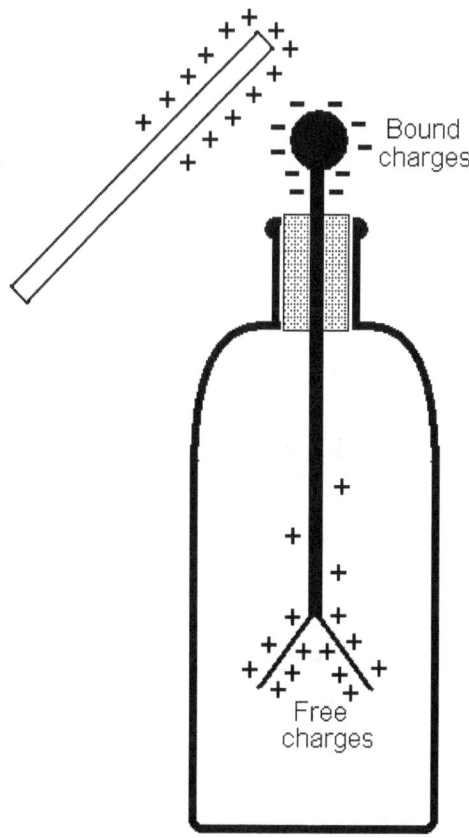

Charging by Induction

Grounding the charges: In the case of charging by induction, as shown above, the charge on the

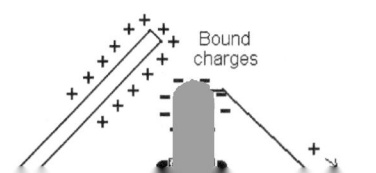

knob is *bound charge* because it is held in place by the attraction from the positive charge on the glass rod. However, the charge at the farther end, that is, on the gold leaves, is *free charge*. If a person touches the knob, the body of the person provides a conducting path for the charges to go to the earth. The positive charge on the knob, being *bound charge*, does not budge. But the *free charge* on the leaves flows to ground through the person's body, the leaves turn neutral and collapse. Now when the rod is removed, the negative charges that were bound become free and move all over the rod and leaves; the leaves diverge again.

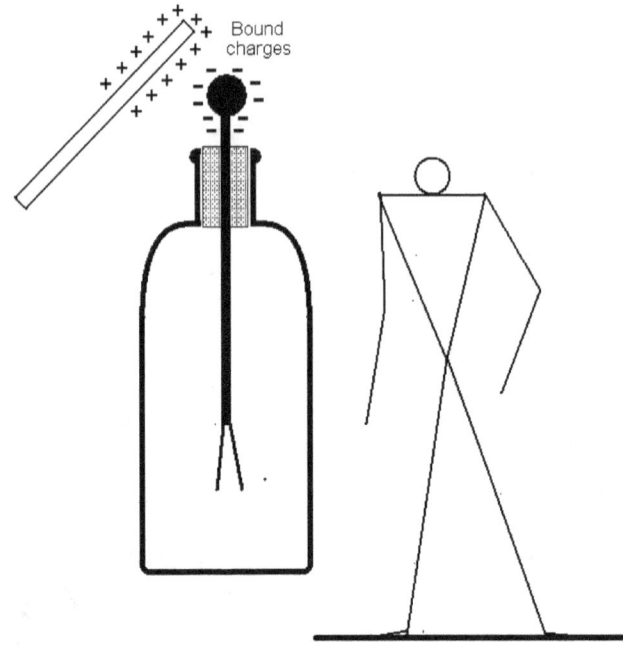

Earth has unlimited capacity to absorb negative as well as positive charges!

Electric charge is quantized: The smallest unit of negative charge is that on an electron i.e. -1.6 X 10^{-19} C. The smallest unit of positive charge is equal and opposite to that on an electron i.e. +1.6 X 10^{-19} C. This is the charge on a proton. The elementary particles called 'quarks' carry 1/3 and 2/3 of these charges. However quarks do not exist as free particles.

Some interesting facts & figures
• Thales' theorem (named after Thales of Miletus) states that if A, B and C are points on a circle where the line AC is a diameter of the circle, then the angle ABC is a right angle.
• Ben Franklin was of the Founding Fathers of the United States.
• Among other inventions and discoveries, Ben Franklin invented the bifocals.

Multiple Choice:

1. The tip of a plastic rod carries a negative charge. It is brought near an initially neutral pith ball. The electrostatic induction causes the closer end of the pith ball to get a positive charge and the farther end a negative charge. Which of the charges is called free charge?
(A) The negative charge on the plastic rod
(B) The positive charge on the pith ball
(C) The negative charge on the pith ball
(D) All of these
(E) None of these

2. An isolated piece of metal is initially neutral. It is given a positive charge. Theoretically, the mass of the metal

piece has
(A) Increased   (B) Decreased   (C) Stayed the same

3. A metal sphere can be charged by addition or removal of electrons from its surface. In order to give a positive charge of 1.6 µC
(A) 10^{19} electrons need to be added.
(B) 10^{19} electrons need to be removed.
(C) 10^{13} electrons need to be removed.
(D) 10^{13} electrons need to be added.
(E) 2.25×10^{-25} electrons need to be removed.

4. A negatively charged rod is brought near an initially neutral gold leaf electroscope. Which of the following diagrams correctly represents the sign of the charges and the labels for the electroscope?

(A)

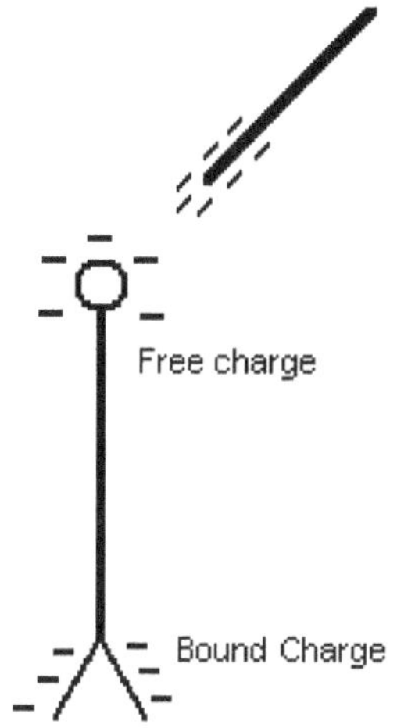

Free charge

Bound Charge

(B)

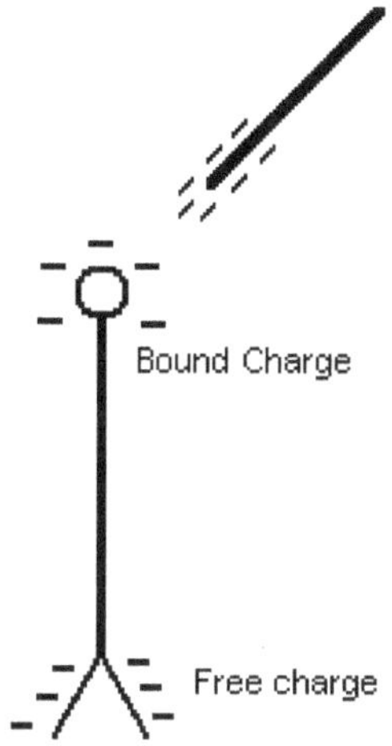

Bound Charge

Free charge

(C)

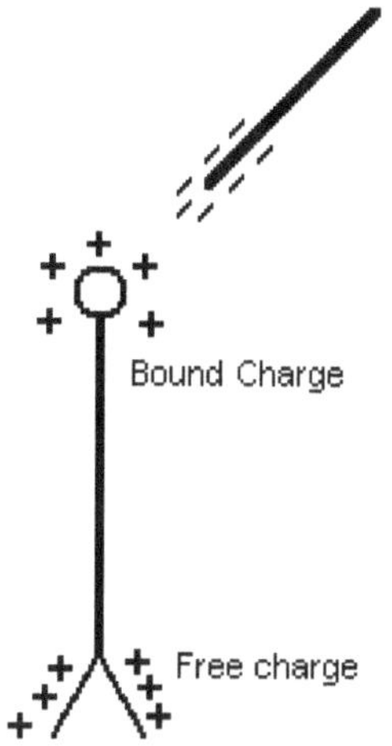

Bound Charge

Free charge

(D)

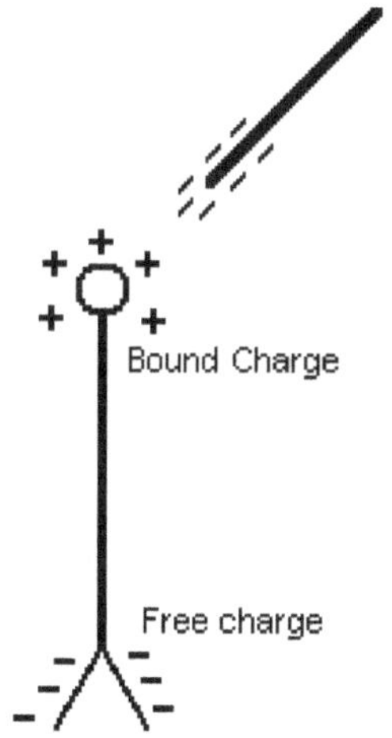

Bound Charge

Free charge

(E)

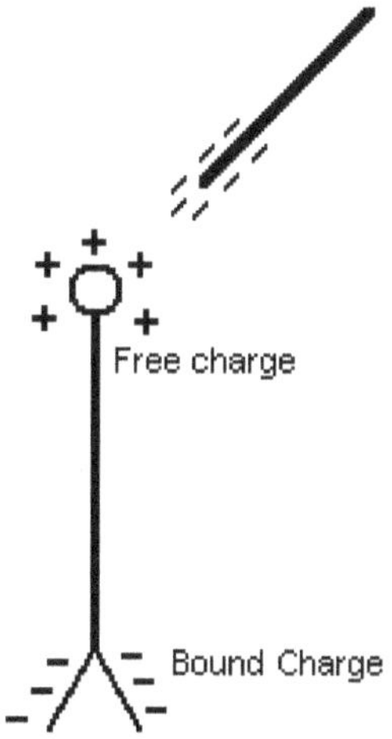

Free charge

Bound Charge

5. Two identical isolated metal spheres X and Y are in contact and initially uncharged. A glass rod is rubbed with silk and brought near X. The sphere Y is then removed away from X while the glass rod stayed near X. The glass rod is then removed. The signs of charges on the two spheres are

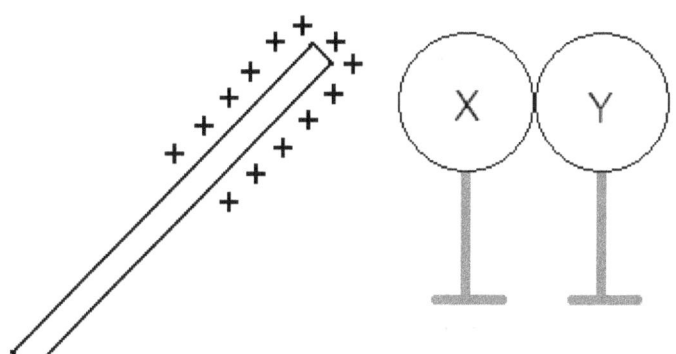

	Sphere X	Sphere Y
A	+	-
B	-	+
C	+	0
D	-	0
E	0	0

6. Two identical isolated metal spheres X and Y are in contact and initially uncharged. A glass rod is rubbed with silk and brought near X. The glass rod is then removed. The sphere Y is then removed away from X. The signs of charges on the two spheres are

	Sphere X	Sphere Y
A	+	-
B	-	+
C	+	0
D	-	0
E	0	0

7. A negatively charged rod is brought near the knob of a gold-leaf electroscope. Then the knob is touched by hand to ground the electroscope. The rod is now moved away and <u>then</u> the hand is removed. Which of the following diagram represents the final state of the electroscope?

(A)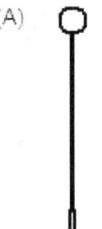

(B)

(C)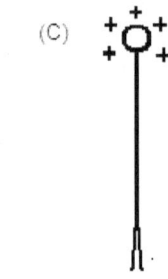

(D)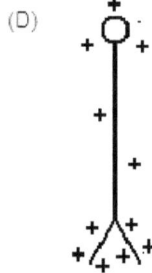

(E)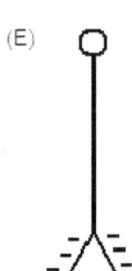

Chapter 32
Coulomb's Law and Electric Field

French Physicist **Charles-Augustin de Coulomb** [14 June 1736 – 23 August 1806] carried out experimental study of the interaction between the charged particles and stated the result, which is now called Coulomb's Law.

In mathematical form the Coulomb's law is,

$$F = k_e \frac{|q_1 q_2|}{r^2}$$

Take absolute value for the product $q_1 q_2$ in the equation above.

F is the magnitude of the force of attraction or repulsion between q_1 and q_2
q_1, q_2 are point charges, measured in coulombs
r is the distance between the point charges
k_e is a constant = **9.0×10^9 N.m^2/C^2**

The Coulomb constant k_e is compressed form of the constant $1/(4\pi\varepsilon_0)$ in which the constant ε_0 is the electric permittivity of vacuum and equals 8.85×10^{-12} C^2/N/m^2.

The equation above provides the magnitude of the force and direction depends on the positions of the interacting charges.

$$F = \frac{k_e |q_1 q_2|}{r^2} \qquad\qquad F = \frac{k_e |q_1 q_2|}{r^2}$$

$$F = \frac{k_e |q_1 q_2|}{r^2} \qquad\qquad F = \frac{k_e |q_1 q_2|}{r^2}$$

$$F = \frac{k_e |q_1 q_2|}{r^2} \qquad F = \frac{k_e |q_1 q_2|}{r^2}$$

For three or more interacting point charges the net force on any charge is the vector sum of all the forces due to the remaining charges.

Electric field is the region around a charge in which it can influence other electric charges.

The electric field strength or simply electric field ($\vec{E}$) at a point is defined as force per unit positive charge placed at that point. Hence,

$$\vec{E} = \frac{\vec{F}}{q}$$

SI unit for E is N/C
Electric field is a vector quantity.

Determination of Force from the Electric Field:

If a point charge q is placed at a point where the electric field is E then the force acting on the particle is given by, $\boxed{F = qE}$

Note: The force F is in the same direction as E if the charge q is positive.
The force F is in the opposite direction to E if the charge q is negative

For a point charge q the electric field at any point a distance r
from it is then take the absolute value for q in the equation above.

$$\boxed{E = k_e \frac{|q|}{r^2}}$$

Direction of Electric Field:
The direction of E at a point is
- away from the charge q if q is positive
- toward the charge q if q is negative

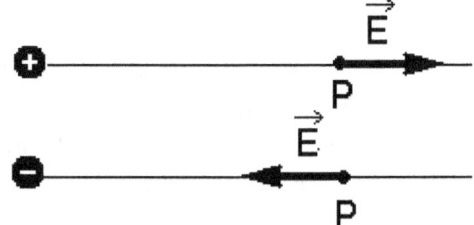

The Net Electric Field due to 2 or more Point Charges
If electric field at a point due to two or more point charges is to be determined then the net electric field at that point is given by the vector sum of all the electric fields due to the given point charges.
If electric field at a point due to two or more point charges is to be determined then the net electric field at that point is given by the vector sum of all the electric fields due to the given point charges.

For example:

In each of the diagrams below, q_1 and q_2 create electric fields E_1 and E_2 respectively at P. The resultant or net field, E_{net}, at P is determined in each case.

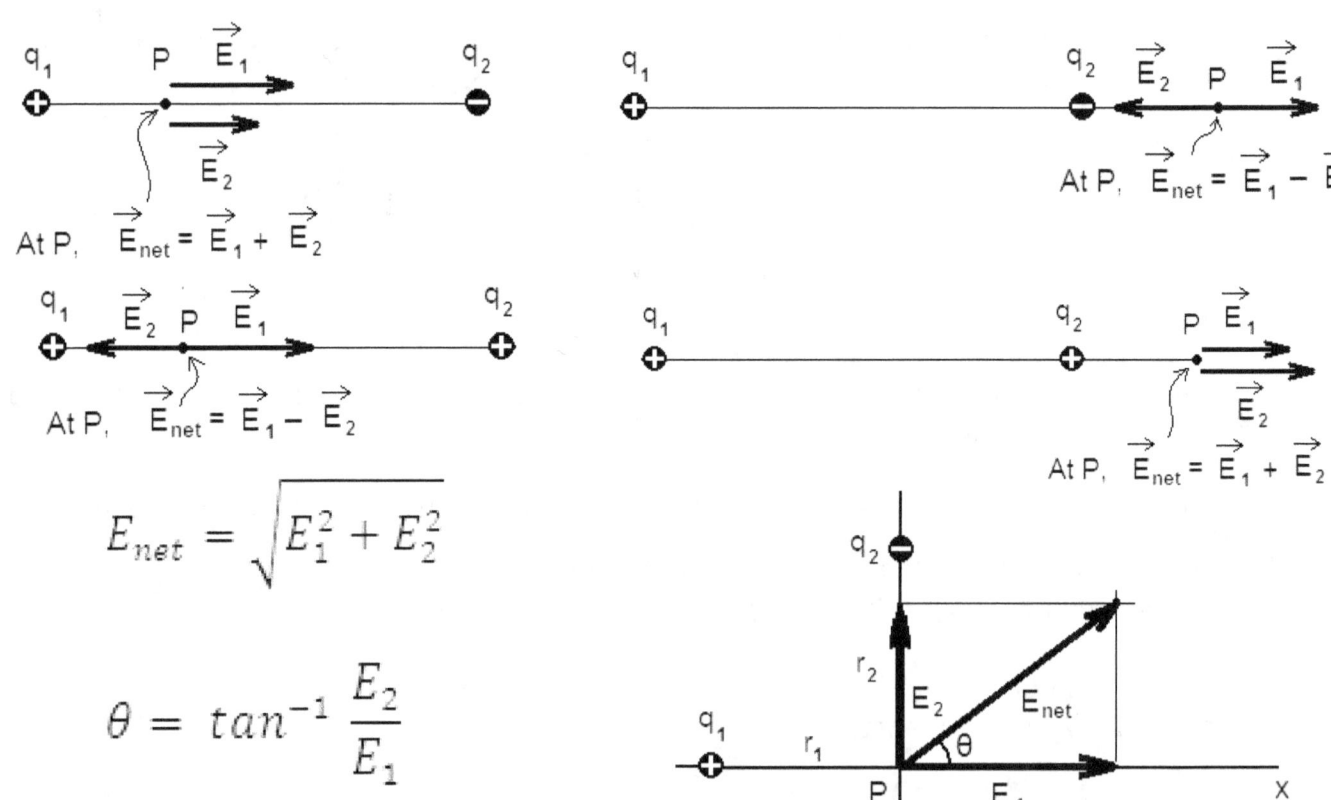

Surface Charge Density (σ): If a charge q is spread uniformly over an area A then the surface charge density is defined as

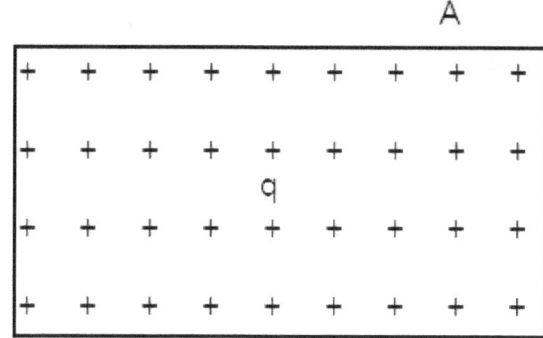

$$\sigma = q/A \quad C/m^2$$

Uniform Electric Field: The electric field in a region is said to be uniform if at every point of the region the electric field is constant in magnitude and direction.

The surface charge density $\sigma = \dfrac{q}{A}$

A uniform electric field is created in the space between two identical metal plates kept close to each other and carrying equal and opposite charge +q and -q respectively. The electric field between the plates is then given by

$$E = \sigma/\varepsilon_0$$

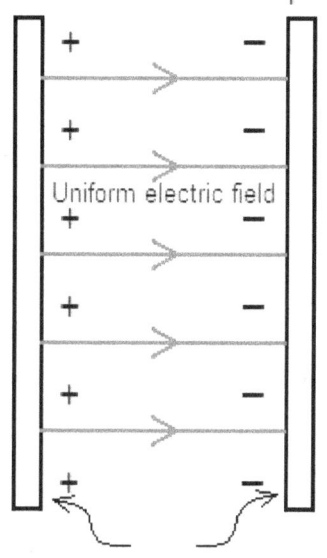

Area of each plate = A

Some interesting facts & figures
• A typical AA battery can supply charge of about 10, 000 coulombs before depletion.
• *Astraphobia* is the irrational fear of lightning and thunder.
• *Fulminology* is the study or science of lightning,
• A *fulminologist* is someone who studies lightning.
• A strong bolt of lightning may transfer as much as 20 C of charge to or from ground

PROBLEMS:

Coulomb's Law Problems

1. How many electrons should be removed from a neutral object to give it a charge of +1 μC?

[6.25x10^{12}]

2. How far apart should two electrons be placed so that the force of repulsion between them is (a) 1 N?

(b) Equal to the weight of an electron?

3. Sketch a graph of F vs. r for the Coulomb Law equation and discuss the characteristics of the graph. *[F∝ 1/r², As r→ 0, F → ∞ and as r → ∞, F→ 0]*

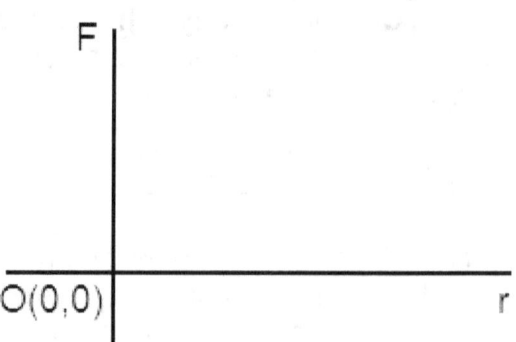

4. Two small metal spheres initially neutral are placed 3.0 m apart. Now one billion electrons are transferred from one sphere to the other. Determine the charge on each sphere and the force between them.

5. The force of repulsion between two point charges placed 6.0 m apart is 5.0×10^9 N. If one charge has twice the magnitude of the other, determine those magnitudes.
[3.16 C, 6.32 C]

6. Two point charges of +6 µC and -4 µC are placed in a plane at the points (6m,-2m) and (0m,4m). Determine the magnitude and the direction of force on each.

**7. How would you divide the charge of +9 C between two particles so that when these particles are kept a distance of 3.0 m apart the force of repulsion between them would be 1.4×10^{10} N?
[2 C, 7 C]

8. The point charges q_1 = +6 µC, q_2 = +5 µC, and q_3 = -8 µC are placed at the corners A, B, and C of an equilateral triangle ABC, respectively. Each side of the triangle ABC is 4.5 m. Determine the magnitude and the direction of the net force acting on the charge q_2 due to the other two charges.

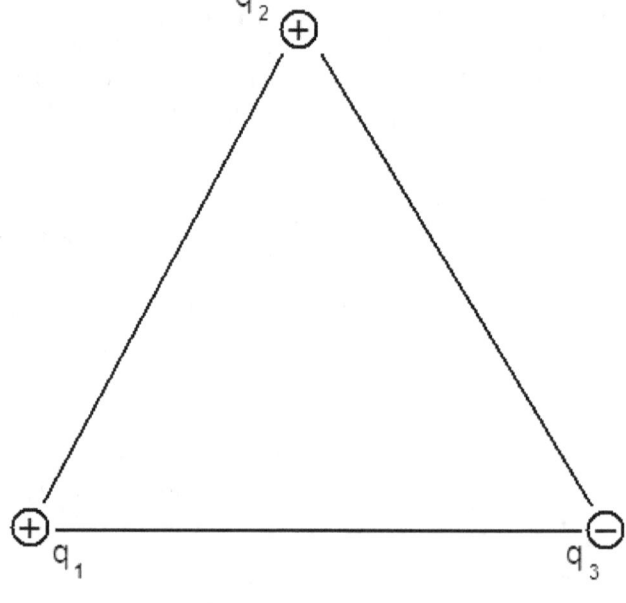

Electric Field Problems

9. A particle carries a charge of +10 µC. At what distance from the charge is the electric field
(a) 10 N/C   (b) 0.001 N/C   (c) 0 N/C   (d) infinity N/C
                                                                    [95 m, 9487 m, ∞, 0]

10. Two point charges q_1= +49 µC and q_2 = +9 µC are
placed 20 m apart. Determine the point(s) at which the net
electric field is zero.

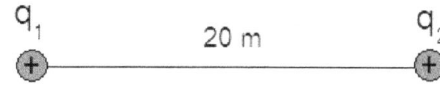

11. Two point charges q_1= +49 µC and q_2 = -9 µC are placed
20 m apart. Determine the point(s) at which the net electric
field is zero. *[15m on the side of q_2 opposite to q_1]*

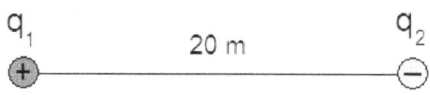

12. The point charges q_1 = +6 µC and q_2= -8 µC are placed at the
corners B, and C of an equilateral triangle ABC, 4.5 m on a side.
Determine the magnitude and the direction of the net electric field
at A.

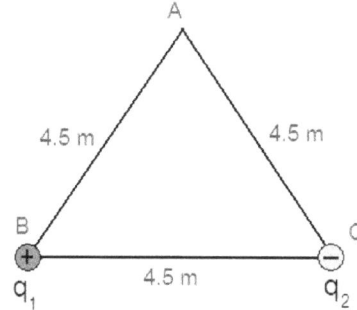

13. Three point charges, q_1 = +7 µC, q_2= -8 µC, and q_3= +5 µC are placed on the x-axis at the
points (-6m,0m), (-1m,0m), and (7m, 0m). Determine the net electric field at (a) the origin, and (b)
at the point (4m,0m).   *[-71168 N/C, -7250 N/C]*

14. Determine the magnitude and direction of an electric field required balance an electron
against its own weight.
[m(electron) = 9.11 x 10^{-31}kg, q(electron) = -1.6 x 10^{-19}C]

15. How far above a proton should an electron be placed so that force of attraction on the proton
due to the electron will balance the pull of gravity on the proton.
{m(proton) = 1.67 x 10^{-27}kg, q(proton) = 1.6 x 10^{-19}C}

                                                                    [0.12 m]

16. Two charge particles, q_1 = +5 µC and q_2= -2 µC are placed at the coordinates (3m, 0m) and (0m, 4m) respectively. Determine the net E at the origin.

17. Two identical metal plates each of area 0.03 m^2 are placed 1 cm apart and are given the charges +6 µC and -6 µC.
(a) Determine the surface charge density on each plate.   *[2x10⁻⁴ C/m²]*

(b) Determine the electric field between the plates.   *[2.26x10⁷ N/C]*

(c) If a particle of mass 2 x 10⁻⁶ kg and charge +1.8 µC is placed near the positive plate, determine its acceleration and its velocity as it hits the negative plate.
[2.03x10⁷ m/s², 638 m/s]

Multiple Choice:

1. The force of attraction between two charged particles is 80 N. If the distance between them is reduced to 1/4th the original distance, the force between them will change to
(A) 5 N   (B) 20 N   (C) 40 N   (D) 320 N   (E) 1280 N

2. Two point charges each +1 C are placed 1 m apart. The force of repulsion between them is
(A) 1 N   (B) 9 x 10⁻³N   (C) 9 x 10⁹N   (D) 1.1 x 10⁻¹⁰N   (E) 81 x 10¹⁸N

3. P is the midpoint between the equal and opposite charge particles. The direction of net electric field E at P is

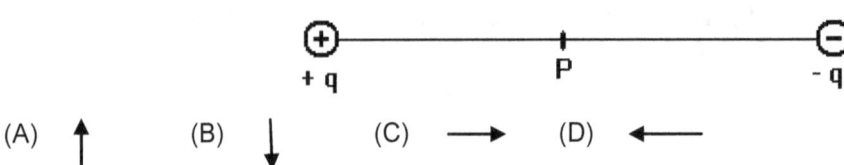

(A) ↑   (B) ↓   (C) →   (D) ←

(E) not defined because it is zero.

4. A particle of negative charge –q experience an upward force F when placed at a point P in an electric field. Which of the following is true of the magnitude and direction of the electric field?

	Magnitude	Direction
A	qF	up
B	qF	down
C	F/q	up
D	q/F	down
E	F/q	down

5. An electron traveling due <u>east</u> enters a region of uniform electric field directed <u>north</u>. The electron will
   (A) deflect toward North   (B) deflect toward South   (C) deflect toward West (D) deflect toward East
   (E) not deflect but accelerate due east

6. A proton traveling due <u>east</u> enters a uniform electric field directed due <u>west</u>. The proton will
   (A) deflect toward North (B) deflect toward South (C) deflect toward West (D) deflect toward East
   (E) not deflect but accelerate due east

Chapter 33
Electric Lines of Force
Motion of Charged Particle in E

The electric lines of force are the imaginary lines representing the intensity and the direction of the electric field in a region.

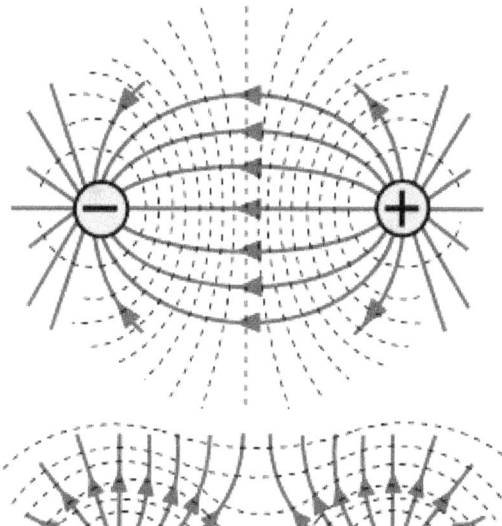

PROPERTIES OF ELECTRIC LINES OF FORCE:

*   A point positively charged particle will trace out an electric line of force if allowed to move slowly (such as through a viscous medium) in the field. At any point, a tangent drawn to the line of force represents the direction of electric field at that point.
*   Any number of lines can be drawn in a region.
*   The lines of force are crowded in the regions of high intensity electric field and are spaced farther apart in the region of low intensity electric fields.
*   No two lines of force intersect each other.
*   In a uniform electric field the lines of force are parallel and evenly spaced.
*   The lines of force originate at the positive charges and terminate at the negative charges.
*   The lines of force start at right angles to the surface of a charged conductor.
*   Conductors are 'opaque' to the electric lines of force.

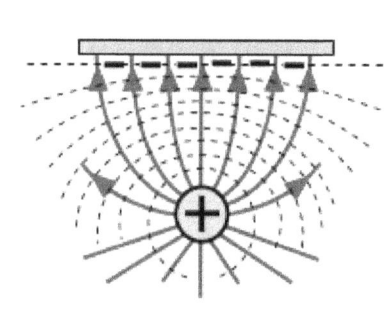

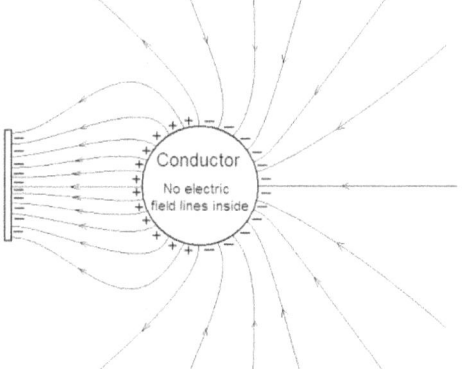

Motion of a charged particle in Electric Field:

The acceleration of a particle of charge **q** and mass **m** in an electric field **E** is given by,

$$a = F/m = qE/m$$

Note:

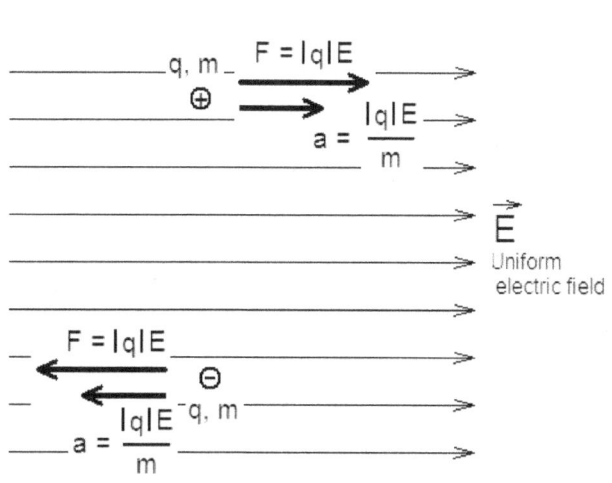

*   A **positively** charged particle accelerates in the **same direction** as the applied electric field E.
*   A **negatively** charged particle accelerates in the **opposite direction** to the applied electric field E.

The following Equations of Motion from Mechanics would be useful to analyze the motion of charged particles in an electric field:

Linear Motion	Projectile Motion
$x = \frac{1}{2}(v_o + v_o)t$	$y = v_{av} t \quad ; \quad x = v_{ox}t$
$v = v_o + at$	$v = v_{oy} + at$
$x = v_o t + \frac{1}{2}at^2$	$y = v_{oy} t + \frac{1}{2}at^2$
$v^2 = v_o^2 + 2ax$	$v^2 = v_{oy}^2 + 2ay$

PROBLEMS:

[q(electron) = -1.6 x 10^{-19} C, q (protons) = +1.6 x 10^{-19} C, m(electron) = 9.11 x 10^{-31} kg
m(proton) = 1.67 x 10^{-27} kg]

1. Determine the acceleration of a proton and an electron due to an electric field of 1000 N/C.
   [9.60x10^{10} m/s^2, 1.76x10^{14} m/s^2]

2. A beam of electrons enters the region between two parallel plates of length 10 cm with a velocity of 8 x 10^6 m/s as shown in the diagram. A uniform electric field of 60 N/C is created between the plates perpendicular to the initial direction of the beam. Determine the transverse displacement (y) of the beam.
[Apply the ideas of projectile motion]

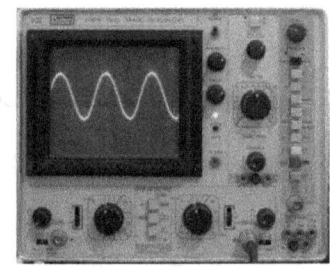

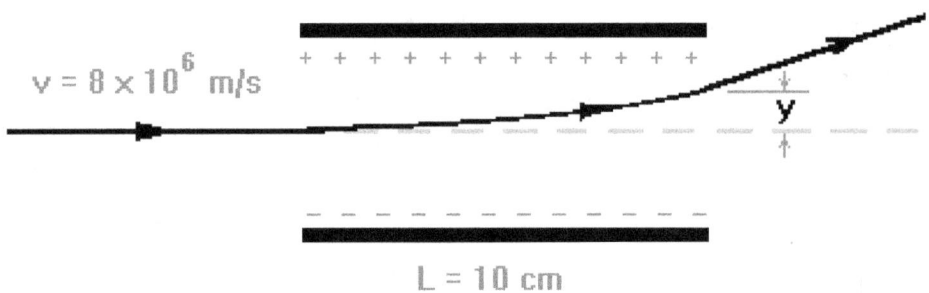

$v = 8 \times 10^6$ m/s

y

L = 10 cm

3. A particle of charge q = +4.0 μC and mass m = 5 x 10^{-5} kg is placed near the positive plate of a parallel plate arrangement which creates a uniform electric field E = 5,000 N/m. The two plates are 5 mm apart.
(a) Determine the direction and magnitude of the acceleration of the charged particle.
   [400 m/s^2 toward neg. plate]

(b) Determine the velocity of the particle just before it strikes the other plate.   *[2 m/s]*

Some interesting facts & figures

- **A charged placed on an electric field line will follow the line if**
  - **- the field line is straight**
  - **or - the charged particle is massless (no such particles though)**
  - **or – the particles moves slowly through a very viscous medium**

Multiple Choice:

1. An electron traveling in the y-direction enters a uniform electric field pointing in the x-direction. Which of the following best indicates the subsequent path taken by the electron?

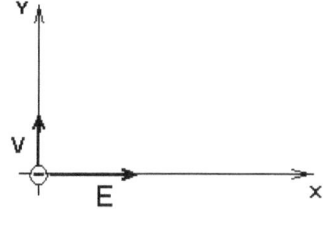

(A)    (B)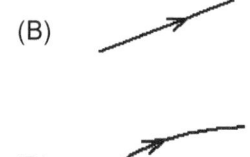

(C)   (D)    (E)

2. A particle of charge q and mass m enter a uniform electric field E. The acceleration of the particle due to **E** is given by
(A) mq**E**   (B) m**E**/q   (C) mq/**E**   (D) q**E**/m   (E) **E**/m

3-5. The diagram below shows points 1,2,3,4, and 5 in an electric field E.   .

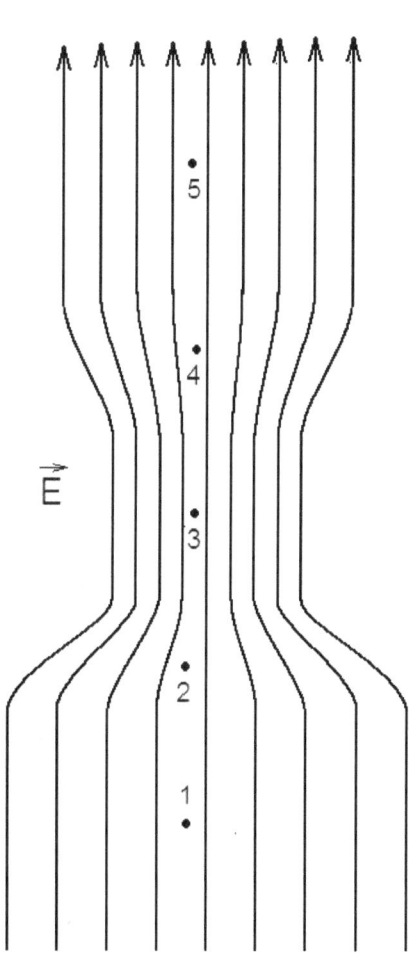

   3. At which point of the 5 points, is the electric field strength the greatest?
         (A) 1   (B) 2   (C) 3   (D) 4   (E) 5

4. If an electron is placed at the point 3
         (A) it wil accelerate to the right
         (B) it will accelerate to the left
         (C) it will accelerate upward
         (D) it will accelerate downward
         (E) it will stay at rest

5. If a proton is placed at the point 3
         (A) it wil accelerate to the right
         (B) it will accelerate to the left
         (C) it will accelerate upward
         (D) it will accelerate downward
         (E) it will stay at rest

Chapter 34
Electric Potential (V) - Electric Potential Energy (U$_E$) -Equipotential Surfaces

Electric Potential Energy (U$_E$) for two point charges q$_1$ and q$_2$:

Electric potential energy (U$_E$) of two point charges: It is the work done in assembling the configuration by bringing the two charges in from infinity.

Alternatively, keep q$_1$ fixed in place and bring q$_2$ from infinity to its position, a distance r from q$_2$. The amount of work done in doing so is the electric potential energy (U$_E$) of the system. The work done can be found by using calculus. The result is

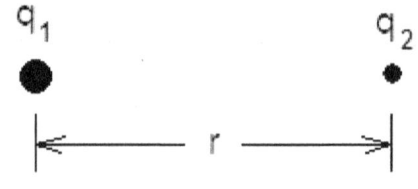

$$U_E = \frac{k_e q_1 q_2}{r}$$

Electric Potential Energy of three point charges:
The total U$_E$ for 3 point charges is the sum of the potential energies for all distinct pairs. Thus,

$$U_E = U_{12} + U_{23} + U_{31}$$

$$U_E = \frac{k_e q_1 q_2}{r_{12}} + \frac{k_e q_2 q_3}{r_{23}} + \frac{k_e q_3 q_1}{r_{31}}$$

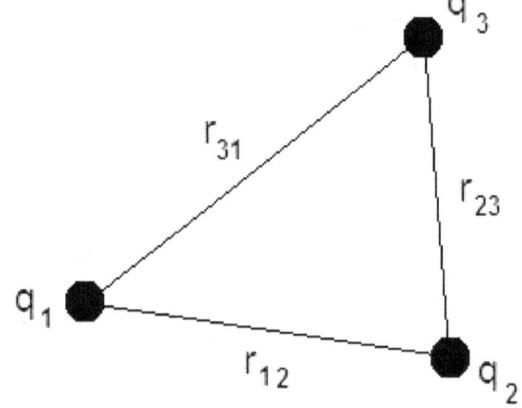

Electric Potential (V):

The electric potential at any point is defined as the work done per unit charge in bringing a positive point charge q from infinity to that point. Thus,

$$V_P = \frac{W_{\infty,P}}{q}$$

The electric potential energy of a point charge q at a point with potential V is $\boxed{U_E = qV}$

Work done in moving q from a point at V$_1$ to a point at V$_2$:

The work done by an <u>external</u> force in moving a charge q from point 1 to point 2 is

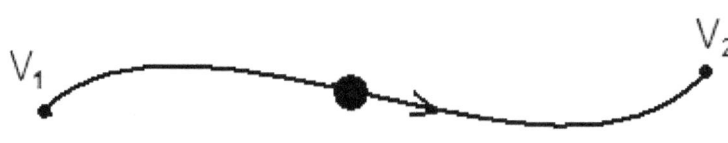

$$W = q(V_2 - V_1)$$

Electric potential energy converting to kinetic energy:

- A positive charge particle accelerates from a higher potential to a lower potential
- A negative charge particle accelerates from a lower potential to a higher potential

If a point charge **q** moves from point **1** at potential V_1 to a point **2** at potential V_2 then, applying the *Conservation of Mechanical energy*

$$KE_1 + q\,V_1 = KE_2 + qV_2$$
$$\Delta KE = q\,(V_1 - V_2)$$

If $KE_1 = 0$, $KE_2 = \frac{1}{2} mv^2$, and $V_1 - V_2 = V$, then the above equation can be written as

$$\boxed{qV = \frac{1}{2} mv^2}$$

Potential V at a point in an electric field:
It is defined as the potential energy per unit positive charge placed at that point. Thus

$V = U_E/q$

V is a scalar quantity and its SI unit is J/C called volt (V)

Potential due to a point charge:
Let P be a point a distance r from a point charge q. The potential V at P is given by

$$\boxed{V = \frac{k_e q}{r}}$$

$$V\ (\text{at P}) = \frac{k_e q}{r}$$

The electric potential V is a scalar quantity. It can be positive, negative, or zero.

Net potential due to two or more point charges:
If there are a number of point charges q_1, q_2, q_3 ... a distance r_1, r_2, r_3... from a point P then the net potential V at P is the sum of the potential V_1, V_2, V_3... due to all the point charges. Thus,

$$V = V_1 + V_2 + V_3 \ldots$$

Or,

$$V = \frac{k_e q_1}{r_1} + \frac{k_e q_2}{r_2} + \frac{k_e q_3}{r_3} + \ldots$$

E and V due to a Charged Spherical Conductor

E and V due to a solid or hollow
spherical charged conductor of radius R

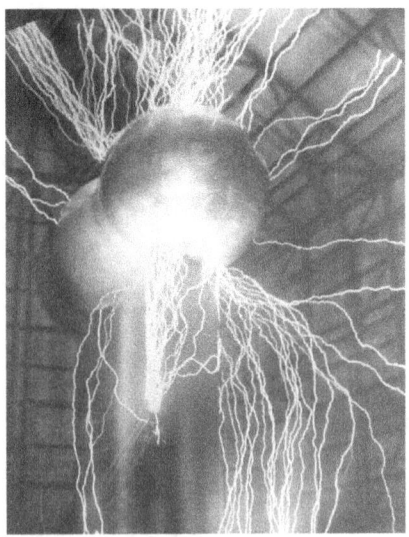

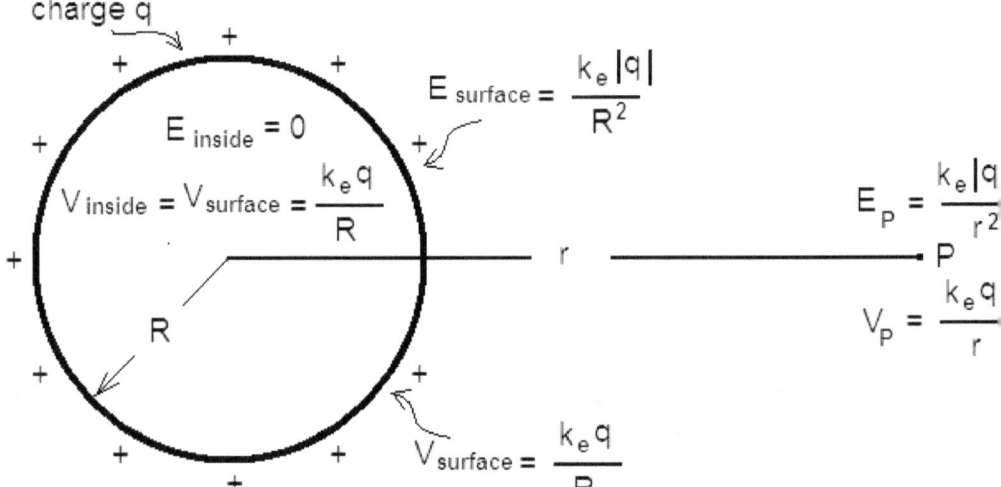

charge q

$E_{inside} = 0$

$V_{inside} = V_{surface} = \dfrac{k_e q}{R}$

$E_{surface} = \dfrac{k_e |q|}{R^2}$

$V_{surface} = \dfrac{k_e q}{R}$

$E_P = \dfrac{k_e |q|}{r^2}$

$V_P = \dfrac{k_e q}{r}$

E and V at a point P a distance r from the center of the circle;

    i. r > R (outside the sphere)
        E and V can be found by regarding the entire charge of the conductor
        to be a point charge at the center of the sphere. Thus,
        $E = k_e |q|/r^2$ and $V = k_e |q|/r$

    ii. r = R (at the surface of the sphere)
        Here too, E and V can be found by regarding the entire charge of the conductor
        to be a point charge at the center of the sphere. Thus,
        $E = k_e |q|/R^2$ and $V = k_e |q|/R$

    iii. r < R (inside the sphere)
        $E = 0$; $V = V$ at the surface $= k_e |q|/R$

Potential difference between two points in a uniform electric field E:

For the points 1 and 2 in a uniform E as shown in
the figure below,

$$V_2 - V_1 = Ed$$

Also note that the point 1 is at a higher potential
compared to the point 2. Why?

Equipotential surfaces and their relationship to electric field lines:

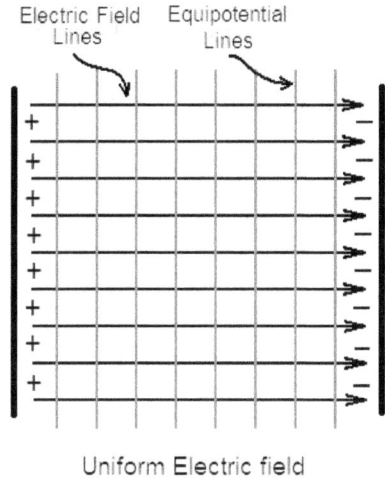

Electric Field Lines   Equipotential Lines

Uniform Electric field

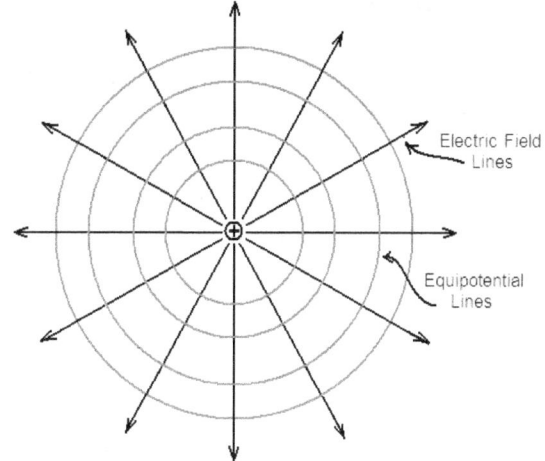

Electric Field Lines

Equipotential Lines

- In an electric filed, an equipotential surface is one, which has all its points at the same potential.

- The electric field vectors are perpendicular to the equipotential surfaces.

- For a point charge, the equipotential surfaces are concentric spherical surfaces with the point charge at the center. The electric field being radial is always perpendicular to these surfaces.

- For a uniform field such as the one shown in the figure above, the equipotential surfaces are the planes intersecting the field lines at right angles.

- If a uniform E field is created between two parallel plates, the equipotential surfaces are planes parallel to the plates.

Two charged spheres connected by a wire: Two spheres of radii R_1 and R_2 carry charges q_1 and q_2 respectively. They are now connected by a wire. If the two spheres have different potentials, positive charge will flow from higher to lower potential until the potential are equal.

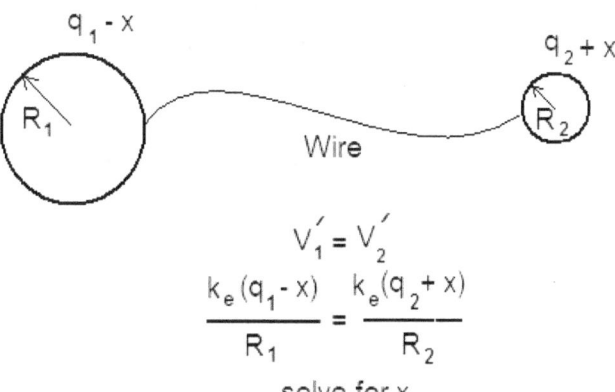

q_1 Charge flow

q_2

R_1 Wire

R_2

$V_2 = \dfrac{k_e q_2}{R_2}$

$V_1 = \dfrac{k_e q_1}{R_1}$

Assume $V_1 > V_2$

$q_1 - x$

$q_2 + x$

R_1

Wire

R_2

$V_1' = V_2'$

$$\frac{k_e(q_1 - x)}{R_1} = \frac{k_e(q_2 + x)}{R_2}$$

solve for x

Some Electrical Properties of a Charged Conductor:

1. Any charge given to a conductor resides on its outer surface under the steady state conditions.

2. The electric charge density on the surface of the conductor is higher at the points of greater curvature.

3. The surface of a charged conductor is an equipotential surface.

4. The electric lines of force are at right angles to the surface of a charged conductor

5. The electric field inside a charged conductor is zero.

6. The electric potential inside a charged conductor is constant and is equal to the potential of the surface of the conductor

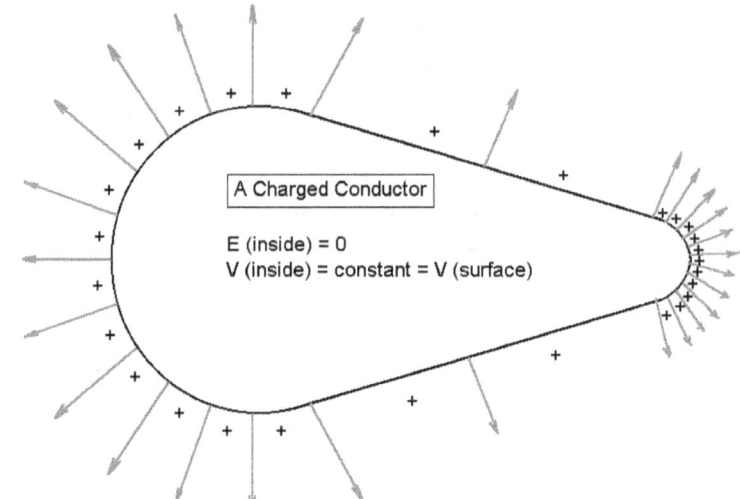

A Charged Conductor

E (inside) = 0
V (inside) = constant = V (surface)

7. For a spherical charged conductor, E and V **outside** the surface can be calculated by assuming the entire charge of the sphere to be concentrated at the center of the sphere.

Electron Volt (eV) as a unit of energy:

One eV is the KE gained by a particle carrying an elementary charge e = 1.6 x 10^{-19} C accelerated through a potential of 1 volt. Thus

$$1 \text{ eV} = q \, \Delta V = (1.6 \times 10^{-19}) (1) = 1.6 \times 10^{-19} \text{ J}$$
$$1 \text{ keV} = 1,000 \text{ eV}$$
$$1 \text{ MeV} = 1,000,000 \text{ eV}$$
$$1 \text{ BeV} = 1 \text{ GeV} = 1,000,000,000 \text{ eV}$$

For example:
- An electron accelerated through a potential difference of 1 V will gain 1 eV of KE.
- A proton accelerated through a potential difference of 1 V will gain 1 eV of KE.
- A proton accelerated through a potential difference of 1 x 10^6 volt will gain 1 MeV of KE.
- A particle carrying 3 times the elementary charge accelerated through a potential difference of 1 V will gain 3 eV of KE.

Conversion between eV and joule:

$$eV \xrightarrow{\text{X } 1.6\times10^{-19}} J$$

$$J \xrightarrow{\div \; 1.6\times10^{-19}} eV$$

Problems:

1. A point charge $q = -7.0 \ \mu C$ is placed at the point (4 m, 0 m) in a coordinate plane. At each of the points P_1 (8 m, 0 m), P_2 = (0 m, 0 m), P_3 = (3 m, 0 m), determine
(a) the electric field (E).   [3938 V/m, 3938 V/m, 63,000 V/m]

(b) the electric potential (V).   [-15750 V, -15750 V, -63000 V]

2. A conducting sphere of radius 9.0 m carries a charge of +5.0 μC. Determine the electric potential and electric field
(a) at the surface of the sphere

(b) inside the sphere

(c) at 12.0 m from the center of the sphere.

3. A point charge q_1 = +5.0 μC is placed 6.0 m to the right of another point charge q_2 = -5.0 μC. Determine the net electric field (E) and the net potential (V) at

(a) a point midway between the two charges.
                    [10000 N/C, 0 V]

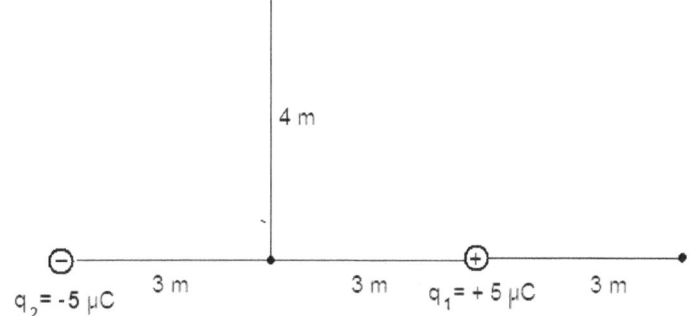

(b) a point 3.0 m to the right of q_1 [4444 N/C, 10000 V]

(c) a point 4.0 m above the point midway between the two charges.   [2160 N/C, 0 V]

4. A particle of mass 0.003 kg and charge -4.0 μC is accelerated from a potential of V_1 = -100 V to a potential of V_2= +100 V.

(a) Determine the KE gained by the particle.

(b) Determine the final velocity of the particle if the particle started from rest.

5. Determine the gain in the KE and the final velocity of
(a) an electron accelerated from rest through a potential of 10,000 V.
$[1.6 \times 10^{-15} J, 5.92 \times 10^7 \ m/s]$

(b) a proton accelerated from rest through a potential of 10,000 V.
$[1.6 \times 10^{-15} J, 1.38 \times 10^6 \ m/s]$

(c) a deuteron accelerated from rest through a potential of 10,000 V.
$[1.6 \times 10^{-15} J, 9.79 \times 10^5 \ m/s]$

6. In problem 5, determine the gain in the KE of the three particles in eV.

7. A particle of charge +6.4 x 10^{-19} C is accelerated to a KE of 1.2 keV. Determine the accelerating voltage.   [300 V]

8. A conducting sphere of radius 30 cm is given a charge q = +3.8 μC. A particle of charge q' = +1.4 μC and mass 1.2 x10^{-6} kg is placed close to the surface. As this particle moves away from the sphere due to repulsion, determine the speed of the particle very far from the sphere.

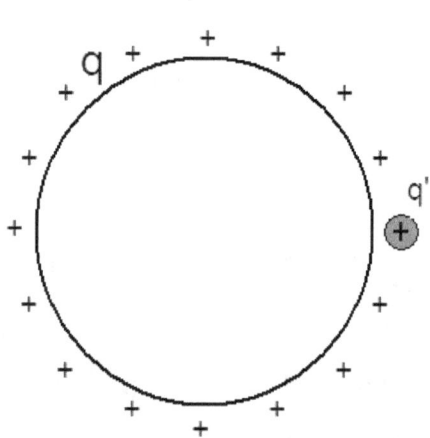

**9. For a point charge q, two equipotential surfaces are 3 m apart and have potentials of 5 V and 3 V respectively. Determine the radius of each potential surface and the magnitude of the charge.
[4.5 m, 7.5 m, 2.5×10^{-9} C]

**10. For a point charge q = 6 nC, two equipotential surfaces are 3 m apart and have a potential difference of 3 V. Determine the radius and the potential of each surface.

11. Two isolated conducting spheres have radii, $R_1 = 0.3$ m and $R_2 = 0.5$ m and carry charges q_1 = +4 μC and q_2 = +6 μC respectively.
(a) Determine the potential of each sphere.
[120000 V, 108000 V]

 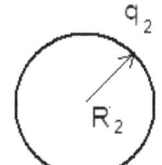

(b) If these two spheres are now connected by a thin wire, determine the amount of charge transferred from one sphere to the other.
[0. 25 μC from the first to the second sphere]

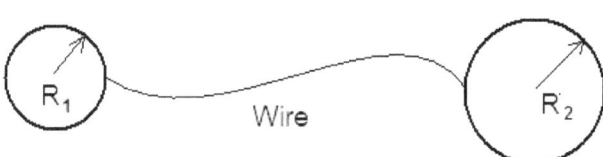

Multiple Choice:

1. An isolated conductor is charged so that it has a potential of 1000 V. If E_{in} and V_{in} are respectively the electric field and electric potential inside the conductor then which of the following is true?
(A) $E_{in} = 0$; $V_{in} = 0$ (B) $E_{in} \neq 0$; $V_{in} = 1000V$ (C) $E_{in} = 0$; $V_{in} = 1000V$
(D) $E_{in} \neq 0$; $V_{in} = 0$

2. The electric field and electric potential at a distance r from a charged particle are E and V respectively. At a distance of 2r, the new electric field and electric potential E' and V' are given by
(A) $E' = \frac{1}{2} E$; $V' = \frac{1}{2} V$ (B) $E' = \frac{1}{4} E$; $V' = \frac{1}{2} V$
(C) $E' = \frac{1}{2} E$; $V' = \frac{1}{4} V$ (D) $E' = \frac{1}{4} E$; $V' = \frac{1}{4} V$
(E) $E' = 4 E$; $V' = 2 V$

3. A proton is to be accelerated from rest to a kinetic energy of 5 keV. The accelerating potential required is
(A) 5 kV   (B) 5 (1.6×10^{-19}) kV   (C) $5/(1.6 \times 10^{-19})$ kV
(D) $\sqrt{[5 (1.6 \times 10^{-19})]}$ kV   (E) $\sqrt{[5/(1.6 \times 10^{-19})]}$ kV

4. In an electric field, a positive charge q coulomb moves from a point at +20 V to a point at -20 V. The work done by the electric field is
(A) + 40q J   (B) -40q J   (C) 0 J   (D) 20q J   (E) -20q J

5. For any charge distribution, the electric field lines are _____ to the equipotential surfaces.
(A) tangential   (B)perpendicular   (C) at 60°   (D) at 45°   (E) at 30°

6. In uniform electric field, equipotential surface are _____ in shape.

(A) spherical   (B) flat planes   (D) conical   (E) toroidal

Chapter 35
Electrostatic Capacitor

Parallel Plate capacitor (PPC)
A parallel plate capacitor consists of two identical conducting plates placed parallel and close to each other.

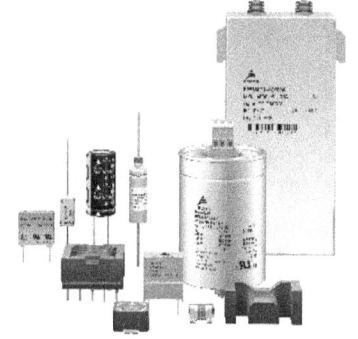

Each of the two plates has an area A. The two plates are connected to a voltage source (say a battery) which creates a potential difference V between the two plates. Thus the two plates get equal and opposite charges +q and -q respectively. The capacitor is said to carry a charge q.

The amount of charge q is proportional to the potential difference V between the two plates. Thus

$$q = CV$$

C is called the capacitance of the capacitor. The SI unit for capacitance is **farad (F)**.

$$1 \text{ microfarad } = 1 \text{ } \mu F = 10^{-6} \text{ F}$$
$$1 \text{ picofarad } = 1 \text{ pF} = 10^{-12} \text{ F}$$

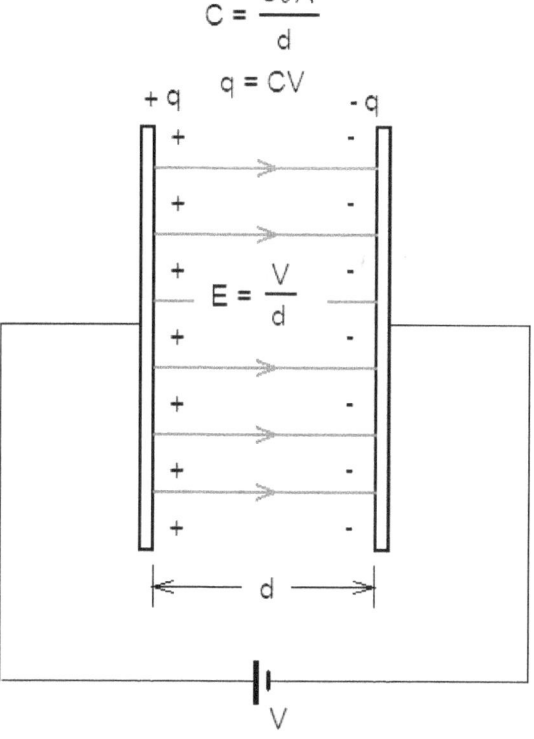

$$C = \frac{\varepsilon_o A}{d}$$

q = CV

$$E = \frac{V}{d}$$

Parallel Plate Capacitor

The value of C depends on the construction of the capacitor and **not** on the applied voltage or the amount of charge on it. The capacitance of the parallel plate capacitor is given by,

A is the area of a plate
d is the separation between the plates
ε_o is the electrical permittivity = **8.85 x 10⁻¹² F/m**

$$C = \frac{\varepsilon_o A}{d}$$

Note: **C** is directly proportional to **A** and inversely proportional to **d**.

Energy Stored in a capacitor (W):

A charged capacitor stores electrical energy W. The stored energy depends on C, V, and q and is given by

$$W = \tfrac{1}{2} qV = \tfrac{1}{2} CV^2 = \tfrac{1}{2} \frac{q^2}{C}$$

Varying the distance, d, between the plates of a charged capacitor:

Case 1:

Battery <u>stays</u> connected as the distance between the plates is varied

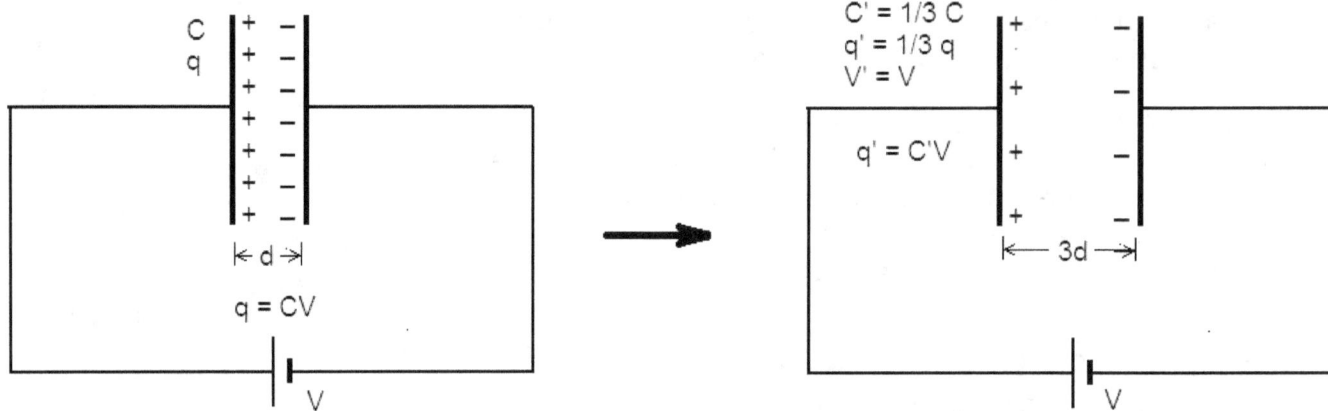

The voltage between the plates stays at V becuase it governed by the battery that stays connected. Some of the charges return the battery.

Case 2:

Battery is <u>disconnected</u> before the distance between the plates is varied

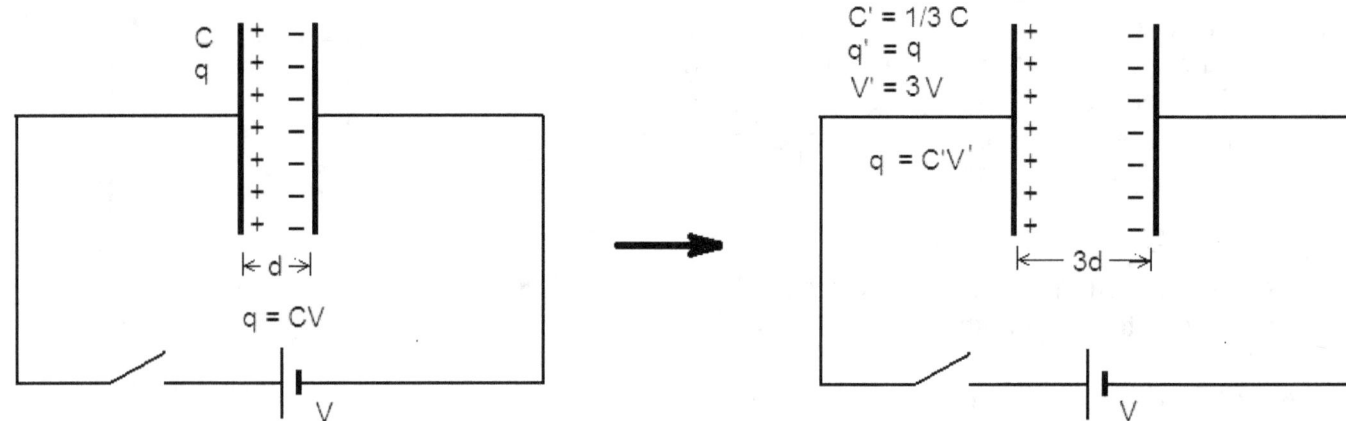

Since the battery has been disconnected, the charges have no place to go. The voltage between the plates is no more governed by the battery and varies in value.

Some interesting facts & figures
• A capacitor can store charges and hence electrical energy. It can be used as a temporary battery
• Capacitor is used in car stereo as a back up to deliver high current in case of large demand during bass.

PROBLEMS:

1. The plates of a PPC (parallel plate capacitor) are circular with a radius of 6 cm each. The separation between the plates is 2 mm. Determine its capacitance. *[5.0x10⁻¹¹F]*

2. The capacitance of a PPC is 10 pF. If the area of its plates is doubled and the separation is

halved determine the new capacitance.

3. How much area should each plate of a PPC have if it is to have a capacitance of 1 µF for a plate separation of 1 mm.   *[113 m²]*

4. A 10-µF PPC is charged by connecting it to a 1.5 V battery. The battery is now <u>disconnected</u> and then the plate separation is doubled. Determine the new potential difference between the plates.

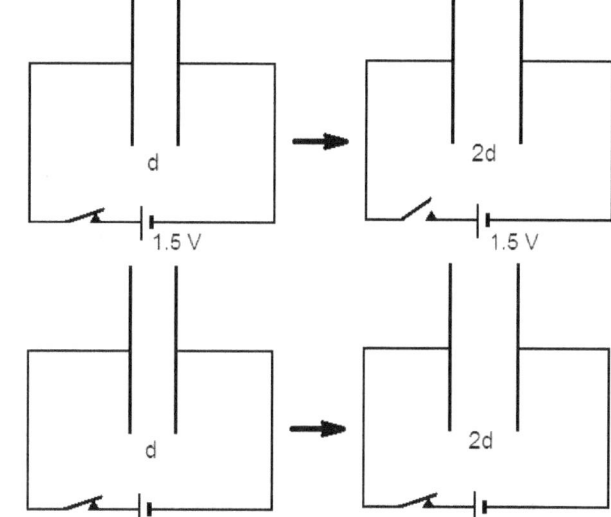

5. A 10-µF PPC is charged by connecting it to a 1.5 V battery. Now the battery remaining connected the separation between the plates is doubled. Determine the new charge on the capacitor.   *[7.5 µC]*

6. A 15-pF capacitor is fully charged by connecting it to a 6-V battery.

(a) Calculate the charge on the capacitor and the energy stored in it.

The battery is now disconnected. The distance between the plates is made 1/3rd the original distance.
(b) Calculate the charge on the capacitor and the energy stored in it now.

Multiple Choice:

1. A capacitor has a capacitance of 24 µF. If the area of its plates is doubled and the distance between its plates is halved the new capacitance will be
   (A) 6 µF   (B) 12 µF   (C) 24 µF   (D) 48 µF   (E) 96 µF

2. When a capacitor is connected to a potential difference of 1.5 V electrical energy of 9 J is stored in it. If the potential difference is increased to 4.5 V the new stored energy will be
   (A) 3 J   (B) 9 J   (C) 27 J   (D) 36 J   (E) 81 J

3. The capacitance of a parallel plate capacitor depends on
   (A) only q   (B) only V   (C) both q and V
   (D) on A and d   (E) on the E-field between the plates

4. The electrical energy stored in a capacitor connected to 1.5 V is 36 µJ. If the 1.5-V battery is replaced by a 4.5-V battery the amount of energy stored must be
   (A) 324 µJ   (B) 6 µJ   (C) 729 µJ   (D) 12 µJ   (E) 108 µJ

Chapter 36
Electric Current, Resistance, emf

Electric Current:

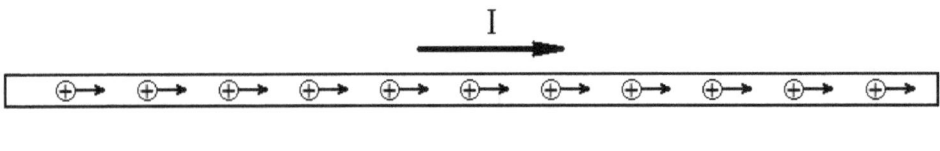

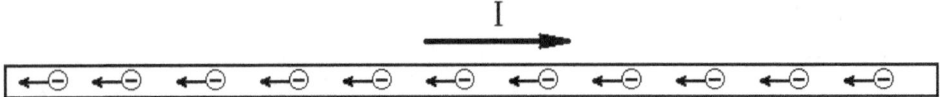

Electric current (I) is defined as the rate of flow of electric charge through a medium. Thus,

The SI unit for the electric current is 'ampere' (A). $1\,A = 1C.s^{-1}$

French Physicist Andre'-Marie Ampere
[20 January 1775 - 10 June 1836]

Ohm's Law: The electric current (I) flowing through a medium is proportional to the potential difference (V) across the conductor.

Hence we may write, $V \propto I$
or,

$$\boxed{V = I\,R}$$

Here, R is the constant of proportionality and is called the 'resistance' of the conductor. The conductor itself may be called a 'resistor'.

The SI unit for the resistance is ohm (Ω). $1\,\Omega = 1\,V/A$

A battery is one of the many common devices that are used to create potential difference across a conductor to create electric current.

German Physicist George Simon Ohm
[16 March 1789 - 6 July 1854]

Electromotive force (emf) and terminal potential difference (V_R) for a battery:
A source of voltage (such as a battery) has an electrical resistance of its own. It is called *internal resistance*, **r**. As a result when it supplies current to an external resistance **R**, a voltage **ir** is "lost" within the battery. If a voltmeter is used to measure voltage across **R** it will show

$$V_R = iR$$

Hence the actual voltage of the battery, called **emf**, is $\varepsilon = V_R + ir$

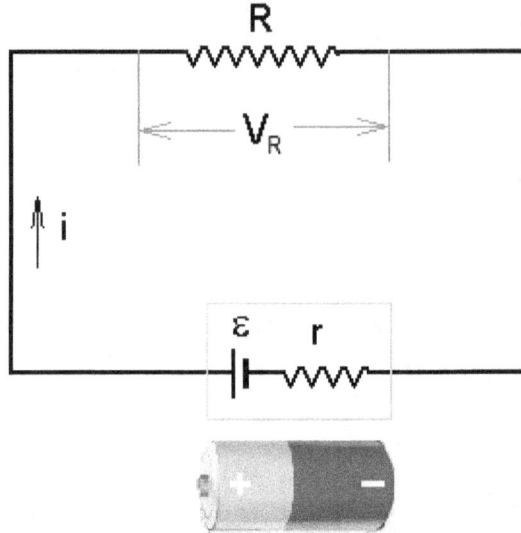

When no current is drawn from the battery the terminal potential difference, V_R, across the terminals of the battery has maximum value equal to the **emf, ε**.

PROBLEMS:

1. When a potential difference of 5 V is applied across a wire a current of 0.2 A flows through it. Determine:

(a) the resistance of the wire.   [25Ω]

(b) the number of electrons/second flowing through a cross-section of the wire.

[1.25x10^{18}]

2. An electrical device has a resistance of 20 ohm and is connected to a 120V supply. Determine the current through the device.

3. A stroke of lightning transfers 4.5 C of charge to the earth in 9.0 microseconds. Determine the average current carried by the lightning stroke.   [500,000 A]

4. A battery has an emf of 8.40 V and an internal resistance of 0.5 Ω. Determine the terminal potential difference and the current drawn from the battery for each of the following external resistors connected to the battery:

(a) 100 Ω   (b) 50 Ω   (c) 4 Ω   (d) 2 Ω   (e) 0.5 Ω   (f) 0.1 Ω

5. A battery has an emf of 1.5 V. When an external resistance of 25 Ω is connected across it the terminal potential difference drops to 1.48 V. Determine the internal resistance of the battery and the current drawn from the battery.   *[0.34Ω, 0.059 A]*

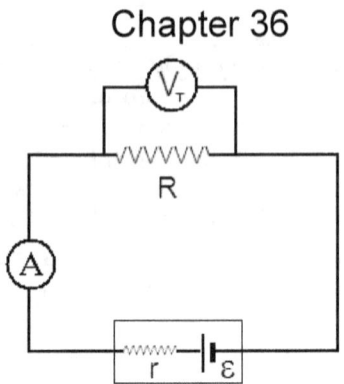

6. A battery delivers 5A when the external resistance is 10 Ω and 3 A when the external resistance is 20 Ω. Determine the emf and the internal resistance of the battery.

7. A voltage source has an emf of ε = 6 V and internal resistance of r = 1.5 Ω. Calculate maximum current that can be obtained from the source.   *[4 A]*

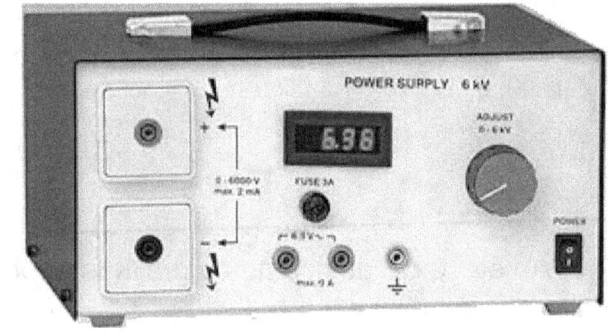

Multiple Choice:

1. Which of the following is the equation for ohm's law?
(A) $V = i/r$ (B) $V = r/i$ (C) $i = Vr$ (D) $i = V/r$ (E) $i = r/V$

2. A battery has an emf of $\epsilon = 9$ V and an internal resistance r. When an external resistance of R = 7 Ω is connected across its terminals a current of 1 A flows through it. If the external resistance were made zero (the terminal are shorted) the current in the circuit would be
(A) zero   (B) infinity   (C) 1/9 A   (D) 9 A   (E) 4.5 A

3. A battery of emf ε and internal resistance r supplies current I through an external resistance R. Which of the following is the correct relationship among these four quantities?
(A) $\epsilon = I(R + r)$ (B) $\epsilon = R + Ir$ (C) $\epsilon = IR + r$ (D) $\epsilon = I(1 + r/R)$ (E) $\epsilon = (R + r)/I$

4. How much voltage is required to send a current of 1 mA through a resistor of 1 kΩ?
(A) 1 V   (B) 1 mV   (C) 1 kV   (D) 1 MV   (E) 1 µV

Chapter 37
Resistivity, Electrical Energy, Electric Power

Electrical resistance of a wire: The electrical resistance of a wire is directly proportional to its length (l) and inversely proportional to its area of

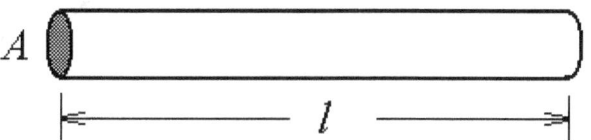

cross-section (A) and depends on the material of the wire. Thus ρ is called the resistivity of the material of the wire. The SI unit of **ρ** is **Ω m**.

$$R = \frac{\rho l}{A}$$

Material	Resistivity (ρ) Ω.m
Aluminum	2.82×10^{-8}
Copper	1.72×10^{-8}
Silver	1.59×10^{-8}
Carbon	3.5×10^{-5}
Teflon	1×10^{16}
Rubber	$10^{13} - 10^{16}$
Maple wood	3×10^{10}
Silicon	20 - 2300

Electrical energy: In many conducting mediums (e.g. metals, carbon) the flow of electric current is due to the flow of free (conduction) electrons.

As the electrons (being negatively charged) move from a lower to a higher potential through the medium, they tend to accelerate. But they loose energy due to collision with other particles in the medium and hence quickly attain *terminal velocity*. The energy loss is their electrical potential energy qV and converts to heat and light as in an electric bulb.

Thus, as charge **q** flows through a conductor due to a potential difference **V** across the conductor the amount of energy dissipated in the conductor is

$$W = q\,V$$

Using q = I t and Ohm's Law V = IR, energy W can be written as

$$W = V\,I\,t$$
$$W = I^2 R\,t$$
$$W = (V^2/R)\,t$$

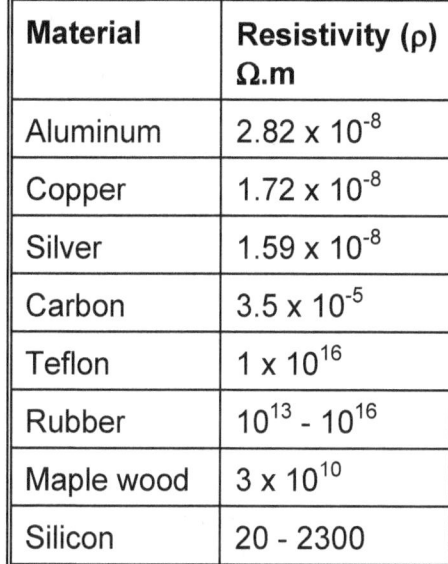

V is measured in volts
I is measured in amperes
t is measured in seconds
W is in joules

Electric power (P): By definition **P = Work/time** or **Energy emitted/time**. Hence the power dissipated in a resistor can be obtained by dividing the expressions for energy **W** above. Thus,

$$P = W/t$$
$$P = V\,I$$
$$P = I^2 R$$
$$P = V^2 / R$$

SI unit of power is **watts**.

Commercial unit of Electrical Energy: To measure the electrical energy consumed, the electric power companies use a unit called **kWh** (kilowatt - hour). Thus energy in kWh is given by

$$\text{Energy (kWh)} = P\ (kW) \times t\ (hours)$$

For example, if a 1200-W hair dryer is used 8 hours in a month, its monthly usage in joules and kWh is as below:

For answer in joules: W = P t = (1200 W) x (8x3600 s) = 3.456×10^7 J (not good for electric bill)
For answer in kWh: W = Pt = (1200/1000 kW)(8) = 9.6 kWh (manageable number for electric bill)

Some interesting facts & figures
• **Superconductor shows zero resistance to the flow of electric current.** • **The temperature below which a material becomes superconductors is called its critical temperature T_C.** • **A sure test for that a material is superconductor is Meissner Effect.** • **Meissner Effect: A magnet can be levitated over a superconductor.**

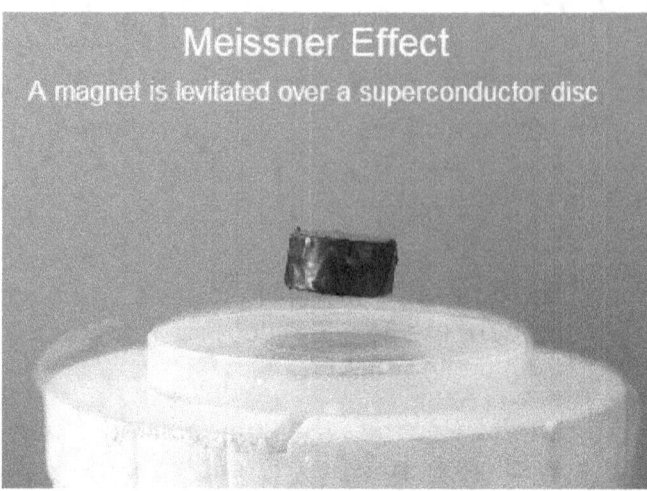

Meissner Effect
A magnet is levitated over a superconductor disc

PROBLEMS:

1. Determine the length of a 20-gauge (A = 5.2×10^{-7} m^2) copper extension cord which would have a resistance of 1 ohm.   *[30.2 m]*

2. A wire has a resistance R. If the wire is uniformly stretched so that its length increases to n times its original length (density remaining constant), determine the new resistance of the wire.

3. A wire has length l, the radius r and the resistance 5 ohm. Determine the resistance of another wire of the same material but length ½ l and radius ½ r.   *[10 Ω]*

4. When a toaster oven is connected to a 120 V supply the hot filament has a resistance of 15 ohm. Determine
(a) the current through the toaster

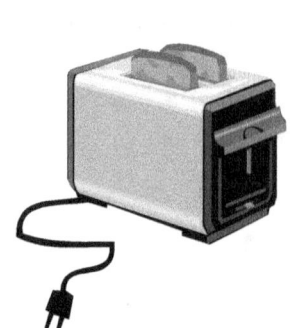

(b) the power consumed by the toaster

(c) the energy consumed by the toaster in 10 min, in joules and kWh.

(d) the cost of running the toaster for 10 min at the rate of 10 cents/kWh.

5. Determine the hot-resistance of an electric bulb which is rated 120 V - 75 W.
[192 Ω]

6. A carbon resistor is rated 2000 ohm - 0.25 W. The resistor will burn off if a power of more than 0.25 W is dissipated in it. Determine the maximum possible current and voltage for this resistor.

7. A battery with an emf of 6 V has an internal resistance of 5 ohm. If the external resistance is 25 ohm determine:
(a) the current drawn from the battery. *[0.2 A]*

(b) the external (terminal) potential difference. *[5 V]*

(c) the internal potential difference. *[1 V]*

(d) the power dissipated in the external resistance. *[1 W]*

(e) the power lost within the battery. *[0.2 W]*

Multiple Choice:

1. A wire has a resistance of 18 Ω. If it is stretched to 3 times its original length without a change of its density the new resistance will be
(A) 2 Ω   (B) 6 Ω   (C) 18 Ω   (D) 54 Ω   (E) 162 Ω

2. When a resistor R is connected across a battery of emf Є and negligible internal resistance the battery delivers a power P. If the resistance of the resistor is doubled the power delivered by the battery would be
(A) still P   (B) ½ P   (C) 2P   (D) ¼ P   (E) 4 P

3-4.
For the 4 wires shown here,:

A [   ]   C [   ]

B [   ]   D [   ]

3. Which wire has the least resistance?
(A) A   (B) B   (C) C   (D) D
(E) It depends on the voltage applied across the wires.

4. Which wire has the most resistance?
(A) A   (B) B   (C) C   (D) D
(E) It depends on the voltage applied across the wires.

5. A voltage V is applied across a resistor R. The resistor dissipates power P. The power dissipated by the resistor will increase by a factor of 2 if the voltage is changed by a factor of
(A) 2   (B) 4   (C) ½   (D) ¼   (E) √2

Chapter 38
Resistors and Capacitors in Series and Parallel

RESISTORS: Any medium that offers resistance to the flow of charges through it is called a resistor.

In most metals, electrons are responsible for the electric current. The resistance is created due to the collision of the electrons with the particles of that medium. Such collisions results in the loss of kinetic energy of the electrons and the electrons achieve 'terminal velocity' which is their drift speed. The loss of the kinetic energy of the electrons appears as dissipation of light and/or heat in the resistor.

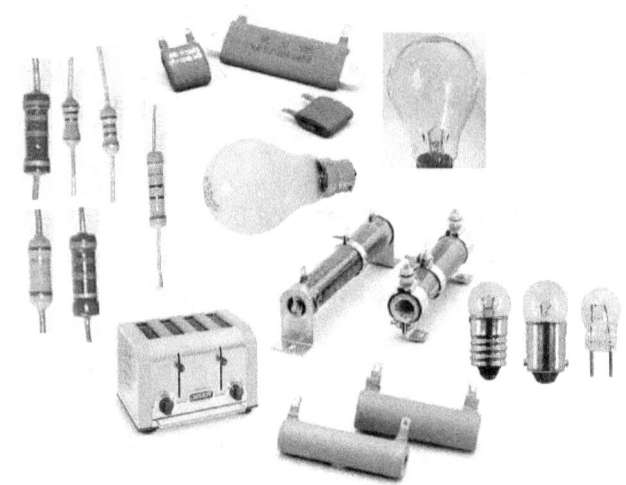

Resistors in Series: For the resistors R_1, R_2, and R_3 connected in series,

1. the effective resistance R_s is given by
$$R_s = R_1 + R_2 + R_3$$

2. the total voltage applied across the circuit
$$V = V_1 + V_2 + V_3$$

3. the current is same through each resistor, i.e. $I = I_1 = I_2 = I_3$

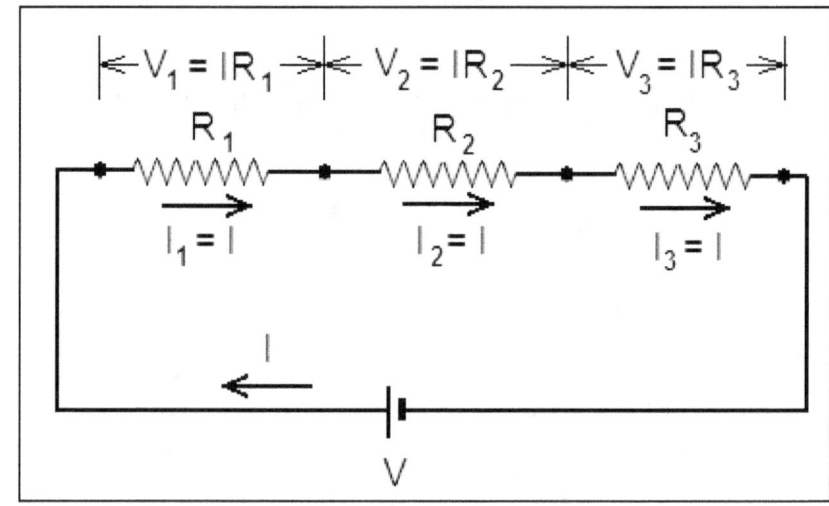

Resistors in Parallel: For the resistors R_1, R_2, and R_3 connected in parallel,

1. the effective resistance R_p is given by
$$1/R_p = 1/R_1 + 1/R_2 + 1/R_3$$
2. the total current drawn from the battery is
$$I = I_1 + I_2 + I_3$$
3. the potential difference across the resistor is same as that of the battery, i.e.
$$V = V_1 = V_2 = V_3$$

$$I_1 = \frac{V}{R_1}$$

$$I_2 = \frac{V}{R_2}$$

$$I_3 = \frac{V}{R_3}$$

Two resistors in Parallel:
If there are only two resistors R_1 and R_2 in parallel, it can be shown that
$$R_p = R_1R_2 / (R_1 + R_2)$$

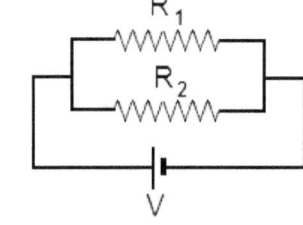

CAPACITORS

Capacitors in Series: For the resistors C_1, C_2, and C_3 connected in series,

1. the effective resistance C_p is given by
$$1/C_s = 1/C_1 + 1/C_2 + 1/C_3$$
2. the charge on each capacitor is same as the total charge drawn from the battery

$$q = q_1 = q_2 = q_3$$
3. the total voltage applied across the circuit is,
$$V = V_1 + V_2 + V_3$$

Capacitors in Parallel: For the resistors C_1, C_2, and C_3 connected in parallel,
1. the effective resistance R_p is given by

$$C_p = C_1 + C_2 + C_3$$

2. the total charge drawn from the battery is

$$q = q_1 + q_2 + q_3$$

3. the potential difference across each capacitor is same as that of the battery, i.e.

$$V = V_1 = V_2 = V_3$$

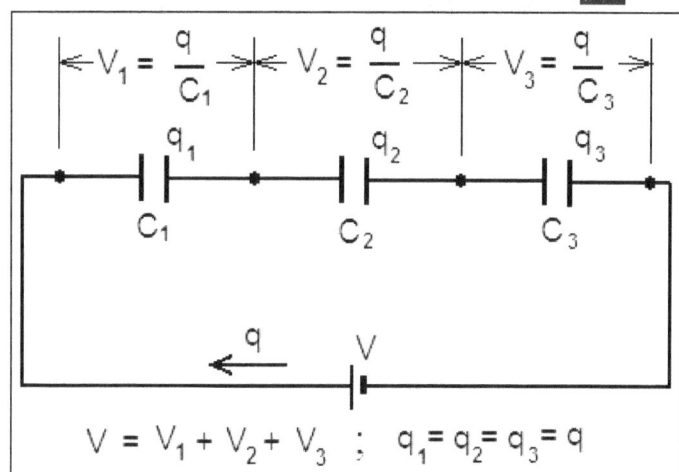

If there are only two capacitors C_1 and C_2 in series, it can be shown that

$$C = \frac{C_1 C_2}{C_1 + C_2}$$

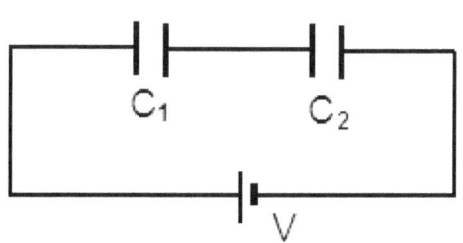

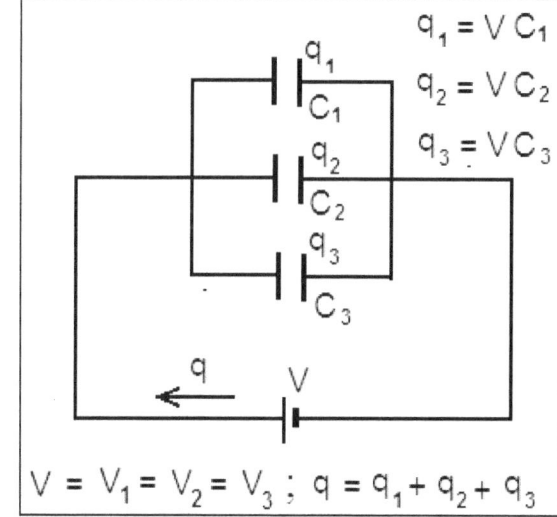

Some interesting facts & figures
• Electrical appliances and devices at home are all connected in parallel for two resons: One, they can be turned on and off independent of each other and two, they can all be designed ofr the same voltage.
• Large capacitor banks (Reservoir) are used as energy sources for the exploding *slapper detonators* in nuclear weapons and other specialty weapons.
• Capacitors and inductors together are used in tuner circuits to select particular frequency bands such as radio receivers
• Thermistor, use as a temperature sensor, is a resistor whose resistance varies with temperature.
• Negative temperature coefficient thermistor has its application in car for monitoring temperatures of coolant and engine oil.

Problems (Circuits with resistors):

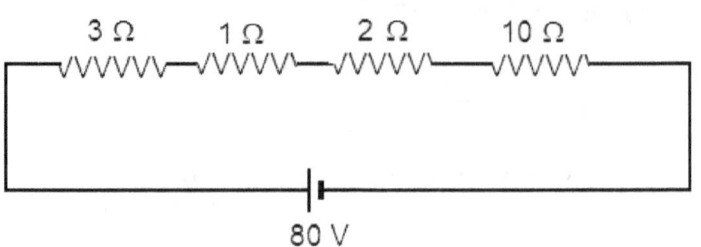

3 Ω   1 Ω   2 Ω   10 Ω

1. A series circuit contains 4 resistors of values 3 Ω, 1 Ω, 2 Ω, 10 Ω in that order. It is connected to a battery of emf 80 V. Determine,
(a) the current through each resistor.   [5 A]

80 V

(b) the total current drawn from the battery.   [5 A]

(c) the potential difference across each resistor.   [15 V, 5 V, 10 V, 50V]

(d) the potential difference across the combination 3 Ω and 1 Ω resistors.   [20 V]

(e) the power dissipated in each resistor.   [75 W, 25 W, 50 W, 250 W]

(f) the power delivered by the battery.   [400 W]

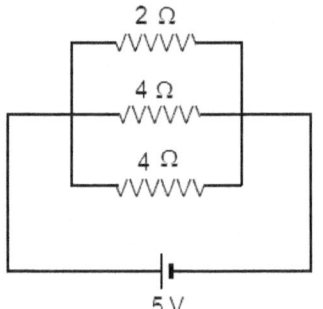

2 Ω

2. A parallel circuit consists of the three resistors: 2 Ω, 4 Ω, and 4 Ω. It is connected to a battery of emf 5 V. Determine
(a) the current through each resistor

4 Ω

4 Ω

(b) the total current drawn from the battery.

5 V

(c) the potential difference across each resistor

(d) the power dissipated in each resistor

(e) the power delivered by the battery

3. The resistors 3 Ω and 6 Ω are connected in parallel and this combination is connected in series with a 7-Ω resistor. The resulting combination is connected across a battery. If the current through the 6- Ω resistor is 1 A, determine
(a) the potential difference across the parallel combination.   [6 V]

(b) the potential difference across the 7-Ω resistor.   [21 V]

(c) the emf of the battery.   [27 V]

(d) the current drawn from the battery.   [3 A]

(e) the power supplied by the battery.   [81 W]

(f) the current through the 3-Ω resistor.   [2 A]

(g) the resistance of the whole circuit.   [9 Ω]

4. Two resistors 8 Ω and 4 Ω are connected in series. This combination is then connected in parallel with a 6-Ω resistor. This circuit is now connected across a battery of emf 24 V. Determine,
(a) the potential difference across the 8-Ω resistor

(b) the current drawn from the battery

5. A wire of resistance 16 ohms is bent into the form of a circle and the two ends are soldered together. Let P be the center of this circular wire. Two points A and B are on this wire such that angle APB = 90°. A battery of emf 36 V is connected to the points A and B. Determine
(a) the total resistance of the circuit.   [3 Ω]

(b) the current drawn from the battery.   [12 A]

6. You are given four resistors each of resistance 6 Ω. How will you connect all of them in simple series, parallel, or their combination to get an effective resistance of

(a) 1.5 Ω   (b) 24 Ω   (c) 6 Ω (two different ways) (d) 8 Ω   (e) 3.6 Ω
(f) 2.4 Ω (g) 15 Ω (h) 4.5 Ω

7. Two electric bulbs, one with resistance R_1 and the other with resistance R_2 are to be connected to a voltage source. $R_1 > R_2$ at all temperatures. Which bulb will be brighter if they are connected in series? and then in parallel?   [First bulb, Second bulb]

8. For each of the three circuits shown below, determine
 (i) the effective resistance

 (ii) the current drawn from the battery

 (iii) the potential difference across each resistor

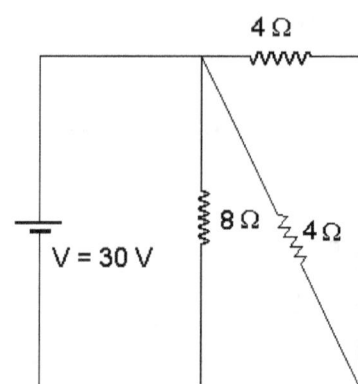

(c)

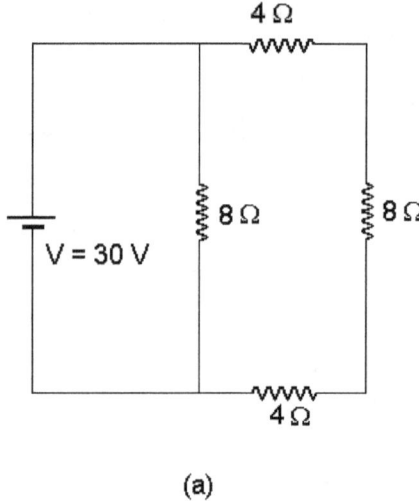

(a)

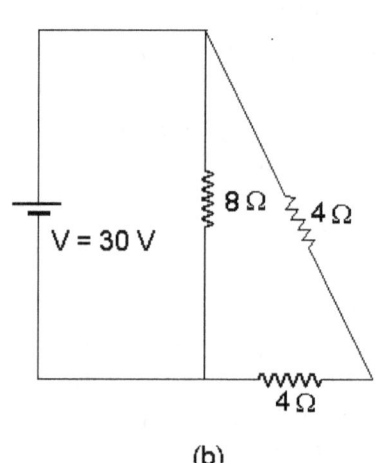

(b)

Problems (Circuits with capacitors):

9. Three capacitors, $C_1 = 12\ \mu F$, $C_2 = 6\ \mu F$, and $C_3 = 4\ \mu F$ are connected in series with a battery of emf 8 V. Determine
(a) the equivalent capacitance of the circuit.   [2 μF]

(b) the charge drawn from the battery.   [16 μC]

(c) the charge on each capacitor. *[16 μC each]*

(d) the energy supplied by the battery in charging the capacitors. *[64 μJ]*

(e) the energy stored in each capacitor. *[10.7 μJ, 21.3 μJ, 32.0 μJ]*

10. Three capacitors, C_1= 5 μF, C_2= 7 μF, and C_3= 3 μF are connected in parallel with a battery of emf 30 V. Determine
(a) the equivalent capacitance of the circuit

(b) the charge drawn from the battery

(c) the charge on each capacitor

(d) the energy supplied by the battery in charging the capacitors

(e) the energy stored in each capacitor

11. Two capacitors, C_1= 8 μF and C_2 = 16 μF are connected in parallel. This combination is connected in series with another capacitor C_3= 12 μF and a battery of emf 16 V.
(a) Determine the equivalent capacitance of the circuit. *[8 μF]*

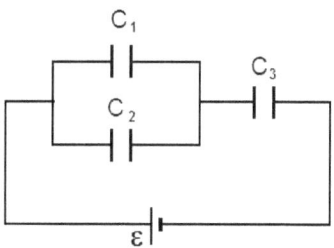

(b) Determine the charge on each capacitor. *[42.6 μC, 85.3 μC, 128 μC]*

(c) Determine the charge drawn from the battery. *[128 μC]*

(d) Determine the energy supplied by the battery in charging all the capacitors. *[1024 μJ]*

12. Two capacitors, $C_1 = 12 \, \mu F$ and $C_2 = 6 \, \mu F$ are connected in series. This combination is connected in parallel with another capacitor $C_3 = 1 \, \mu F$. The entire combination is connected to a battery of emf 10 V.
(a) Determine the equivalent capacitance of the circuit.

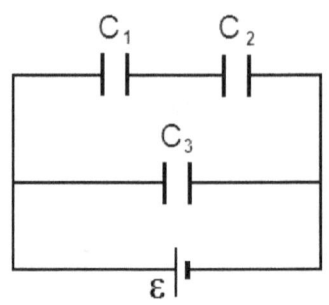

(b) Determine the charge on each capacitor.

(c) Determine the charge drawn from the battery.

(d) Determine the energy supplied by the battery in charging all the capacitors.

13. For each of the three circuits shown below, determine

(i) the effective capacitance.

[(a): 9.6 μF, (b): 6 μF, (c):15 μF]

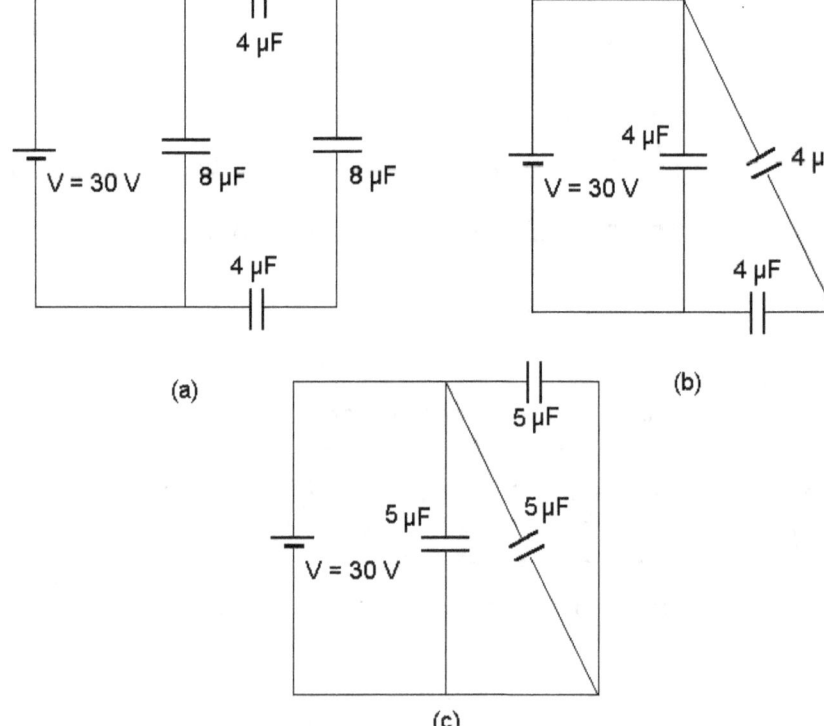

(ii) the total charge drawn from the battery   *[(a): 288 μC, (b): 180 μC, (c): 450 μC]*

14. For the circuit shown below, the charge on C_2 is 15 μC.

14. For the circuit shown below, the charge on C_2 is 15 μC.

(a) Calculate the total capacitance C.

(b) Calculate q_1.

(c) Calculate V_1 and V_2, across C_1 and C_2 respectively.

(d) Calculate the emf ε.

(e) Calculate q_3.

(f) Calculate the total charge drawn from the battery.

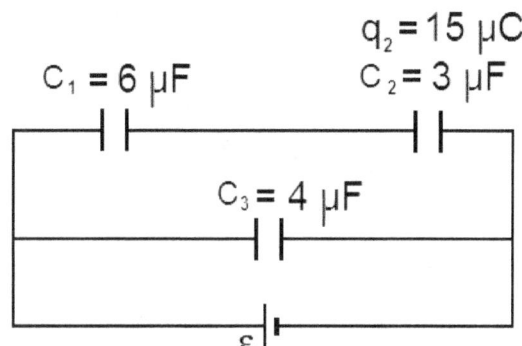

Multiple Choice:

1. One thousand identical resistors, each of resistance 5 Ω are connected in series. The effective resistance of the circuit is
(A) 5,000 Ω   (B) 1/200 Ω   (C) 200 Ω   (D) 1005 Ω   (E) None of these

2. One thousand identical resistors each of resistance 5 Ω are connected in parallel. The effective resistance of the circuit is
(A) 5,000 Ω   (B) 1/200 Ω   (C) 200 Ω   (D) 1005 Ω   (E) None of these

3. One thousand identical capacitors, each of capacitance 5 μF are connected in series. The effective capacitance of the circuit is
(A) 5,000 μF   (B) 1/200 μF   (C) 200 μF   (D) 1005 μF   (E) None of these

4. One thousand identical capacitors, each of capacitance 5 μF are connected in parallel. The effective capacitance of the circuit is
(A) 5,000 μF   (B) 1/200 μF   (C) 200 μF   (D) 1005 μF   (E) None of these

5- 6. Consider the circuits below:

5. Which of the circuits have a resistance of 3 Ω between the points P and Q?
(A) A
(B) B
(C) C
(D) D
(E) None of these

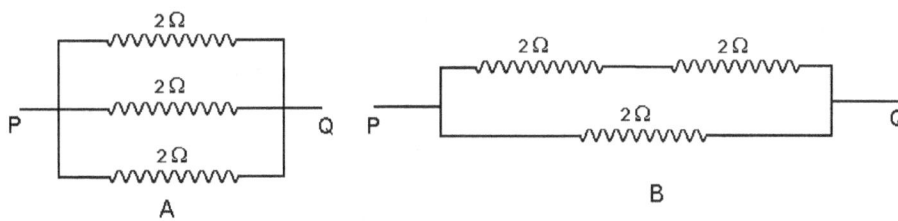

6. Which of the circuits have a resistance of 2/3 Ω?
(A) A
(B) B
(C) C
(D) D
(E) None of these

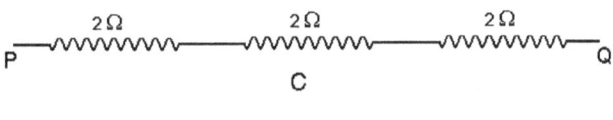

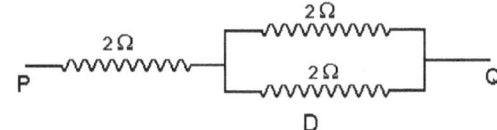

Chapter 39
Kirchoff's Rules and RC Circuits

Gustav R Kirchoff

Kirchoff's Rules:

Some circuits that are neither series, nor parallel, nor series-parallel combinations can be solved using the two Kirchoff's rules.

1. Junction Theorem (or Current Rule): In any circuit the total current entering a junction is equal to the total current leaving the junction. Therefore,

$$\Sigma \ i_{in} = \Sigma \ i_{out}$$

Example: As in the diagram, five wires in a circuit meet at a junction P.

By the Junction Theorem, $I_1 + I_2 + I_4 = I_3 + I_5$

2. Loop Theorem (or Voltage Rule): For any closed loop chosen in a circuit, total voltage of all the batteries plus the total potential difference across all the resistors is equal to zero (when proper sign convention is used!). Therefore,

$$\Sigma \ V + \Sigma \ IR = 0$$

Sign Convention for the Kirchoff''s Loop Rule:

Start from any point on the loop and 'walk' along the loop (CW or CCW).
SIGN OF EMF: Whenever you meet the positive terminal of the battery, take its emf V (or ε) to be positive and vice versa.
SIGN OF iR: Sign of iR is decided by the way current is flowing in the resistors. When you 'walk' through a resistance R, take the potential difference across the resistor to be +iR if the current i is flowing in the same direction as you are walking. The potential difference is -iR otherwise.

Note: The direction of current does not affect the sign of the emf V of the battery.

Example:
Consider the closed loop here. Any circuit components outside the loop are irrelevant for applying the loop theorem for this loop. Let us start from A and go clockwise around the loop and return to A.

Adding all IR's and V's by the convention above, we get

$$I_1R_1 + V_1 - V_2 + V_3 - I_3R_2 - I_3R_3 - I_4R_4 - V_4 = 0$$

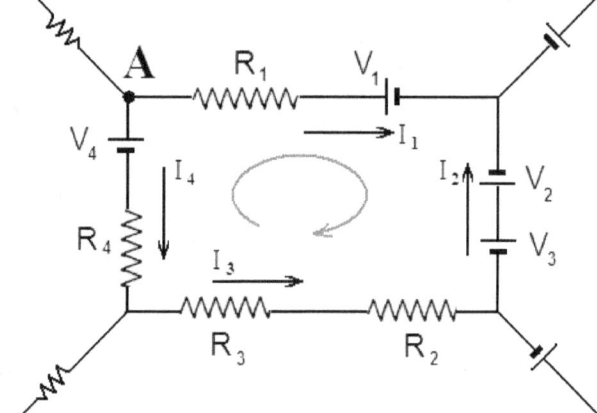

Note that the equation does not have any term involving the current I_2 because in this loop I_2 is not passing through any resistance.

RC-Circuits: This circuit consists of a resistor and capacitor in series.   There may or may not be a battery in the circuit.

Charging:

In the circuit below the switch is initially open and there is no charge on the capacitor.

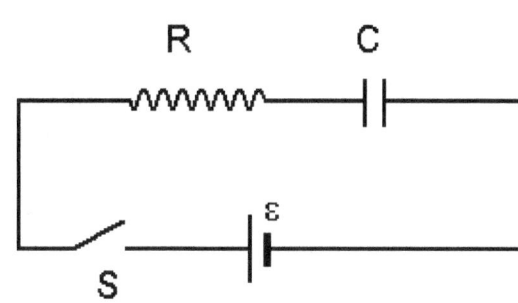

- **At time t = 0:** When the switch is closed at time t = 0 the charges flowing to the plates of the capacitor find the plates empty and no resistance to flow. Hence, <u>momentarily</u> the current flows as if the circuit is complete with only the battery and the resistor in the circuit. Hence, i_o (at t = 0) = Ɛ/R

- As the time progresses, the voltage opposite to that of the battery is building up on the capacitor. The voltage on the capacitor reaching the voltage of tha battery and the current is decaying exponentially.

- **At t = infinity (after a long time):** A long time after the switch is closed, the capacitor is charged and has voltage equal and opposite to that of the battery. Hence no more current flows through the circuit.

i (at t = infinity) = 0

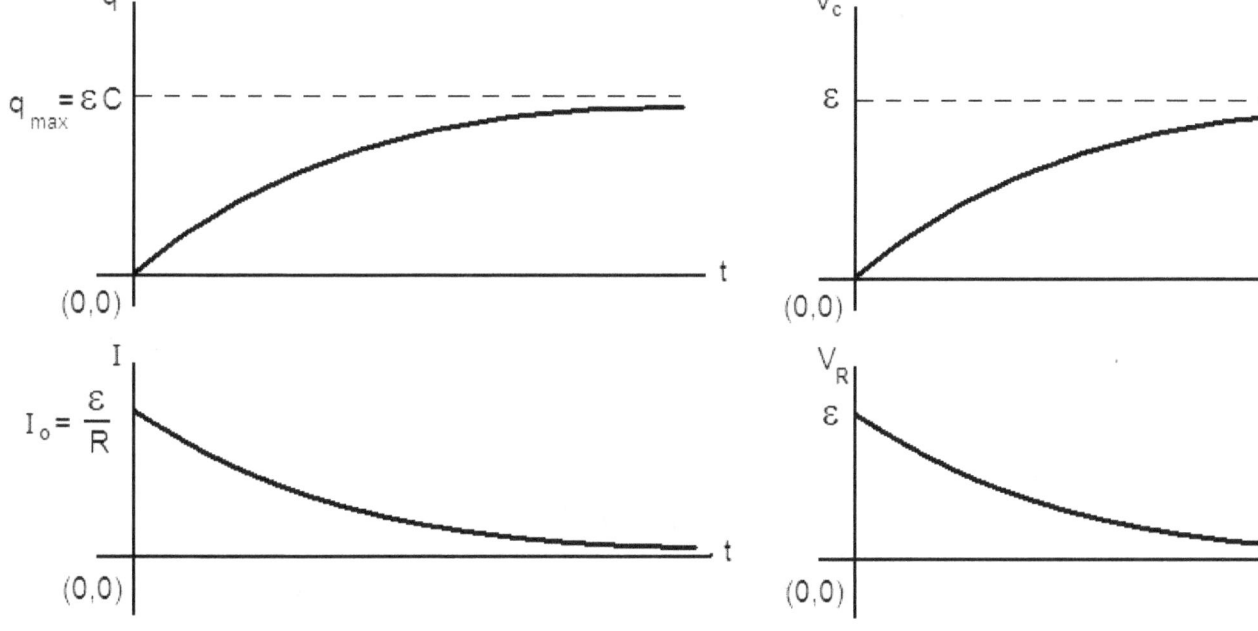

Discharging:

A capacitor is initially charged to a voltage of V_o and carries a charge q_o. It is then connected to a resistor (R) and an open switch S.

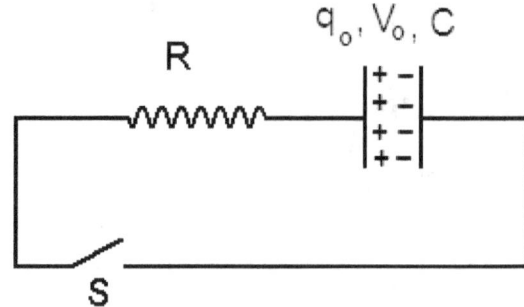

- **At time t = 0:** When the switch is closed at time **t = 0** the charges flow from one plate of the capacitor to the other plate through the resistor. The capacitor, momentarily acts like a battery with emf V_o. Hence, I_o (at t = 0) = V_o/R

- As the time progresses, the capacitor is depleting of its charge exponentially. The capacitor voltage ($V_C = q/C$), the current I and the voltage across the resistor ($V_R = IR$) are also dropping off exponentially

- **At t = infinity (after a long time):** A long time after the switch is closed, the capacitor is discharged completely and no more current flows through the circuit. The circuit is dead!

$$i \text{ (at } t = \text{infinity)} = 0$$

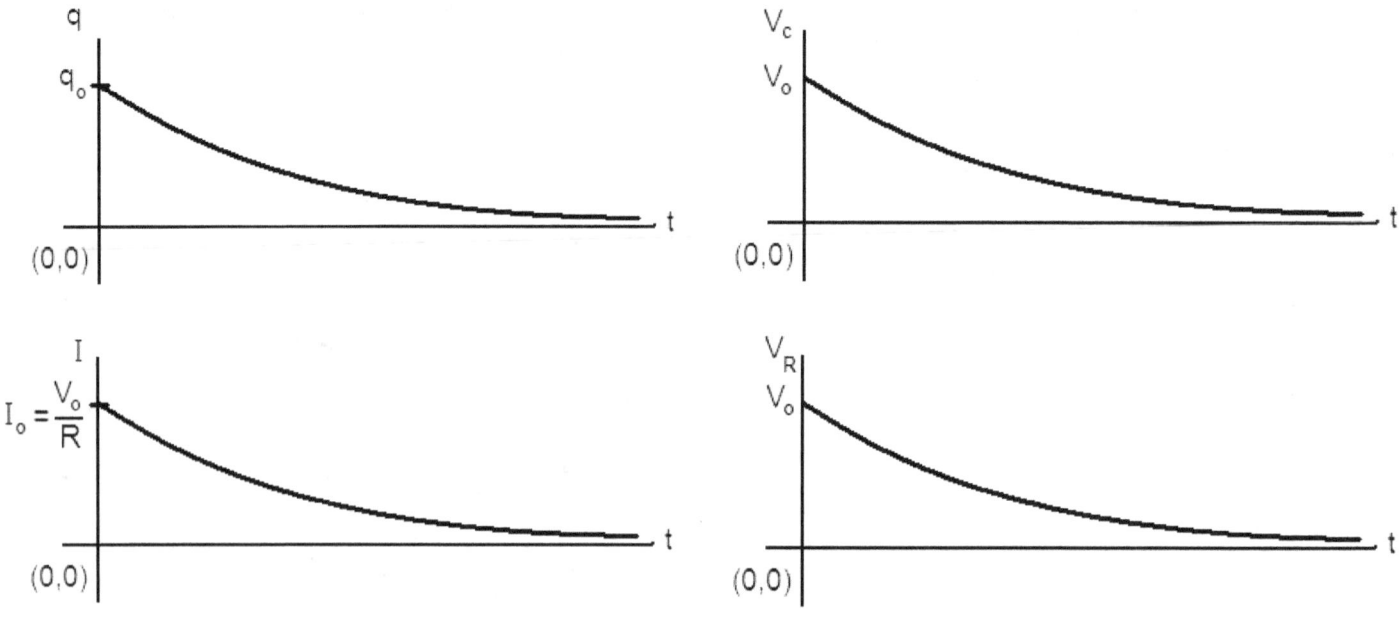

Some interesting facts & figures
• RC circuits can be used to filter a signal by blocking certain frequencies and passing others.   • Continuous repayment mortgage and charging of RC circuit are treated the same way mathemtically.   • RC circuit can be used to store energy in the capacitor and trigger a flash by dumping the entire energy stored in capacitor in the bulb,

Problems:

1. In the circuit shown below, use Kirchoff's Rules to determine the current in the resistor R_1.

[0.04 A ←]

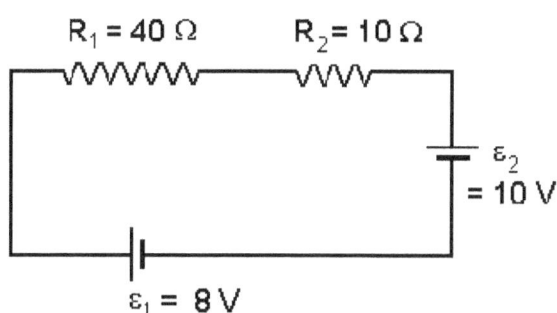

2. In the circuit shown below, use Kirchoff's rules to determine the emf ε_2.

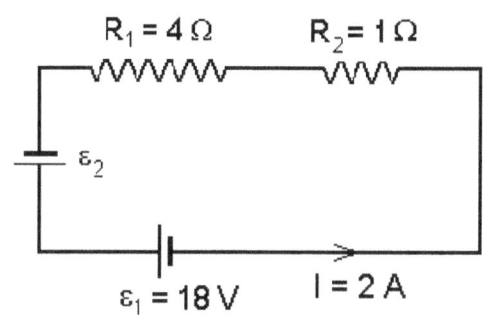

3. In the diagram below determine the charge on the capacitor and the current in the circuit,

(a) immediately as the switch is closed (t = 0). *[2 A]*

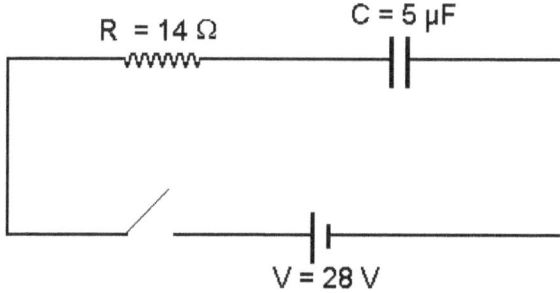

(b) after a long time (t = ∞) .   *[0 A]*

4. In the circuit below,

(a) **Initially, the switch S_1 is closed and S_2 is left open.**

Determine:
(i) The current drawn from the battery.

(ii) The charge on the capacitor

(iii) The current through the resistors R_1 and R_2

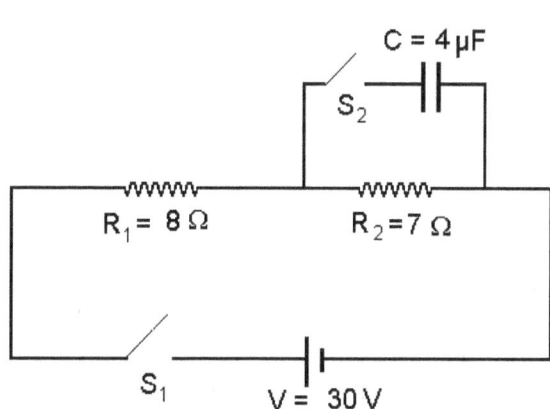

(b) **Now the switch is S$_2$ is also closed**

At time t = 0, determine:
(i) the current through R$_1$

(ii) the current through R$_2$

(iii) the charge on the capacitor

(iv) the voltage across the capacitor
After a long time (t = ∞) determine,
(v) the current through R$_1$

(vi) the current through R$_2$

(vii) the charge on the capacitor

(viii) the voltage across the capacitor

5. Two resistors R$_1$ = 6 Ω & R$_2$ = 4 Ω, a battery of emf 30 V, a capacitor C = 25 μF , and switches S$_1$, S$_2$, and S$_3$ are connected in a circuit as shown in the diagram above.

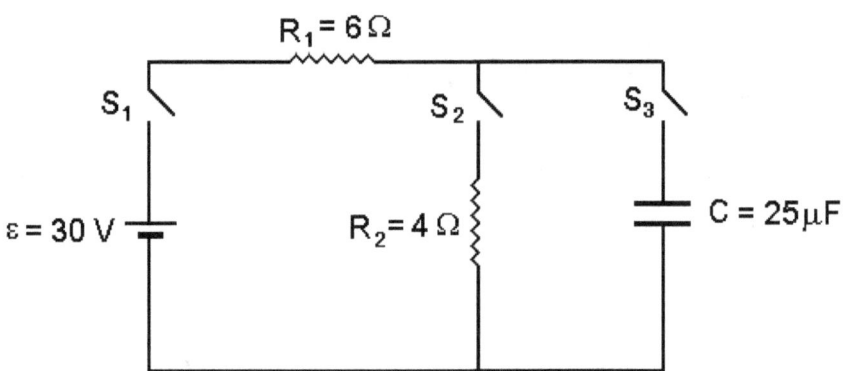

Initially the capacitor is uncharged and all the switches are open. Now switches S$_1$ and S$_2$ are closed.

(a) Calculate the currents through R$_1$ and R$_2$. *[Both 3 A]*

Now switch S$_2$ is opened and the switch S$_3$ is closed

(b) Calculate the currents through R$_1$ & R$_2$ and charge on C
<u>immediately</u> after the switch S$_3$ is closed.   *[5 A, 0 A, 0 C]*

(c) Calculate the currents through R$_1$ & R$_2$ and charge on C
a <u>long time </u>after the switch S$_3$ is closed.   *[0 A, 0 A, 750 μC]*

(d) Sketch a current vs. time graph for R$_1$ <u>over</u> a long period of time after S$_3$ was closed.

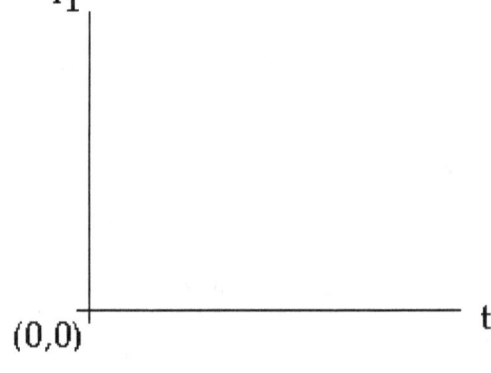

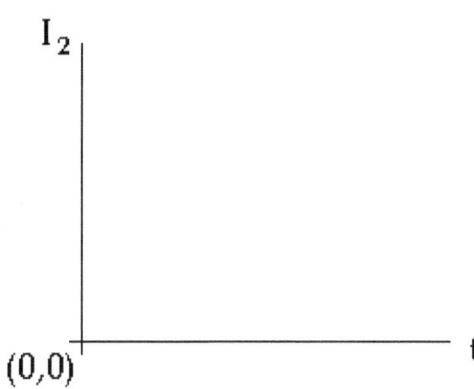

A long time after the switch S_3 has stayed closed the switch S_1 is opened and then S_2 is closed.

(e) Calculate the currents through R_1 & R_2 <u>immediately</u> after the switch S_2 is closed. *[0 A, 7.5 A]*

(f) Calculate the currents through R_1 & R_2 and charge on C a <u>long time</u> after the switch S_2 is closed. *[0 A, 0 A, 0 C]*

(g) Sketch a current vs. time graph for R_2 <u>over</u> a long period of time after S_2 was closed.

Multiple Choice:

1. At a junction P in a circuit a number of branches carry currents in the directions shown. Which of the following is the correct application of the junction rule at the point P?

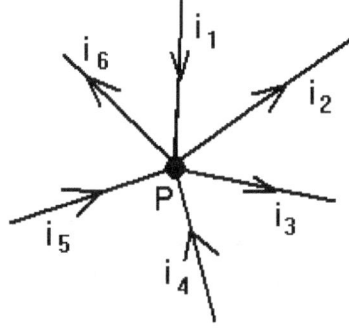

(A) $i_1 + i_2 + i_3 = i_4 + i_5 + i_6$
(B) $i_1 + i_2 + i_3 + i_4 + i_5 + i_6 = 0$
(C) $i_1 + i_4 + i_5 = i_2 + i_3 + i_6$
(D) $i_5 + i_2 + i_3 = i_1 + i_4 + i_6$
(E) $i_1 + i_2 + i_3 + i_4 + i_5 + i_6 = 360$

2. The current in the circuit shown above is
(A) 0 A
(B) 0.6 A
(C) 1.2 A
(D) 1.67 A
(E) 3.33 A

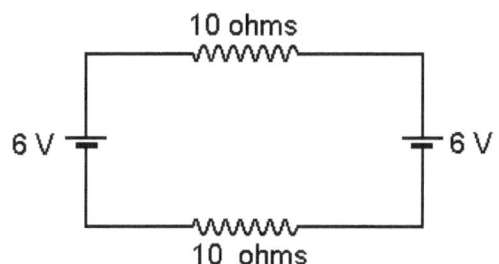

3-4. In the circuit below,

       3. what is the reading in the ammeter a long time after the switch has been closed?
(A) 0 A
(B) 10 A
(C) 12 A
(D) 20 A
(E) 30 A

       4. What is the reading in the ammeter immediately after switch is opened.
(A) 0 A
(B) 10 A
(C) 12 A
(D) 20 A
(E) 30 A

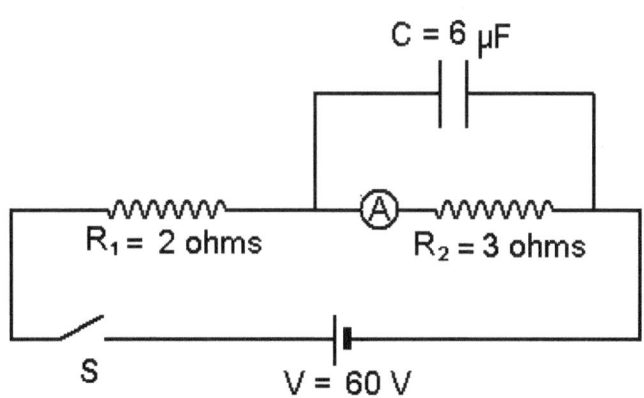

Chapter 40
Magnetic Fields

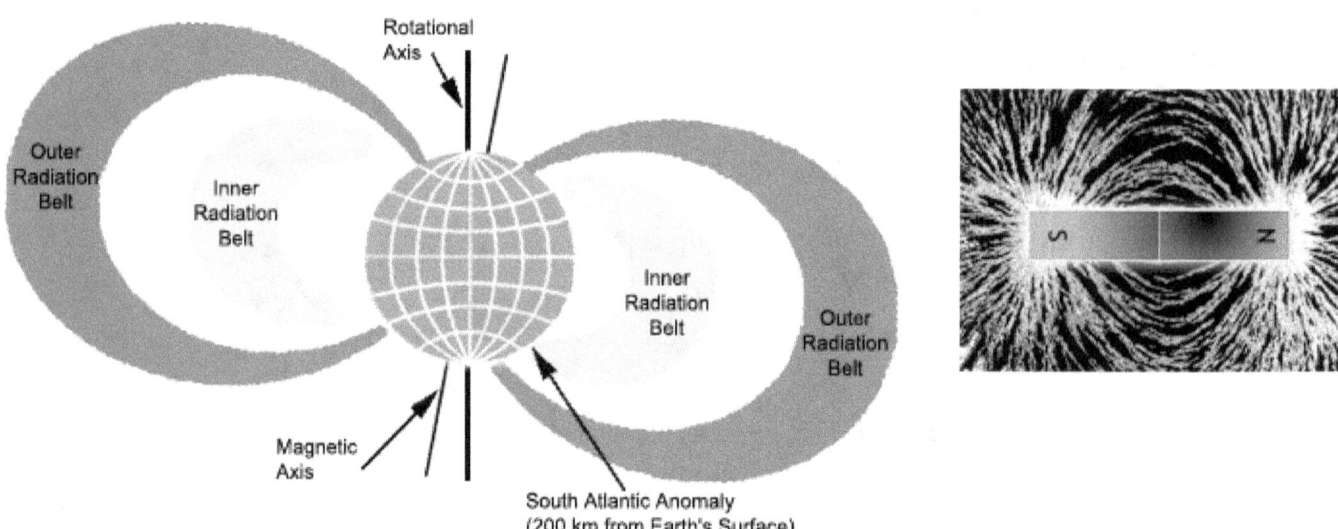

Magnetism is one of the fundamental forces in nature. Toward the end of last century, Maxwell showed that the electric and magnetic forces are the manifestations of a single force. This force is called electromagnetic force.

The SI Unit for the magnetic field is **tesla (T)**. $1 \text{ T} = 10^4$ gauss

The Force on a Moving Charged Particle due to Magnetic Field B:

A charged particle moving with a velocity **v** in a magnetic field **B,** experiences a force **F.** For this force the MAGNITUDE is

$$F_B = |q|vB \sin\theta$$

where θ is the angle between the vectors **v** and **B**.
$F_B = 0$ if $\theta = 0°$ or $180°$

The DIRECTION of the magnetic force on a **positive charge** is given by **right-hand-palm rule**
The DIRECTION of the magnetic force on a **negative charge** is given by **left-hand-palm rule**

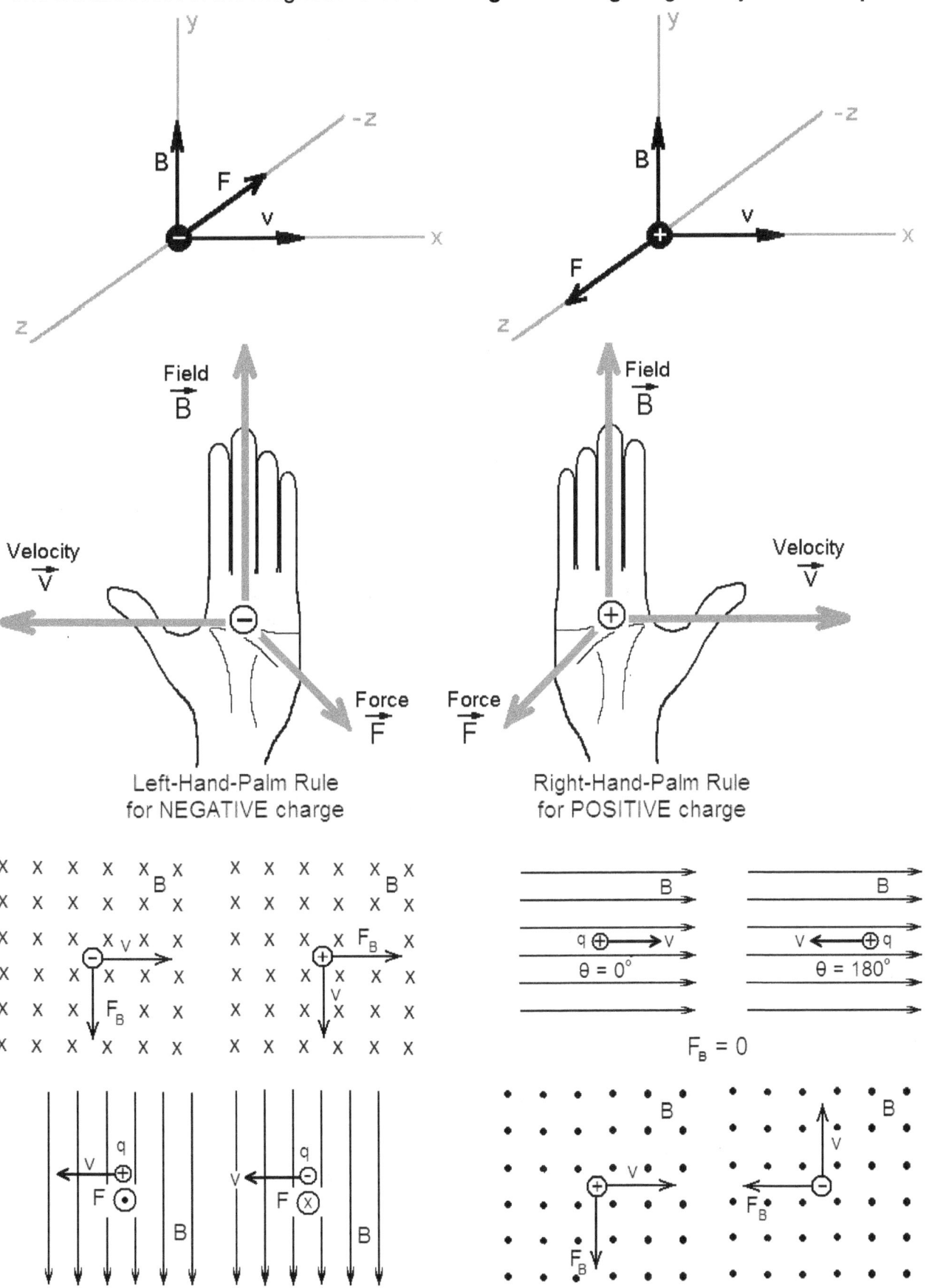

Left-Hand-Palm Rule
for NEGATIVE charge

Right-Hand-Palm Rule
for POSITIVE charge

Circulating charge in a uniform magnetic field:

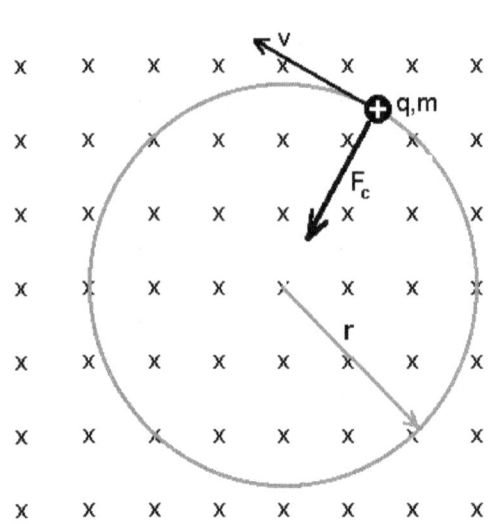

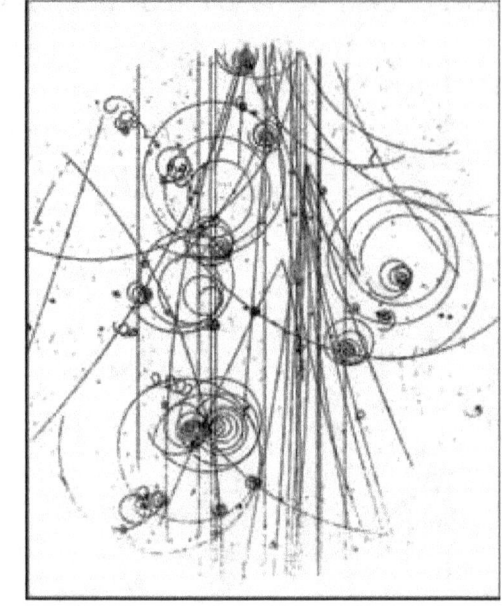

Cloud chamber particle tracks

The charged particles tracks are curved becasue of the strong applied magnetic field

Let a particle of charge q and mass m enters a uniform magnetic field **B** with a velocity **v** perpendicular to **B**. The particle falls into a circular trajectory because the force due to the magnetic field (qvB) is always acting at right angles to v – a characteristic of the uniform circular motion.

*Radius (**r**) of the Circular Trajectory:*

$$r = \frac{mv}{qB}$$

*The time period (**T**) for the circular motion:*

$$T = \frac{2\pi r}{v}$$

Magnetic force does not do any work: The magnetic force is always perpendicular to the velocity of the particle hence it does not do any work on the particle.

The speed and kinetic energy are not affected by the magnetic field.

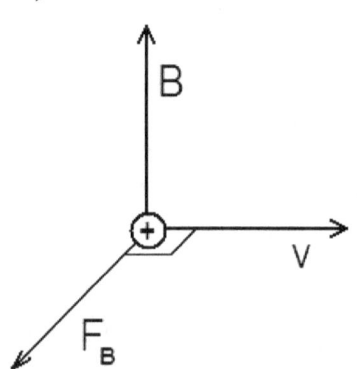

F$_B$ is always perpendicular to v hence no work is done by it on the particle.

Velocity selector:

A uniform electric field **E** is created between two equal
and oppositely charged parallel plates. A beam of charged
particles (say positive) moving with velocity **v** will be deflected
toward the negative plate due to the electric force **F$_E$ = qE**. A
magnetic field **B** can be created at right angles to the electric
field **E** to undo the deflection and make the beam straight. The
field **B** should be such that it causes a force **F$_B$ = qvB** equal

and opposite to **F$_E$**. Hence

F$_B$ = F$_E$ ➔ qvB = qE ➔ $$v = \frac{E}{B}$$

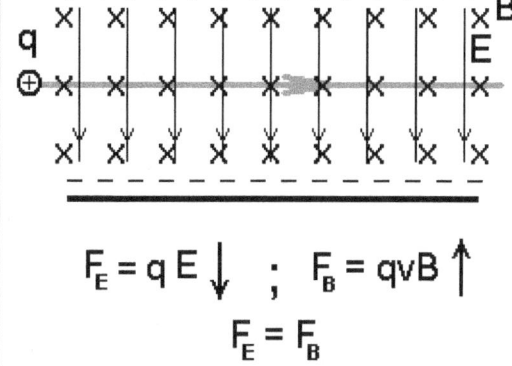

$$F_E = qE$$

$$F_E = qE\downarrow \quad ; \quad F_B = qvB\uparrow$$
$$F_E = F_B$$

Mass Spectrometer and Isotope Separator
Charged particles of different kinds in a beam can be
separated by allowing the beam to enter a uniform magnetic
field, usually at right angles. The particles will follow curved
paths of radius r given by the equation

$$r = \frac{mv}{qB}$$

For example, the three
isotopes of hydrogen can be
separated by this method.

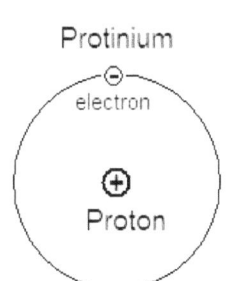
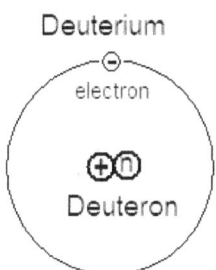
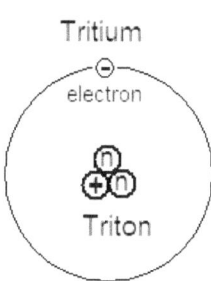

The three isotopes of HYDROGEN

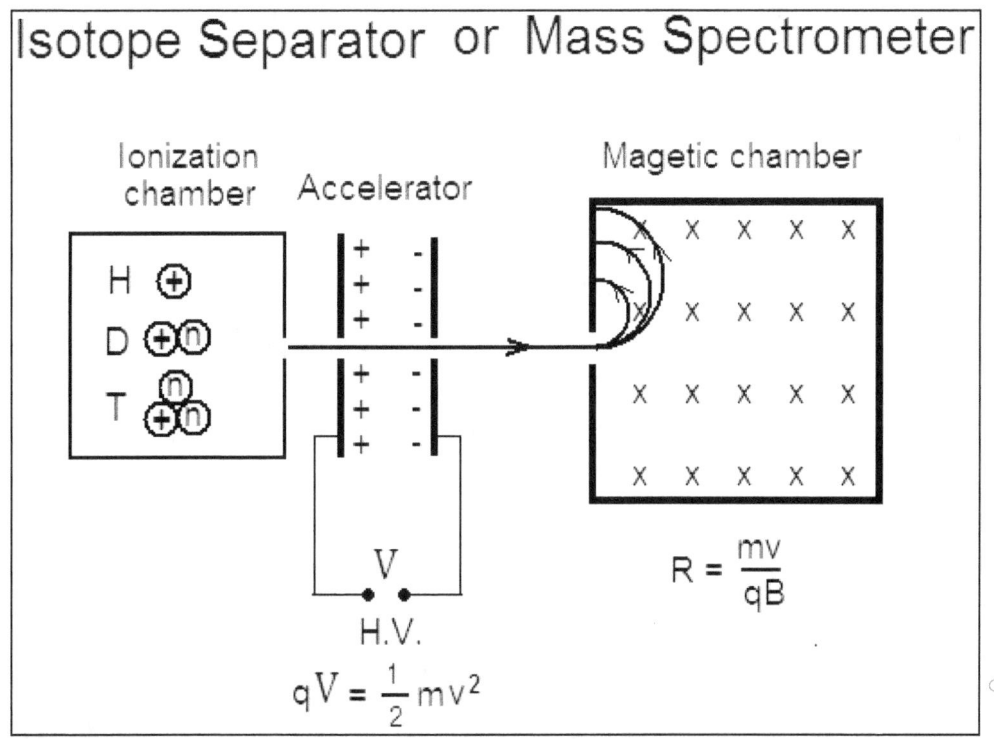

Isotope Separator or Mass Spectrometer

$$R = \frac{mv}{qB}$$

$$qV = \frac{1}{2}mv^2$$

Aurora - northern and southern lights - are caused by the exciation of air molecules by the electrons and protons of solar wind trapped in the earth's magnetic field.

Some interesting facts & figures
<ul><li>Magnetic monopoles have been theorized but never observed.</li><li>Powerful superconducting magnets are used to bend highly energetic proton beams and contain the beam over a circular path in accelerator such as at Fermi National Lab near Chicago and Large Hadron Collider in CERN, Europe.</li><li>Magnetic force bends charged particles in in detectors such as cloud and bubble changers. The circular trajectories of the charged particles can be analyze the mass, charge and energy of the particles</li><li>magnetic force is used in Hall Effect to measure the magnetic field strength or to determine dominant charge carriers in semiconductors.</li></ul>

Problems: Force on a moving charge particle

1. In each of the diagrams below, determine the magnitude and direction of the force acting on the charged particle; q = 4 μC, v = 500 m/s, B = 0.08 T.

   [1.6x10^{-4} N, +z; 1.6x10^{-4} N, -x; 1.6x10^{-4} N, -

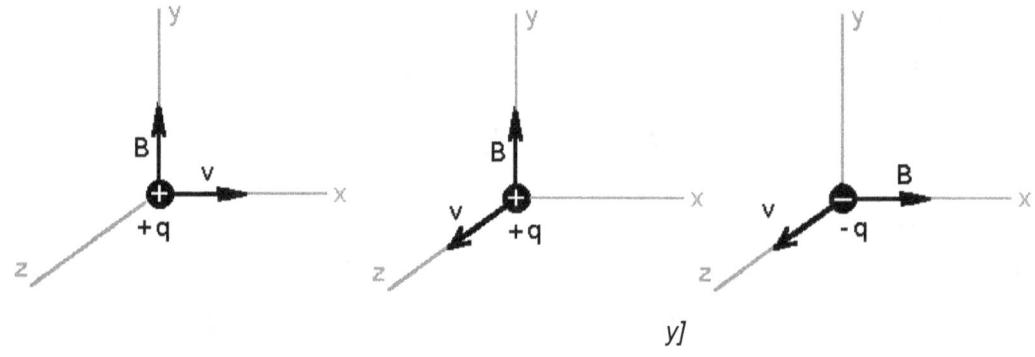

y]

2. In a region a uniform magnetic field **B** = 0.05 T is applied in the positive x-direction. A charged particle (q = 1.2 µC) enters the magnetic filed with a velocity **v** = 800 m/s. The velocity of the charged particle is in the xy-plane and makes and angle of 40° with the direction of **B**. Determine the magnitude and the direction of the force acting on the particle.

3. In a region a strong uniform magnetic field of 1.5 T is directed east to west. An electron moving horizontally toward north with a speed of 4 x 10⁶ m/s enters the magnetic field.
(a) Determine the magnitude of the force acting on the electron.   *[9.6x10⁻¹³ N]*
(b) In which direction will the electron be deflected?   *[Downward]*

4. In the figure below determine the magnitude and the direction of the force due to the magnetic field on each of the charged particles.

**B = 0.8 T,   q = 2 µC,   v = 800 m/s
lying in the plane of the diagram**

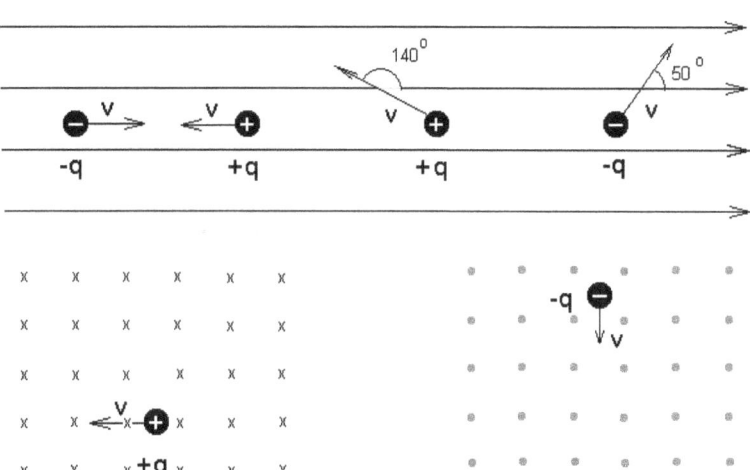

5. The electron beam in a TV set is accelerated through a potential difference of 1kV. They enter a uniform 0.01-T magnetic field. Determine the maximum possible force acting on the electrons.
[3.0x10⁻¹⁴ N]

6. A proton moving horizontally northward with v = 2 x 10⁵ m/s enters a uniform horizontal magnetic field and is deflected downward due to a force of 6.4 x 10⁻¹⁵ N. Determine the magnitude and direction of minimum possible magnetic field.

7. A beam of electrons moving with a velocity of 5x10⁷ m/s enters the space between the parallel plates as shown in the figure below. The plates are 10 cm long and 5 mm apart. Now a uniform electric field E is created by connecting the plates to a potential difference of 60 V.

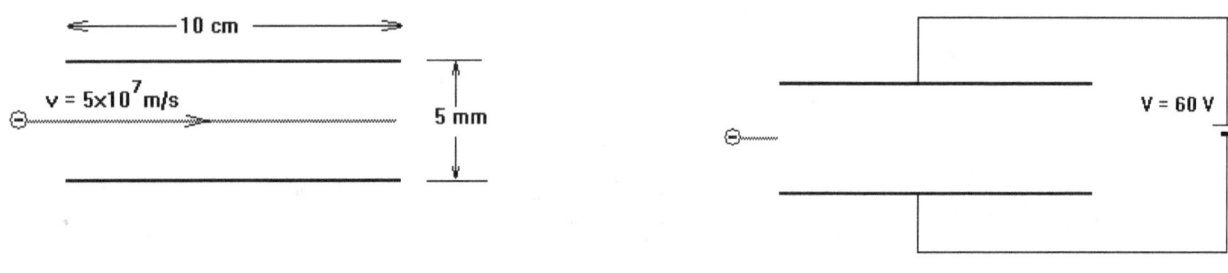

(a) Calculate the electric field E between the plates.   *[12,000] V/m*

(b) Sketch the direction of the electric field between the plates.

(c) Calculate the deflection of the beam from its original position as it exits (at the farther end of the plates) the field E. Sketch a diagram for the path of the beam.   *[0.0042 m]*

(d) The deflection of the beam can be cancelled and the beam can be made straight again by applying a magnetic field of appropriate intensity perpendicular to the electric field. Calculate the magnitude of the magnetic field and show in the diagram its direction.   *[2.4x10⁻⁴ T]*

Problems: Circulating Charge

8. An electron beam is moving with a velocity of 4 x 10⁶ m/s in a region. A 2 x 10⁻⁴-T uniform magnetic field is turned on at right angles to the beam. The beam falls into a circular path. Determine
(a) radius of its circular path.

(b) frequency of revolution along the circular path.

(c) the period of the uniform circular motion.

(d) the centripetal force acting on the particle.

9. A mass spectrometer is used to separate protons and deuterons in a beam. The particles are first accelerated through a potential of 3 kV and they enter the spectrometer at right angles to a magnetic field of strength 0.5 T.

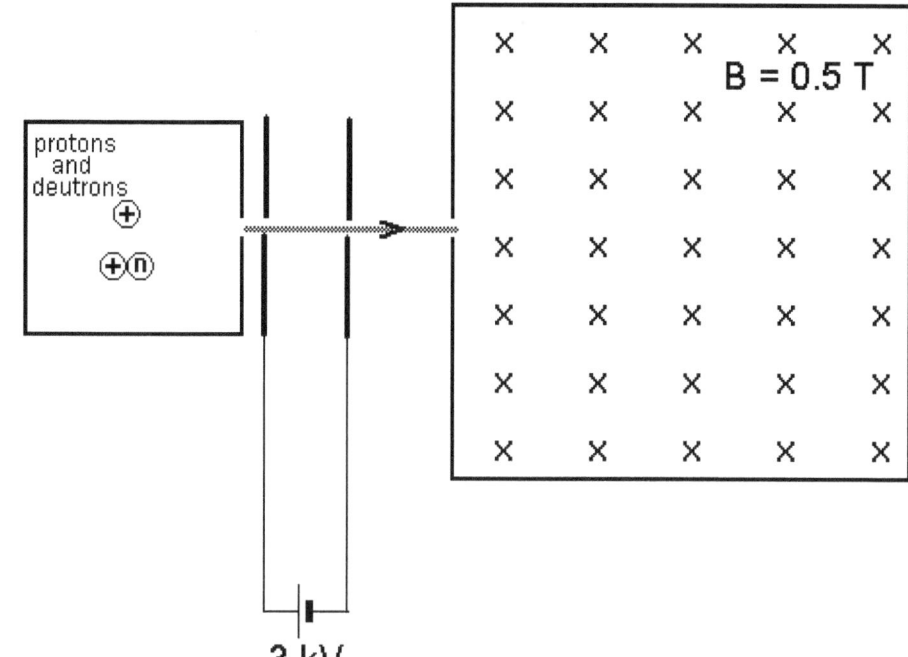

(a) Calculate the radii of curvature of the path taken by protons and deuterons in the beam.   *[R_p = 0.016 m, R_d = 0.022 m]*

(b) Sketch the trajectories of the two beams in the magnetic field above.

Multiple Choice:

1. An electron is traveling horizontally toward north. It enters a magnetic field, which is directed horizontally toward east. The electron will be deflected in which direction?
(A) South   (B) East   (C) West   (D) Up   (E) Down

2. An proton is traveling horizontally toward east. It enters a magnetic field, which is directed horizontally toward south. The proton will be deflected in which direction?
(A) South   (B) East   (C) West   (D) Up   (E) Down

3-4. A particle of charge q and mass m is moving at a speed v along a circular trajectory of radius r due to the force created by a uniform magnetic field B.

  3. The radius r is proportional to
  (A) q   (B) m only   (C) v only   (D) mv   (E) B

  4. The work done by the magnetic field on the charged particle in one revolution is
  (A) Zero   (B) B x 2πr   (C) qvB x 2πr   (D) q x 2πr   (E) 2πr/v

Chapter 41
Force due to B on a Current Carrying Conductor

Let a straight conductor be placed in a uniform magnetic field B. The conductor is carrying a current i in a direction which makes an angle θ with the field **B**. A length **L** of the conductor then experiences whose magnitude is given by

$$F_B = i \, L \, B \sin \theta$$

- F_B is always perpendicular to both **L** and **B** and hence to the plane formed by **L** and **B**.
- F_B is maximum at **θ = 90°**
- F_B is **0** at **θ = 0°** and **90°**
- Its direction is given by the **Right – Hand - Palm Rule**

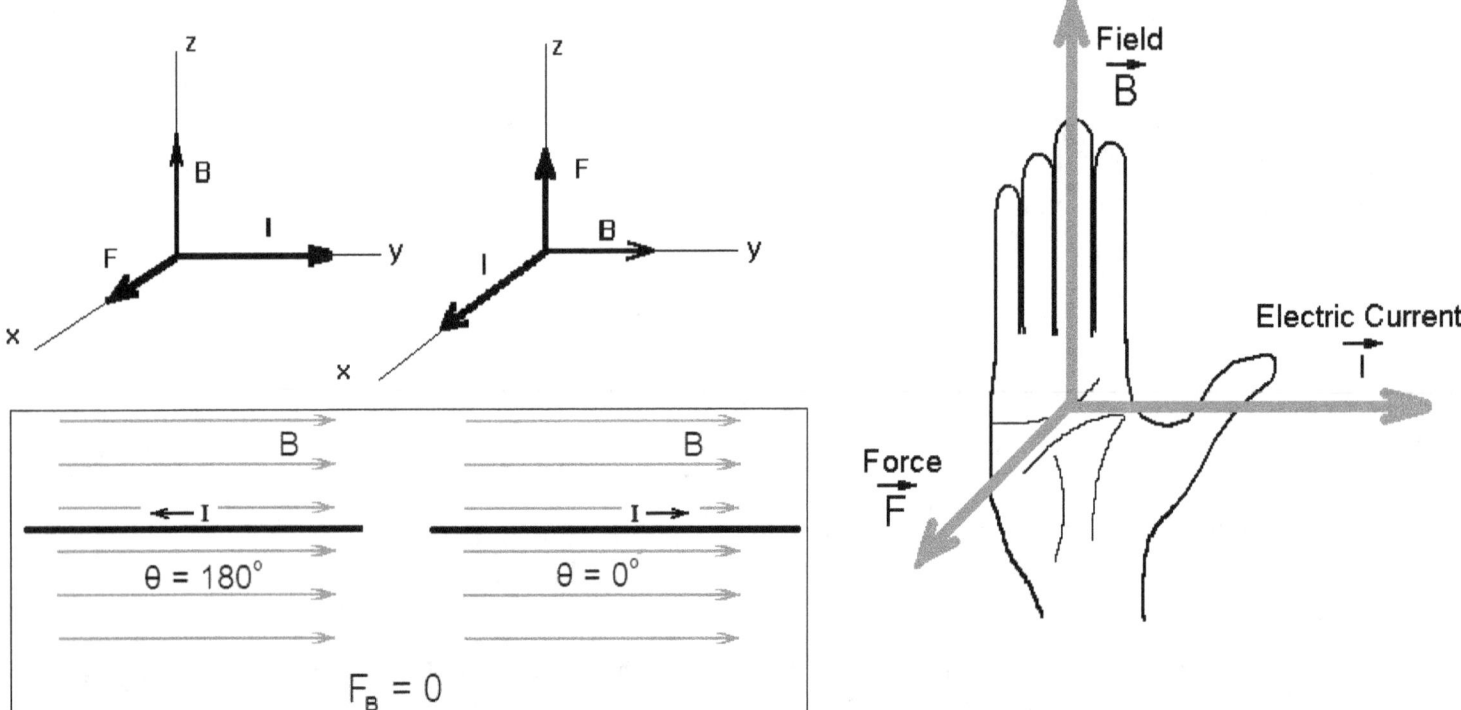

Examples of application of Right-Hand-Palm rule

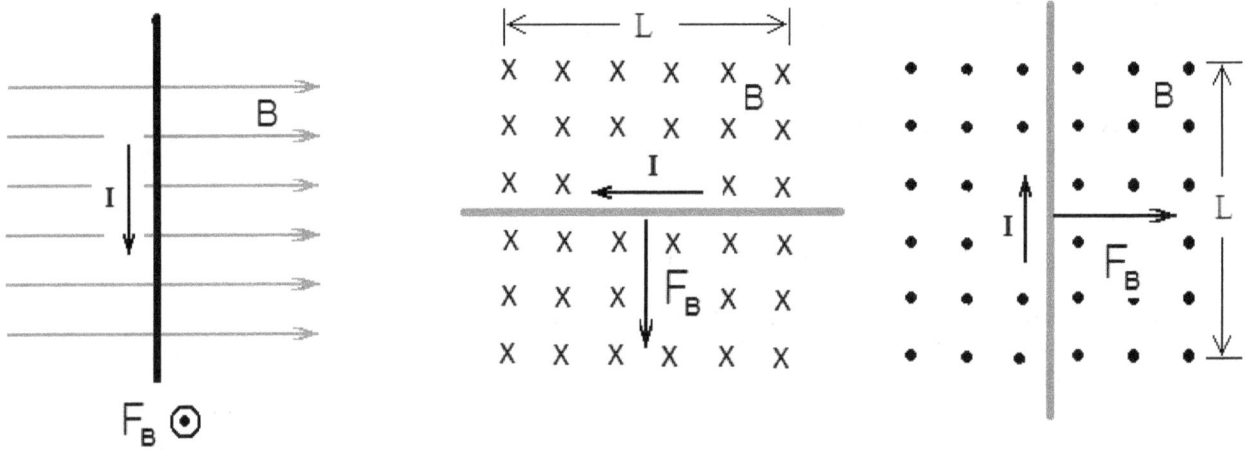

$$F_B = ILB$$

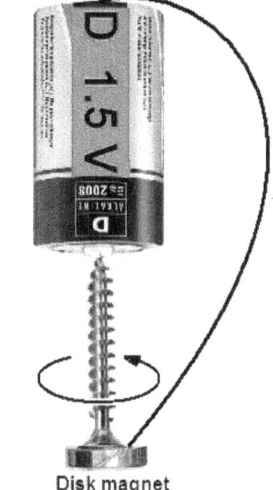

Disk magnet

Some interesting facts & figures
• **Magnetic force on a current carrying wire is used to define the Ampere which is the SI unit of electric current.**
• **Magnetic force on a current carrying wire can be used to design very sensitive weighing balance.**
• **Electric motor works due to this phenomenon of magnetic force on a current carrying wire.**

Problems:

1. A nearly straight power transmission line at a place is carrying a current of 1000 A. The earth's magnetic field at that place is 2×10^{-5} T and is at angle of $30°$ to the transmission line. Determine the force per unit meter on the transmission line due to the earth's magnetic field.

[0.01 N]

2. A thin straight wire of length 40 cm is suspended from a spring balance. The scale reads 2 g. The horizontal component of the earth's magnetic field at that place is $B_H = 3 \times 10^{-5}$ T. The wire is held horizontal and at right angles to B_H. If a current of 500 A is passed through the wire determine the two possible readings on the scale.

3. In each of the diagrams below, a straight wire of length l = 3.0 m is placed in a uniform magnetic field B = 0.2 T. The wire carries a current I = 0.5 A. Determine the magnitude and the direction of the force on the wire in each case.

[(a): 0.3 N, ⊗; (b): 0.3 N, ←; (c): 0 N; (d) 0 N; (e) 0.3 N, ↑; (f): 0.3 N ↑]

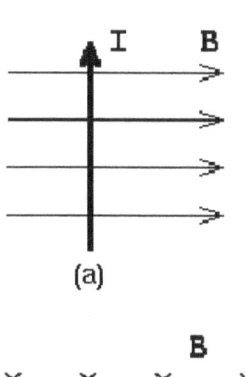

(a)

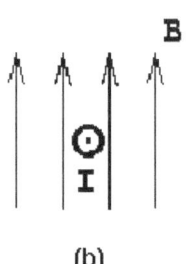

(b)

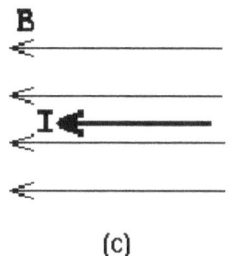

(c)

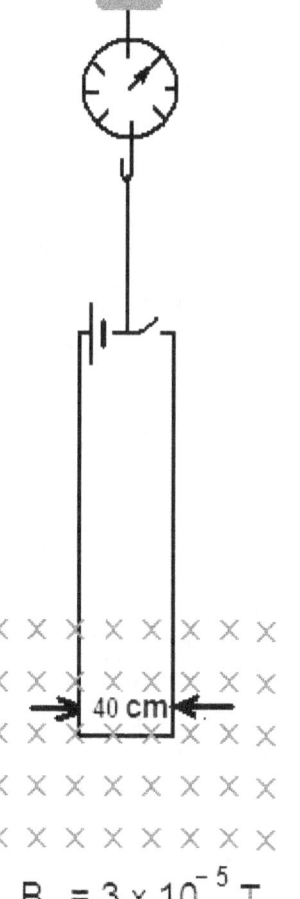

$B_H = 3 \times 10^{-5}$ T

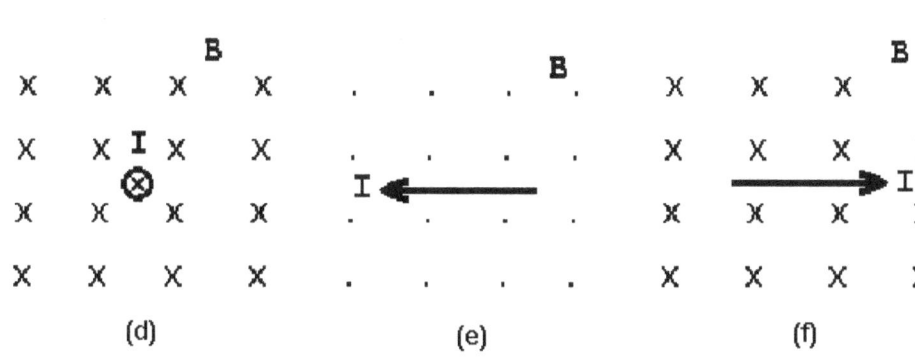

(d)   (e)   (f)

4. In each of the following diagrams below, the directions of two of the three quantities B, i, and F is given. Determine the direction of the missing quantity.

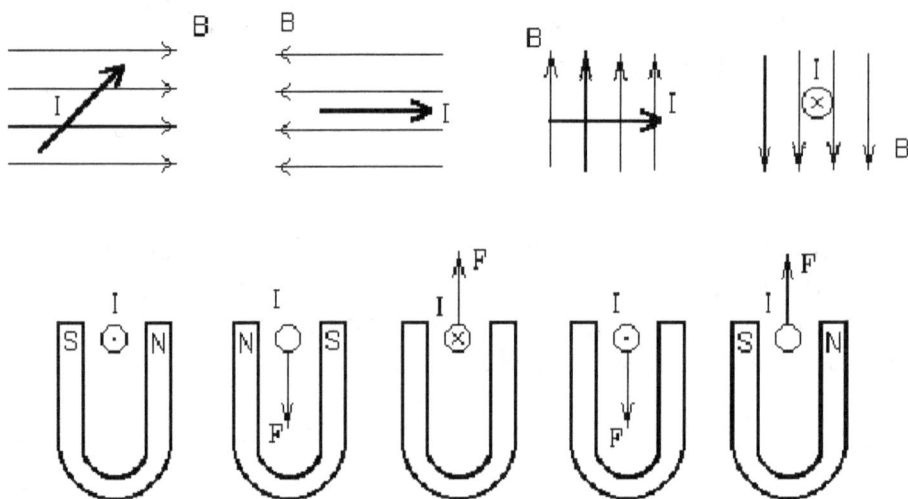

Multiple Choice:

1. A wire is carrying a current Northward in a uniform magnetic field which is directed westward. The direction of force due to the magnetic field on the wire is thus directed
(A) downward   (B) upward   (C) Southward   (D) Westward   (E) none of these

2. A wire is carrying a current Northward in a region in which the uniform magnetic field is directed southward. The force due to the magnetic field on the wire is

(A) directed East   (B) directed West   (C) directed downward
(D) directed upward   (E) none of these

3. A straight current carrying wire is placed in a uniform field as shown in the diagram. The wire lies perpendicular to the magnetic field. The magnetic force acts closest to which of the following?

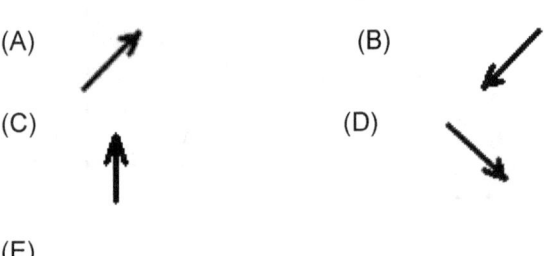

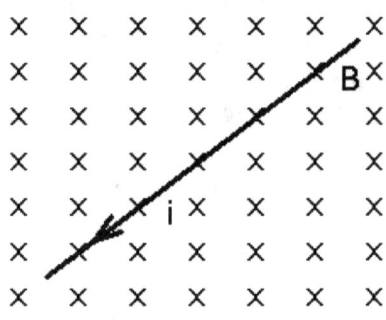

4. A current carrying wire is placed horizontal in a uniform magnetic field as shown in the figure. However, the magnetic force on the wire is zero. The magnetic field must be pointing

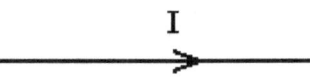

(A) • or ✕   (B) ↑ or ↓   (C) ← or →
(D) ↑ only   (E) ↓ only

Chapter 42
Magnetic Field due to the Electric Current

The phenomenon of magnetic field created by electric current was first observed by the Danish physicist and chemist Hans Christian Oersted [4 August 1777 - 9 March 1851].

Magnetic Field due to a Long Straight Current Carrying Conductor:

A long straight conductor carrying current creates a magnetic field B around it.

The magnitude of **B** at any point a distance **d** from the conductor is given by

$$B = \frac{\mu_o i}{2\pi d}$$

Here
i is the current flowing through the conductor
d is the perpendicular distance of the point from the conductor
μ_o a constant is called the magnetic permeability of vacuum

$$\mu_o = 4\pi \times 10^{-7} \quad \text{T.m/A}$$

The direction of **B** at that point is given by the **right-hand-thumb rule** as shown in the adjoining figure.

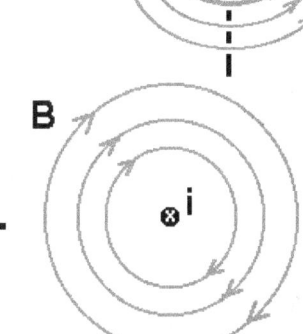

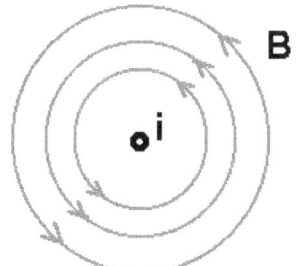
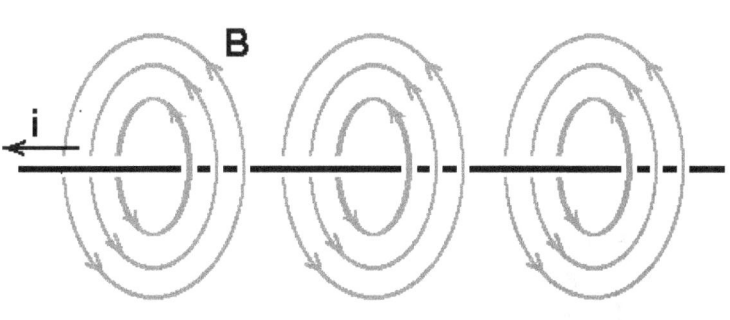

A Circular Current Loop:

A circular loop of wire containing n number of turns and carrying a current acts like a magnetic dipole. One face acts like a North Pole and the other a South Pole.

The magnetic field strength B at the center of the loop is given by

MAGNITUDE:

$$B = \frac{\mu_o ni}{2r}$$

Here
i is the current flowing through the loop
r is the radius of the loop.

Magnetic field direction: Right-Hand-Thumb Rule
Curl the fingers of the right hand along the current loop so that fingers point along the direction of the current. The magnetic field at the center of the loop then points along the direction of the outstretched thumb.

The north and south poles for this loop can be identified by the right hand rule or by the method illustrated in the lower too loops in the diagram below.

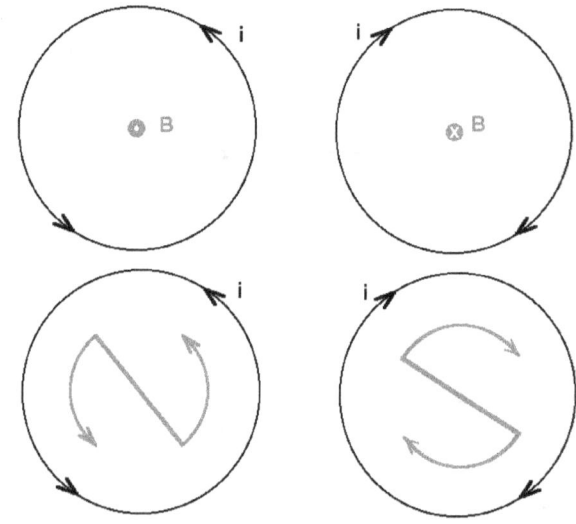

Statue of Hans Christian Oersted
in Copenhagen, Denmark
He first observed magnetic field produced
by electric current

A 100-kroner (1950-1970)note honoring Oersted

A Solenoid (Electromagnet):

A current carrying solenoid creates a magnetic field similar to a bar magnet -- with one end acting like a north pole and the other a south pole.

Here too, the north and south poles can be identified by the **Right-Hand-Thumb rule** mentioned above.

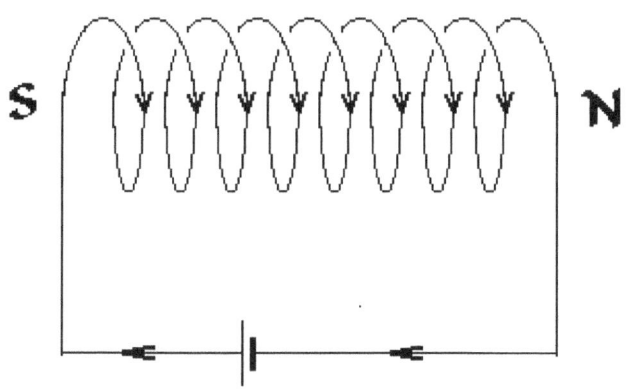

The Force between Two Current Carrying Parallel Wires:

If two parallel wires carry the currents i_1 and i_2 each wire interacts with the magnetic field created by the other. This causes:

- a force of **attraction** between the wires if i_1 and i_2 in the **same** direction
- a force of **repulsion** between the wires if i_1 and i_2 in the **opposite** directions.

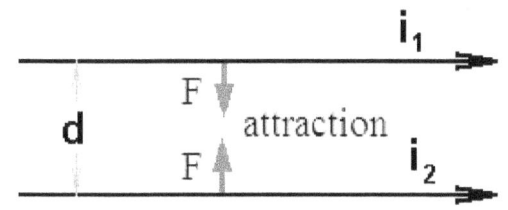

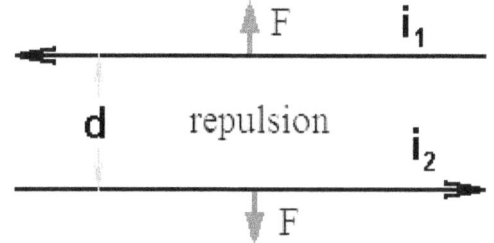

The magnitude of the force on a segment **L** of a wire is given by

$$F = \frac{\mu_o i_1 i_2 L}{2\pi d}$$

where **d** is the distance between the parallel wires.

Some interesting facts & figures
• Hans Christian Ørsted [14 August 1777 - 9 March 1851] was a Danish physicist and chemist first to observe the phenomenon of magnetic field produced by electric current. • Danish government issued a 100 kroner note honoring Oersted. It was in use 1950-1970. • Magnetic field by ferromagnetic materials is limited to 1.6 T. For stronger magnetic fields superconducting electromagnets are used. The coil is cooled toliquid helium temperature and turns superconductor allowing enormous amount of electric current and creating magnetic fields as strong as 10-20 T. The record as of 2009 is 33.8 T

Problems:

1. Determine the magnitude and the direction of the magnetic field created by the long straight current carrying conductor at the points shown in the diagrams below.

(a)

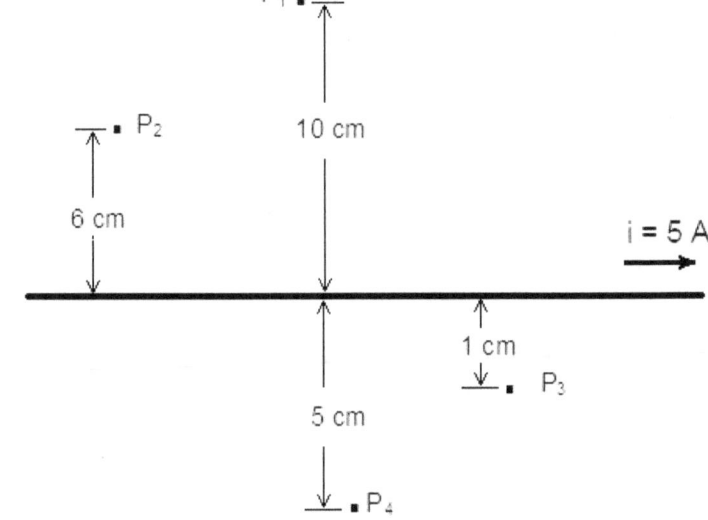

$[1.0 \times 10^{-5}$ T, •; 1.67×10^{-5} T, •; 1.0×10^{-4} T, ⊗; 2.0×10^{-5} T⊗]

(b)

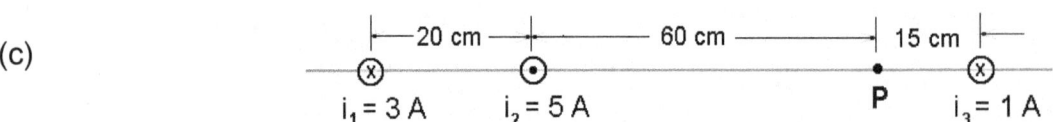

$[5.62 \times 10^{-5}$ T, ⊗]

(c)

[2.25x10⁻⁶ T ↑]

2. As shown in the figure below, two long straight conductors are carrying current at right angles. Determine the net magnetic field B at the point P.

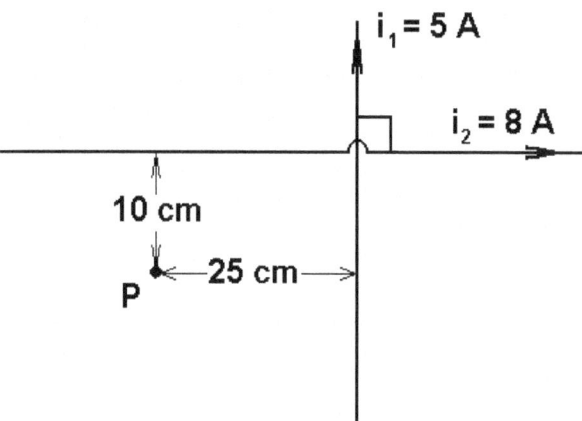

3. A long straight wire carrying of 10 A is held in the plane of a 50 cm X 8 cm rectangular current loop carrying a current of 5 A as shown in the diagram below. Determine the net force acting on the current loop.   [6.2x10⁻⁵ N ↑]

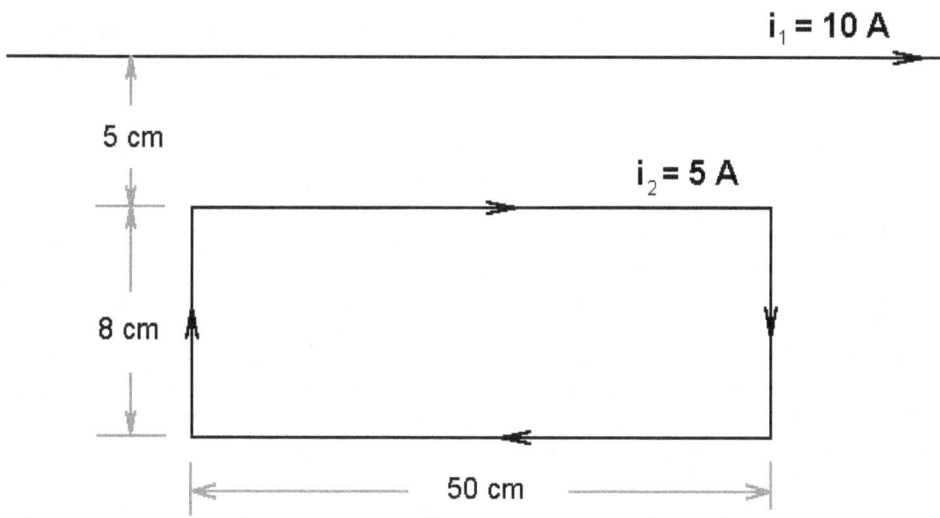

4. An electron is traveling parallel to a long straight wire and 20 cm away from it at 3.0×10^5 m/s as shown in the fig. below. Now if a current of 4.5 A is turned on in the same direction as the electron velocity determine the magnitude and direction of the magnetic force acting on the electron.

I = 4.5 A

20 cm

$v = 3.0 \times 10^5$ m/s

electron

5. As shown in the diagram below, a long straight wire and a circular current loop containing 4 turns lie in the same plane. Determine the radius r of the loop (in terms of d) to make the net magnetic field at its center zero. *[0.83 d]*

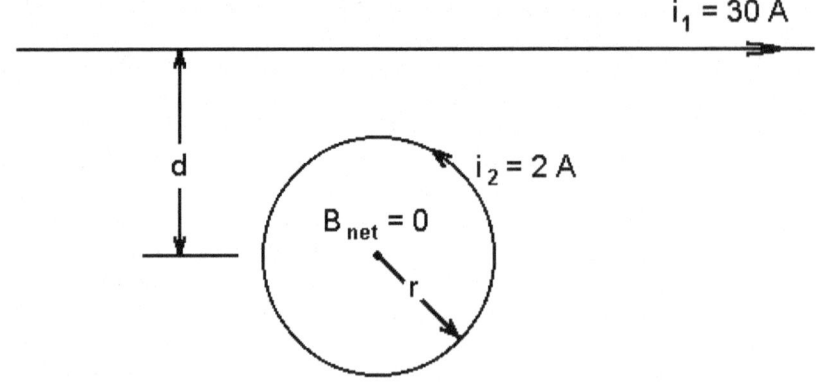

6. In the diagrams shown below, a long straight wire is bent to form a loop in two different ways. Determine the net magnetic field at the center of the loop in each case.

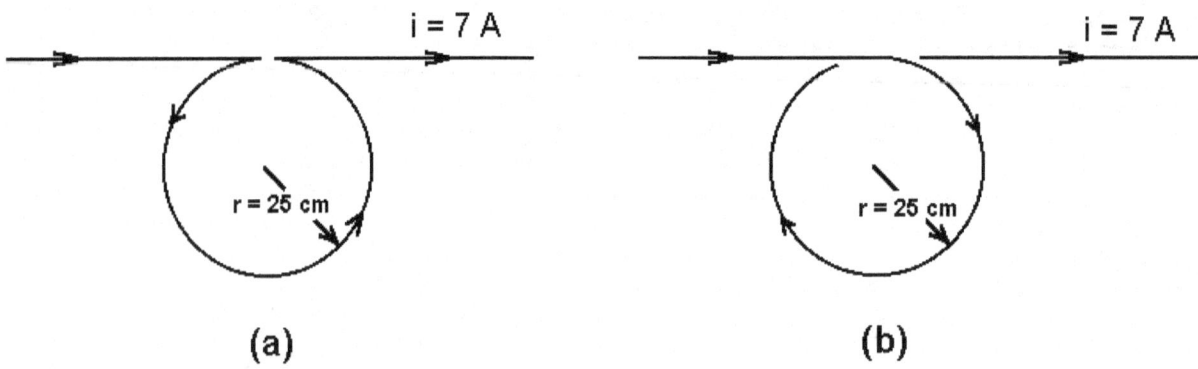

(a)   (b)

Multiple Choice: Choose the best alternative.

1. As shown in each of the diagrams above, a long straight wire is carrying a current from left to right in the pane of the diagram. Which diagram correctly depicts the magnetic field created by the wire?

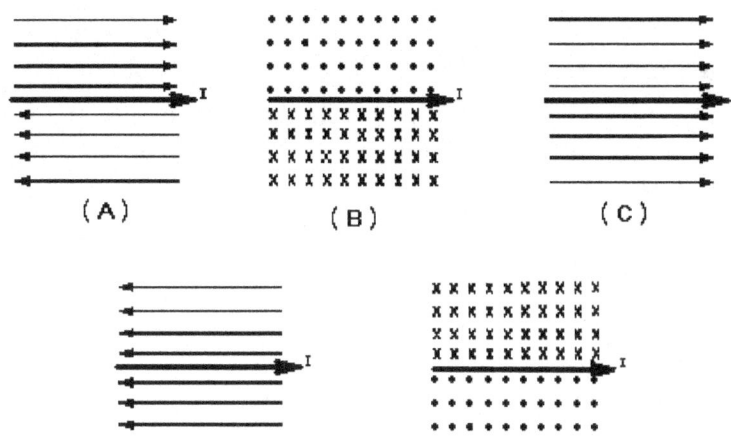

2. Two straight parallel wires carry <u>unequal</u> amounts of current in the same direction. Which of the following is true about their interaction?
(A) They repel each other with equal forces
(B) They repel each other with unequal forces
(C) They attract each other with equal forces
(D) They attract each other with unequal forces
(E) They do not interact.

3. A point P is at a distance d from a long straight current carrying wire. Which of the following is the correct statement of the dependence of the magnetic field at P on the distance d?

(A) $B \propto d$ (B) $B \propto 1/d$ (C) $B \propto d^2$ (D) $B \propto 1/d^2$
(E) B is uniform around the wire and independent of the distance d

4.
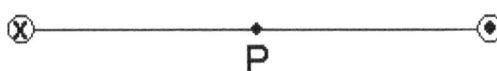

Two long straight wires are parallel and carry equal currents in opposite direction as shown in the diagram above. The net magnetic field midway between the wires is
(A) Zero   (B) ↑   (C) →   (D) ↓   (E) ←

5. Two long straight current carrying wires are at right angles to the xy – plane as shown in the diagram. The direction of the net magnetic field at the origin is closest to

(A) ↘   (B) ↗   (C) ↙   (D) ↘   (E) →

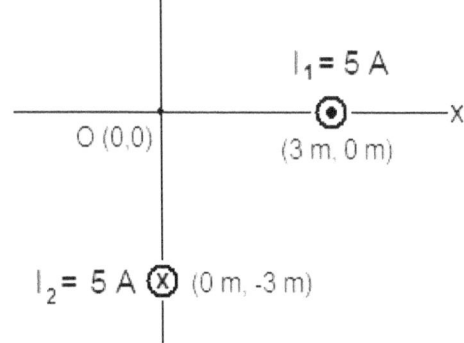

Chapter 43
Magnetic Flux

Area Vector (A): For a given surface S of area **A**, the area vector is defined as a vector at the center of S with a magnitude **A** and directed perpendicular to the surface S. The area vector **A** makes it convenient to define the orientation of the surface with respect to a given direction (e.g. in relation to the direction of applied magnetic field **B**

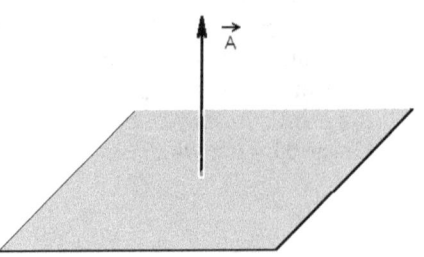

If the surface is **spherical** then the area vector is directed out of the **convex surface**.
If the surface is **plane** there are **two** possible directions for the area vector.

The Concept of Magnetic Flux: Magnetic flux is an abstract quantity. One can conceptualize the idea of the flux in two ways:
- The magnetic flux tells you how many magnetic field lines in a given magnetic field are passing through the surface.
- The magnetic flux tells you how "effectively" the magnetic field vector is penetrating the surface.

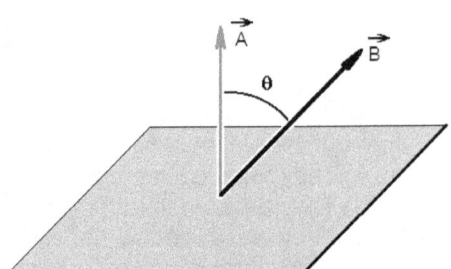

Magnetic Flux ,Φ, Quantitative: For a real or imaginary surface S of area A in a uniform magnetic field B the magnetic flux through the surface is defined as

$$\Phi = BA \cos\theta$$

The SI Unit for the magnetic flux is weber (Wb); $1 \text{ Wb} = 1 \text{ T.m}^2$

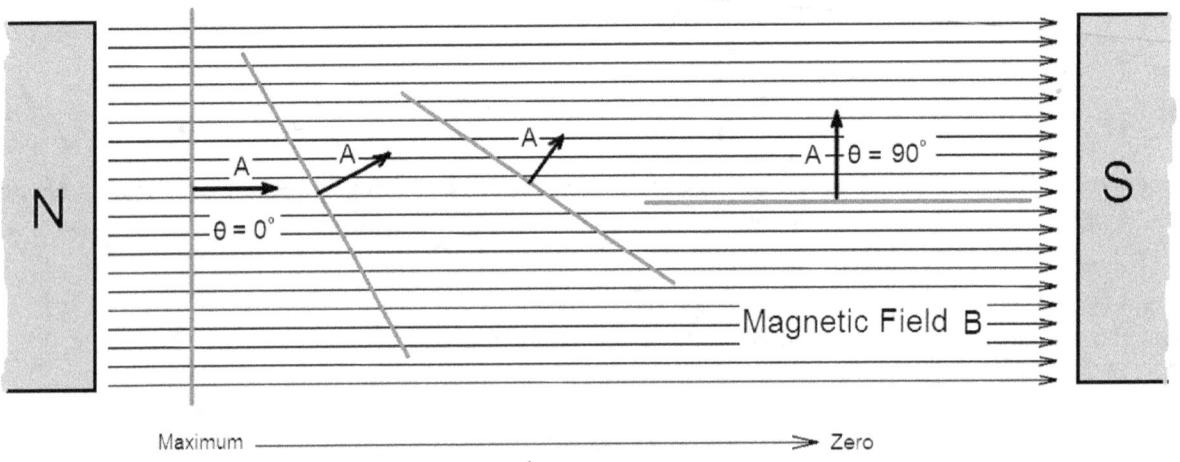

Magnetic Flux ϕ through area A

The magnetic flux through a flat and thin coil of n turns: If the coil has **n** number of turns then the total flux through the coil is the sum of the flux through each turn. Hence,

$$\Phi = nBA \cos\theta$$

Some interesting facts & figures
• **Wilhelm Eduard Weber [24 October 1804 – 23 June 1891] was a German physicist. He and Carl Friedrich Gauss, invented the first electromagnetic telegraph.**
• **The 'magnetic flux quantum' Φ_0 = ½ h/e is the quantum of magnetic flux passing through a superconductor. Applicable only to superconductors.**
• **The magnetic flux quantum Φ_0 can be measured by Josephson Effect.**

PROBLEMS:

1. A flat coil of wire of 15 turns and area 3 m^2 is placed in a uniform magnetic field of 1.2 T. Determine the magnetic flux through the coil for the following orientations of the area vector with respect to the magnetic filed.
(a) 0° (b) 30° (c) 45° (d) 60° (e) 90° (f) 130° (g) 180° (h) 270°
[54.0 Wb, 46.8 Wb, 38.2 Wb, 27.0 Wb, 0 Wb, -34.7 Wb, -54 Wb, 0 Wb]

2. In the problem 1, if the orientation of the coil is initially 0° and then it is flipped so that the orientation of the area vector is now 180°, determine the change in flux (ΔΦ) through the coil.

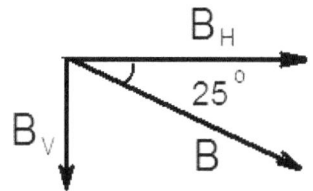

3. The Earth's magnetic field at a place is B = 4 x 10^{-5} T and is directed 25° below the horizontal.
(a) Determine the horizontal and the vertical components (B_H and B_V)of the magnetic field at that place.   *[3.6x10^{-5} T, -1.7x10^{-5} T]*

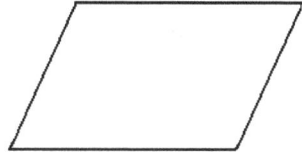

(b) Determine the magnetic flux through a coil of 25 turns and area 0.2 m^2 with its surface placed
(i) horizontal   *[-8.5x10^{-5} Wb]*

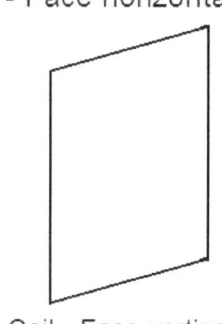

Coil - Face horizontal

(ii) vertical   *[1.8x10^{-4} Wb]*

Coil - Face vertical

Multiple Choice:

1. A coil with area vector A is placed in a uniform magnetic field B. The magnetic flux through the coil is zero if the angle between A and B is
(A) 45°   (B) 90°   (C) 60°   (D) 180°   (E) 0°

2. A coil is turned in a magnetic filed and the flux changes from –10 Wb to +10 Wb. The net change in the flux through the coil is thus,
(A) 0 Wb   (B) +5 Wb   (C) –5 Wb   (D) +20 Wb   (C) -20 W

3. Each of the above diagrams shows a rectangular area A and the associated uniform magnetic field B. The area lies in the plane of the diagram and the magnetic field is either perpendicular to the diagram or in the plane of the diagram.

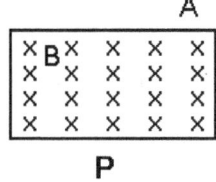

P

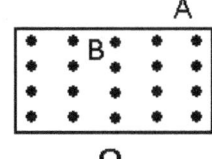

Q

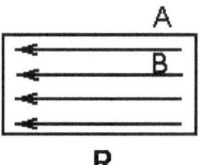

R

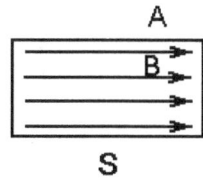

S

In which of the diagram(s) is the flux through the rectangular area zero?
(A) P & Q only   (B) R & S only   (C) R only   (D) All of them   (E) None of them

4. A rectangular loop of area A is placed in a uniform magnetic field B and oriented to make area vector **A** point in the direction of **B**. If the loop is now turned so that **A** is at 90° to **B** the change in the magnetic flux through the loop is
(A) 0   (B) +½ AB   (C) -½ AB   (D) +AB   (E) - AB

Chapter 44
Electromagnetic Induction

Laws of Electromagnetic Induction: Discovered by
Michael Farady

1. FARADAY'S LAW:
The rate of change of flux through a circuit is proportional to the induced emf in the circuit.
Hence

$$\varepsilon = -\frac{\Delta\Phi}{\Delta t}$$

The minus sign is introduced for the direction of the induced emf.
Note that the magnetic flux (**Φ = ABcosθ**) by itself cannot generate induced emf; magnetic flux must change to generate induced emf. The above equation can also be written as

$$\varepsilon = -\frac{\Delta(AB\cos\theta)}{\Delta t}$$

Michael Faraday
[22 September 1791 - 25 August 1867]
An English Physicist and Chemist

If the flux through a circuit changes from Φ_1 to Φ_2 in a time interval Δt, the above equation can be used to obtain the magnitude of the induced emf as below:

$$\varepsilon = -\frac{\left|\Phi_2 - \Phi_1\right|}{\Delta t}$$

The minus sign can be dropped to obtain the magnitude of the emf and direction can be found by using the Lenz's Law below.

2. LENZ'S LAW:

The direction of induced emf (or induced current) is such that it creates a field which opposes the change that creates it.

Induced emf due to relative motion between a solenoid and a magnet

If there is no relative motion between the coil and the magnet no current or emf is induced

As the S-pole of the magnet departs from the solenoid, the solenoid can oppose it by creating a N-

pole at the right end. This happens when the current through the solenoid flows in the direction shown.

The induced current will flow in the opposite direction if the S-pole of the magnet were to approach the coil.
Induced EMF due to Varying Magnetic Field:

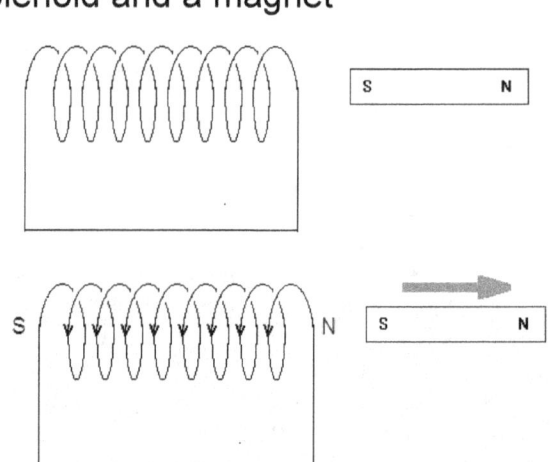

Induced emf due to varying B

A coil of area **A** and **n** number of turns is placed with its plane at right angles to uniform magnetic field **B**. If **B** is varying with time the induced emf is given by

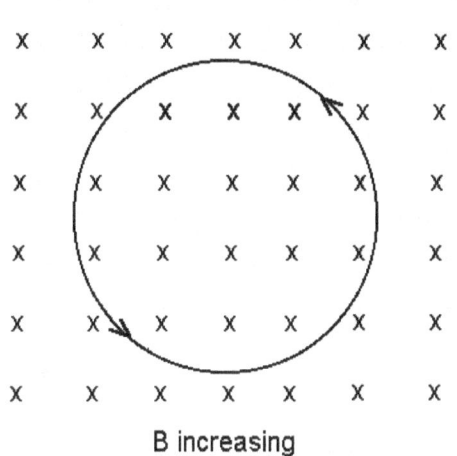

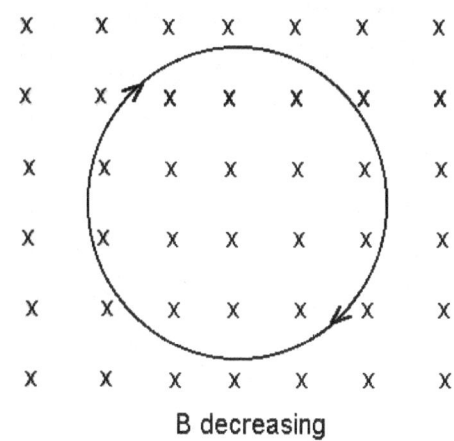

B increasing   B decreasing

$$\varepsilon = nA\frac{\Delta B}{\Delta t}$$

B increasing: As **B** increases through the coil, the induced current must oppose it by creating its own field in the opposite direction (out of the page). Hence the direction of current is as shown in the diagram.

B decreasing: As B decreases through the coil, the induced current must oppose it by creating its own field in the same direction (into the page). Hence the direction of current is as shown in the diagram.

Simple electric generators:

A metal rod of length **L** is moving with velocity **v** at right angles to a uniform magnetic field **B**. The + and − charges (also moving to the right with velocity v) in the rod separate out (as shown) by the magnetic forces given by right-hand-palm rule. This creates a potential difference between the tips of the rod given by $\varepsilon = BLv$ No current is induced here because the circuit is not complete.

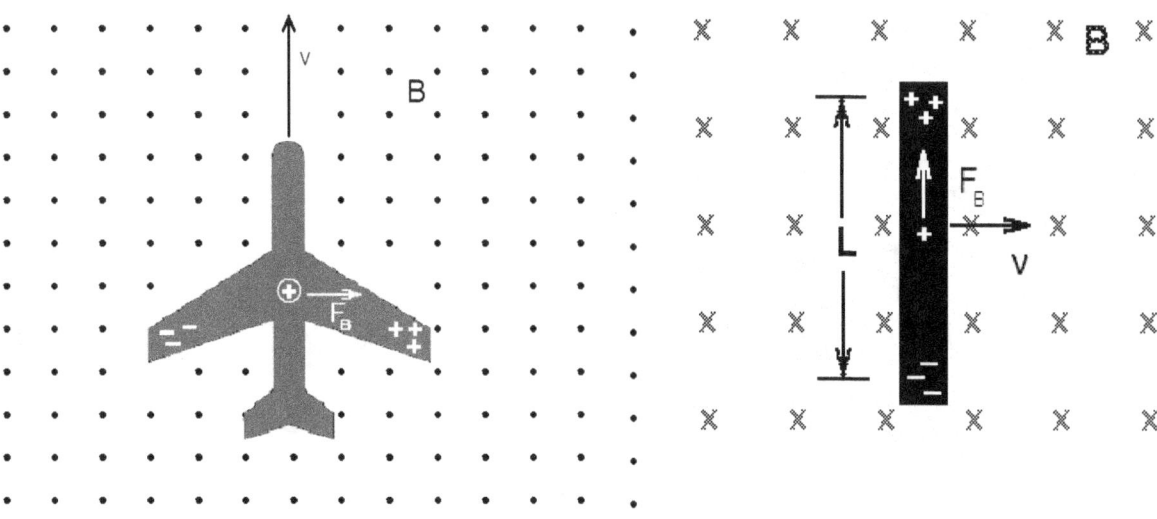

Rectangular Loop:

If a rectangular loop of width L and resistance r is pulled out of a uniform B-field at constant velocity v , the induce emf is again given by $\boxed{\varepsilon = B L v}$

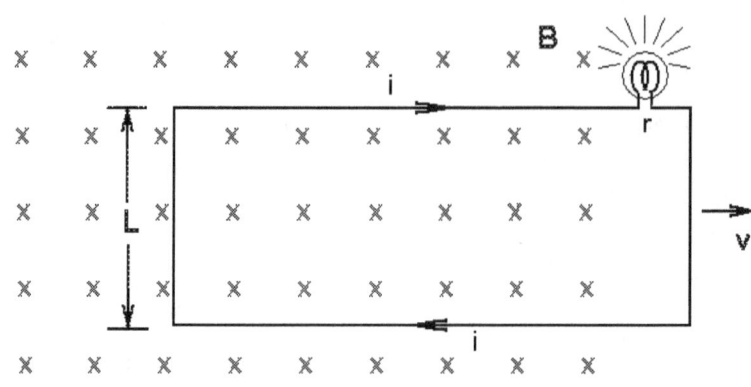

$i = \varepsilon/r$

$P = i^2 r$ or ε^2/r or εi

$F_B = iLB$ to the left

$F_{ext} = F_B$ to maintain the velocity v constant

Electric power output = Mechanical Power input → $P = F_{ext} v$

Direction of current in this loop:

The direction (clockwise or counter clockwise) of the induced current in the circuit above can be determined in any of the following ways:

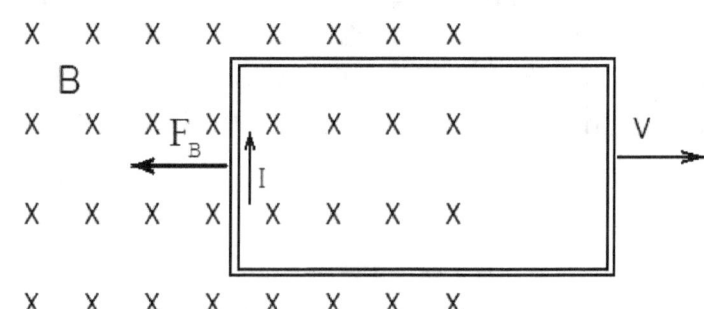

- Since the area of the coil immersed in the magnetic field is decreasing, the number of x's in it decreasing. Using the Lenz's Law, the induced current must undo this by creating more **x**'s through it. By the right-hand-thumb rule, the current must flow CCW.
- Imagine a positive charge in the vertical side of the coil inside the magnetic field. As the coil moves to the right so does this charge. This is a case of a positive particle moving in a magnetic field and will experience a force **upward** given by the right-hand-palm rule. The motion of such positive particles thus constitutes a ccw current in the coil.
- Since the coil is moving to the right, using the Lenz's Law, the current in the coil must be such that the vertical side in the magnetic field will experience the magnetic force to the left. Thus, by the right-hand-palm rule the current must flow upwards and ccw around.

An AC Generator:

Let a coil having n turns and area A be rotating in a uniform magnetic field B with a frequency f. The axis of rotation is perpendicular to B. The continuously changing angle θ is now responsible for flux change through the coil. The induced emf is given by

$$\varepsilon = 2\pi fnAB \ \sin(2\pi ft)$$

Notice that the induced emf is a sinusoidal function of time. The maximum emf induced is $\varepsilon_0 = 2\pi fnAB$

Schematic of a Hydroelectric Power Plant

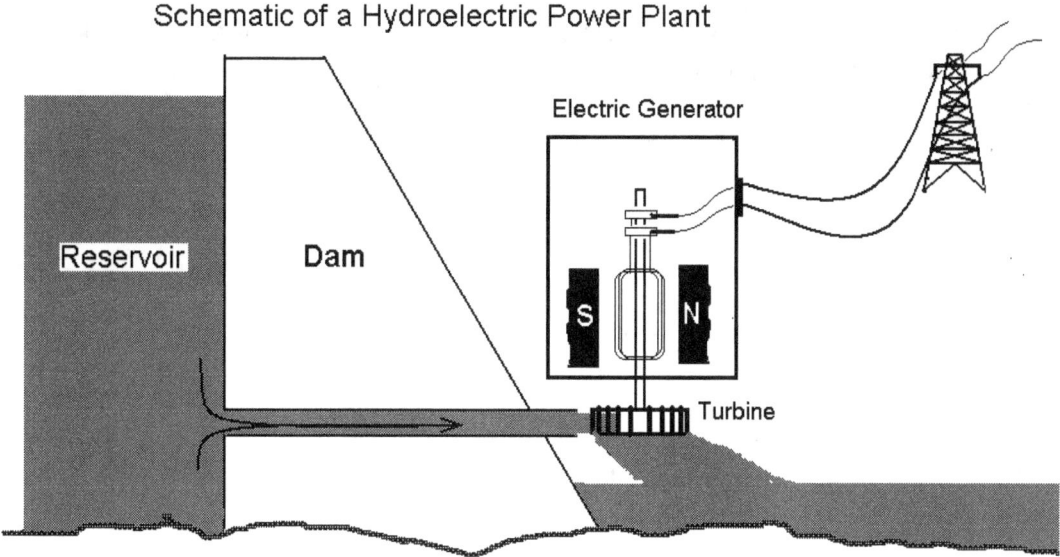

Some fun facts & figures
• Electromagnetic induction is used to hermetically seal plastic and glass containers. A variety of industries including, pharmaceutical, chemical, sports, and food use this process. • The power generated by large wind turbine is of the order 1 to 2 Megawatts. • The largest wind farm in the world is in Taylor county, Texas with capacity of 735 Megawatts • The Three Mile Island Nuclear Generating Station in the USA has a rated capacity of 802 Megawatts. • The coal-fired Ratcliffe-on-Soar Power Station in the UK has a rated capacity of 2 Gigawatts. • The Aswan Dam hydro-electric plant in Egypt has a capacity of 2.1 Gigawatts. • The Three Gorges Dam hydro-electric plant in China, still under construction, will have a capacity of 22.5 Gigawatts when complete.

PROBLEMS:

1. A coil has an area A = 0.8 m^2 and 20 tightly wound turns. It is initially kept with its face at right angles to a uniform magnetic field B = 0.2 T. It is then flipped through 180° about an axis perpendicular to the magnetic field in t = 0.5 seconds.
(a) Determine the average induced emf in the circuit.
[12.8 V]

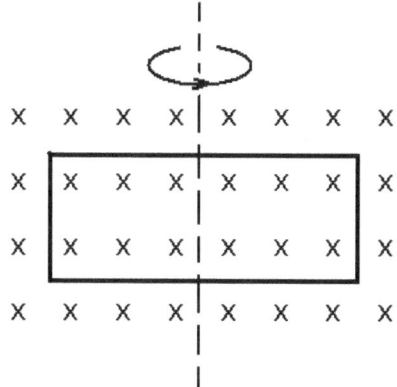

(b) Determine the average induced current in the circuit if the coil has a resistance of 10 ohms.
[1.28 A]

(c) Determine the amount of electrical energy dissipated in the circuit. *[8.2 J]*

(d) Determine the work done in flipping the coil. *[8.2 J]*

2. A plane with a wingspan of 30 m is flying north horizontally with a speed of 500 km/h at a place where the upward vertical component of the earth's magnetic field is
$B_v = 2.0 \times 10^{-5}$ T.
(a) Determine the induced potential difference created between the tips of the plane's wings.

(b) Which wing will be positively charged - west or east?

3. For an AC generator, the coil has an area of 0.05 m^2 and has 200 turns. It is spinning with a frequency of 25 Hz in a uniform magnetic field of 0.05 T. Determine
(a) the peak emf.   [78.5 V]

(b) the peak current drawn from the generator if connected to a resistor of 10 ohms.
                                                                                 [7.85 A]

4. Two long parallel metal wires are placed perpendicular to a uniform magnetic field
   B = 0.08 T as in the figure below. The distance between the wires is L = 10 cm. A metal rod is placed on the rails and moved at a speed of 15 m/s. A resistor r = 10 ohm is introduced in the closed loop.

(a) Determine the induced emf in the circuit.

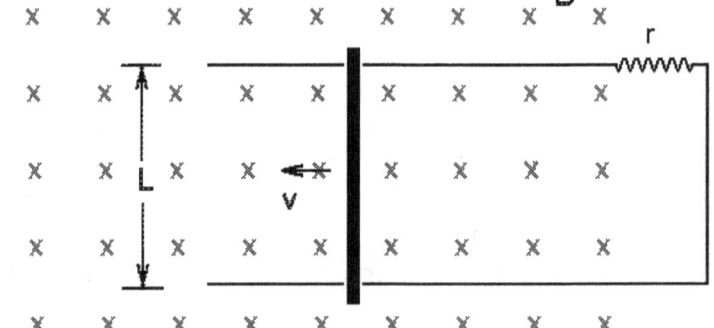

(b) Determine the induced current in the circuit.

(c) Determine the power dissipated in the resistor.

(d) Determine the direction of the induced current in the circuit.

(e) Determine the magnetic force acting on the sliding rod (due to the induced current flowing in it)

(f) Determine the mechanical power input if the frictional losses are negligible.

(g) Determine the force required to maintain the velocity of the rod.

5. A coil of 60 turns and an area of 0.8 m^2 has a resistance of 25 ohms. It is held at right angles to the uniform magnetic field B, which is varying with time as in the graph below.

Determine the induced emf and the induced current for each segment of the graph.

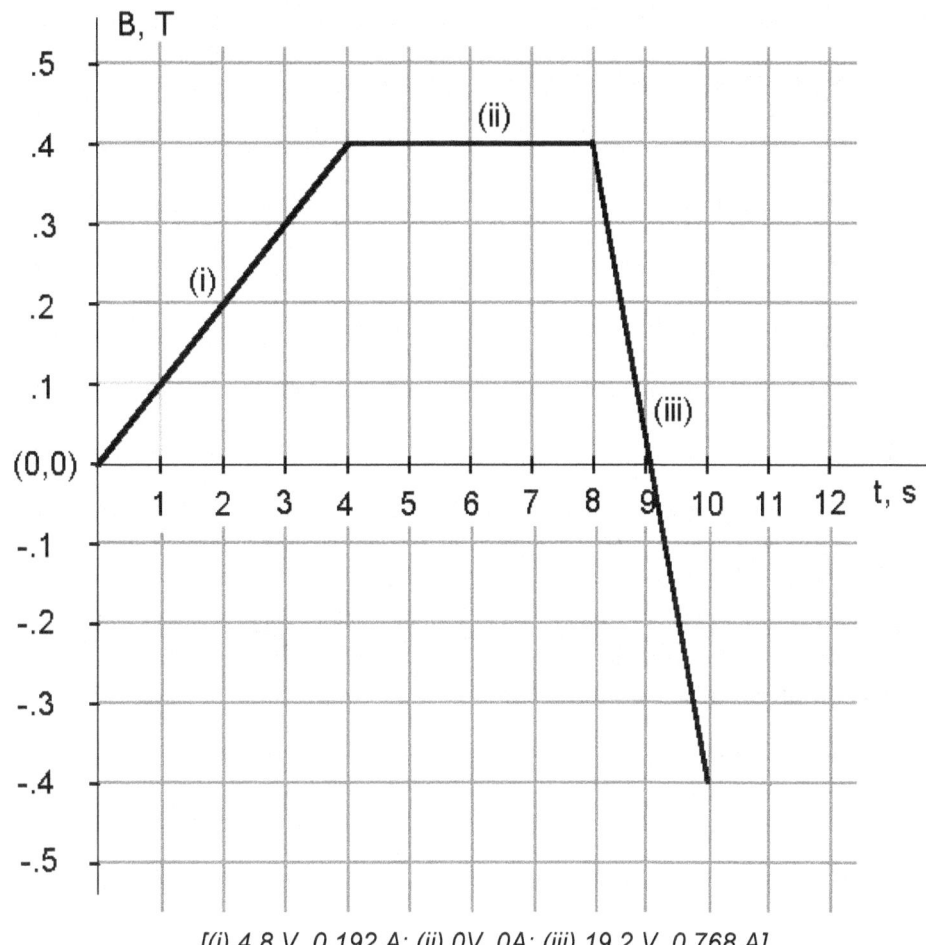

[(i) 4.8 V, 0.192 A; (ii) 0V, 0A; (iii) 19.2 V, 0.768 A]

6.

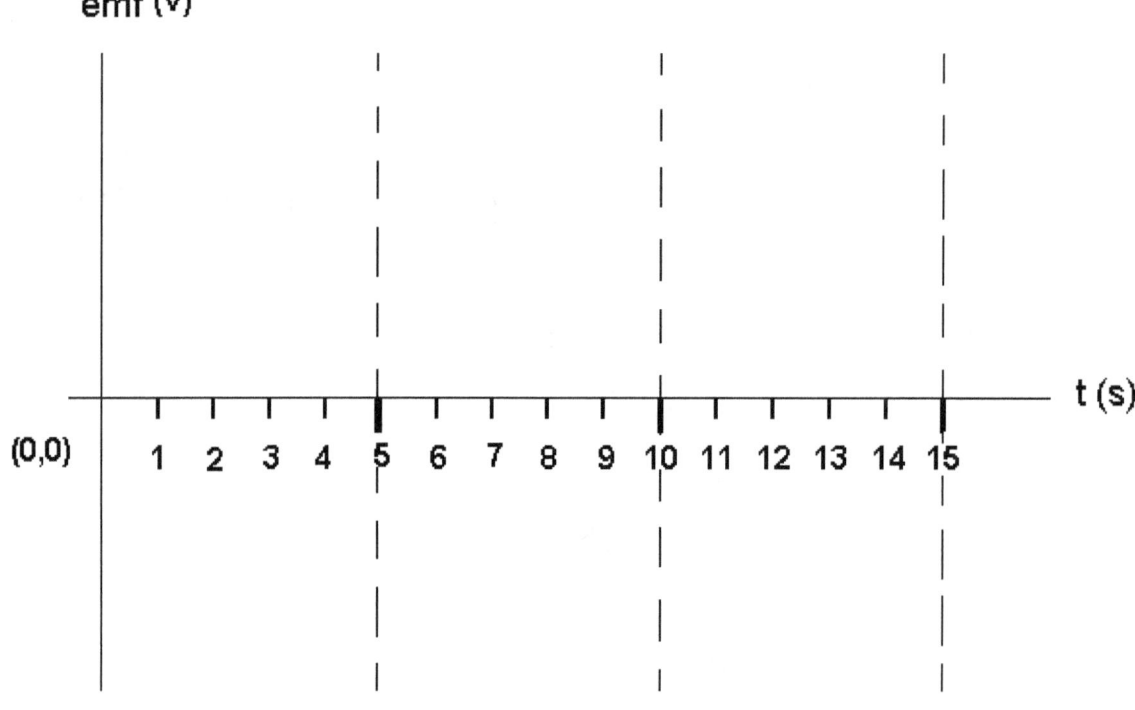

B = 0.8 T

V = 0.2 m/s

← 0.4 m →← 0.4 m →←──────── 1.2 m ────────→

As shown in the figure above, a square loop of wire is moved to the right with a constant speed of 0.2 m/s through a uniform magnetic field that is 1.2 m across and perpendicular to the diagram. The loop is at the position shown, at t = 0 s.

Draw below a graph of induced emf vs. time in the loop over 12 seconds. Assume the emf that appears initially to be positive.

emf (V)

t (s)

(0,0)   1   2   3   4   5   6   7   8   9   10 11 12 13 14 15

199

Multiple Choice:

1-3.

Each of the above diagrams shows a rectangular area A and the associated uniform magnetic field B. The area lies

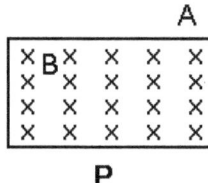

P

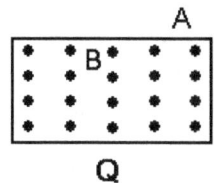

Q

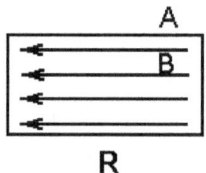

R

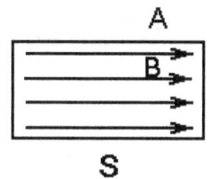

S

in the plane of the diagram and the magnetic field is either perpendicular to the diagram or in the plane of the diagram. In each case the magnetic field B is <u>increasing</u>.

1. In which case will the induced current be clockwise?
   (A) P   (B) Q   (C) R   (D) S   (E) All of them

2. In which case will the induced current be counterclockwise?
   (A) P   (B) Q   (C) R   (D) S   (E) All of them

3. In which case will there be no induced current?
   (A) P & Q   (B) R & S   (C) only R   (S) only S   (D) All of them

4. The magnetic flux through a loop of wire changes from -3.0 Wb to +3.0 Wb in 5.0 seconds. The magnitude of the average induced emf in the loop over the 5.0 seconds interval is
   (A) 0.6 V   (B) 1.2 V   (C) 15.0 V   (D) 30 V   (E) 0V

Chapter 45
Wave Motion

Traveling Wave: A mechanical wave on a stretched string, sound wave, and light wave are the examples of traveling waves. The traveling waves are periodic in space and time. Below is a snapshot of a traveling wave. The intersection **P** of the wave with the y-axis performs simple harmonic motion (SHM) (up and down) along the y-axis..

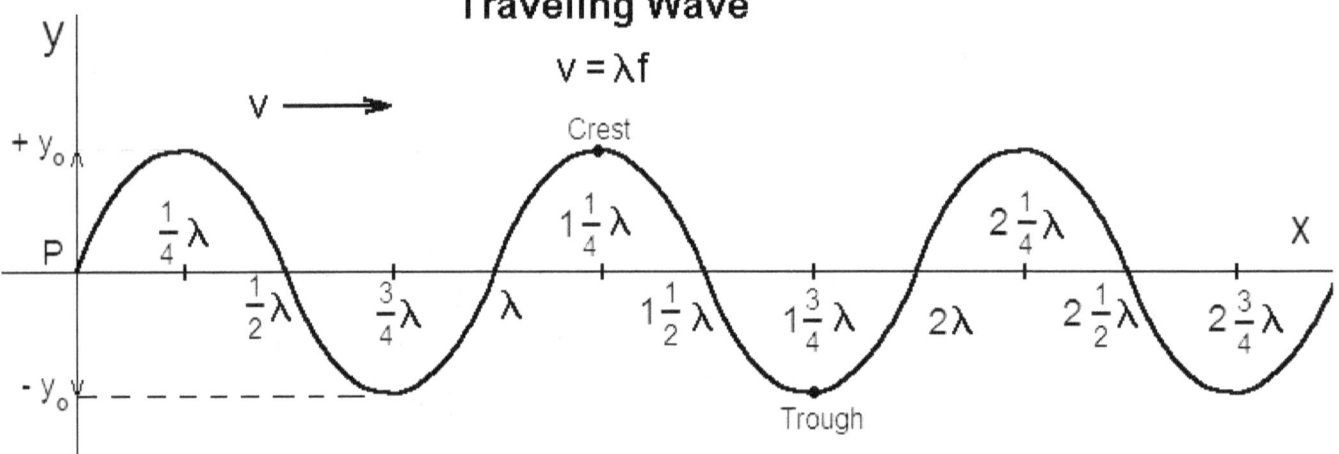

Wavelength of the wave (λ): It is the distance traveled by the wave in one period of the SHM by the point P. It is also the distance between two consecutive troughs or waves or points in the same phase (state of vibration).

Period of the wave (T): It is the same as the period of the SHM of the point P. It is also the time the wave takes to travel a distance equal to its wavelength (λ)

Frequency of the wave (f): It is same as the frequency of the SHM of the point P. It is also the number of 'waves' (the length of the wave generated in one cycle) passing a point per second, for example, the number of crests passing an observer per second. $f = 1/T$

Amplitude (y_o) of the wave: It is the same as the amplitude of the SHM of the point P. It is also half the vertical distance between a crest and trough.

Speed of the wave (v). It is the speed of propagation of the wave. In the diagram above it is the speed of, say, a crest moving to the right.

Note: The actual speed of vibrations of the particles that produce the wave pattern is different and independent of the speed v of the wave.

Relationship between f, λ, and v
For a traveling wave its frequency **f**, wavelength λ, and the speed **v** are related by

$$\boxed{v = f\,\lambda}$$

Two Simple Waves – Longitudinal and Transverse:

1. Longitudinal wave:

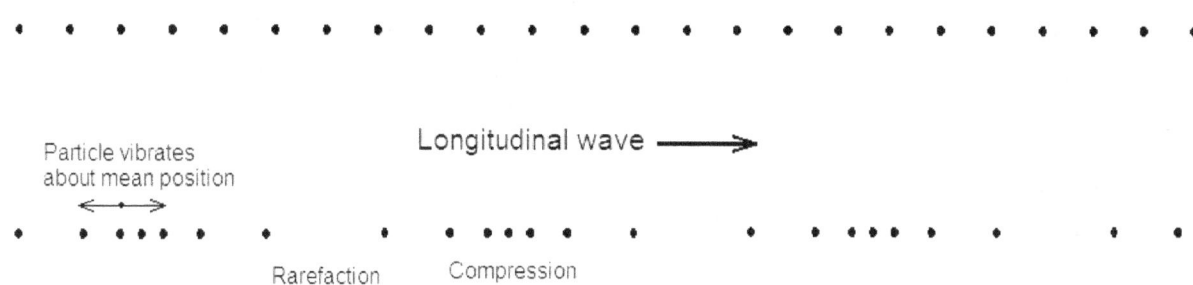

Undisturbed medium

Particle vibrates about mean position

Longitudinal wave ⟶

Rarefaction   Compression

For this kind of wave, the particles of medium vibrate along the line of propagation of the wave. Sound wave is an example of this kind of wave.

Transverse wave: For this kind of wave, the vibrations take place at right angles to the line of propagation of the wave. Light waves and transverse wave on a string are examples of this kind of waves

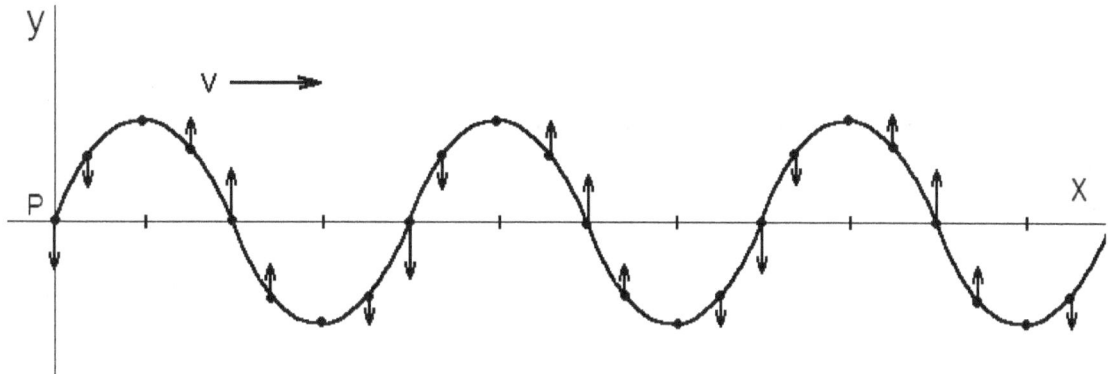

Phase change upon reflection and transmission of waves:

Wave on a string:
When the wave is reflected from a <u>fixed end</u> there is a phase change of π i.e. $\Delta\phi = \pi$. A crest is reflected as a trough and a trough is reflected as a crest.

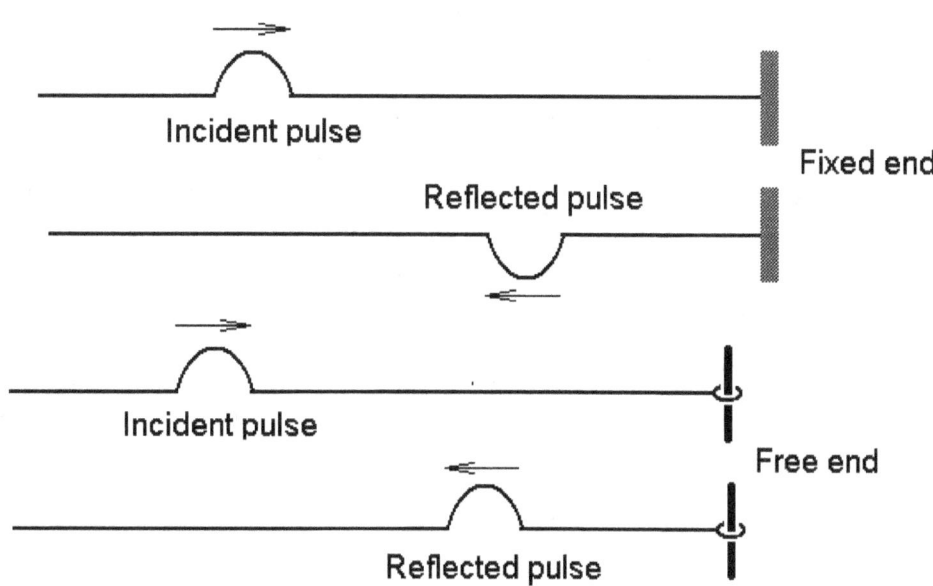

Incident pulse

Reflected pulse

Fixed end

When the wave is reflected from a <u>free end</u> there is no phase change i.e. $\Delta\phi = 0$. A crest is reflected back as a crest and the trough is reflected back as trough.

Incident pulse

Reflected pulse

Free end

Thick and thin strings:

A pulse traveling on a thin string is incident on a thick string. The reflected pulse has a phase change of $\Delta\phi = \pi$. The transmitted pulse has no phase change.

Thinner string — Incident pulse

Thicker string

Reflected pulse — Transmitted pulse

A pulse traveling on a thick string is incident on a thin string. The reflected as well as transmitted pulses have no phase change.

Thicker string — Incident pulse

Thinner string

Reflected pulse — Transmitted pulse

Sound waves:
When the sound wave is reflected from a <u>denser medium</u> there is a <u>phase change of π</u> i.e. $\Delta\phi = \pi$.
A compression is reflected back as a rarefaction and a rarefaction is reflected back as a compression.

When the sound wave is reflected from a <u>rarer medium</u> there is <u>no phase change</u> i.e. $\Delta\phi = 0$.
A compression is reflected back as a compression and a rarefaction is reflected back as a rarefaction.

Light waves:
When the light wave is reflected from an <u>optically denser</u> medium there is a phase change of π i.e. $\Delta\phi = \pi$.
A crest is reflected back as a trough and a trough is reflected back as a crest.

When the sound wave is reflected from an <u>optically rarer</u> medium there is <u>no phase change</u> i.e. $\Delta\phi = 0$.
A crest is reflected back as a crest and a trough is reflected back as a trough.

*NOTE: A transmitted wave or pulse **never** suffers a phase change in transmission*

Speed of wave on a string: The speed (**v**) of a transverse wave on a string depends on
- the tension (**F$_T$**) in the string and
- the linear mass density (**μ**) of the string

and is given by

$$v = \sqrt{\frac{F_T}{\mu}} \qquad \mu = \frac{mass.of.string}{length.of.the.string}$$

Some interesting facts & figures

- According to some claims, *stadium wave* was created, by chance, at the 1976 Montreal Olympics.
- Rogue waves (also known as freak waves, monster waves, killer waves, and extreme waves) are relatively large and spontaneous ocean surface waves that are a threat even to large ships and ocean liners.
- First rogue wave confirmed with scientific evidence was Draupner wave in North Sea, in 1995. It had a maximum height of 25.6 meters (84 ft).
- The speed of seismic waves (due to earthquake) can range from 3-8 km/second in earth's crust to 13 km/second in the earth's mantle.
- Seismic waves consist mainly of 'Body Waves' traveling through the interior of the earth and 'Surface Waves' traveling over the earth surface.
- Body Waves consist of P-Wave (primary waves) that are longitudinal and S-wave (secondary wave) that are transverse waves.

PROBLEMS:

1. Below is the snapshot of a transverse traveling wave, at time t = 0.
(a) Determine the wavelength. *[40 m]*

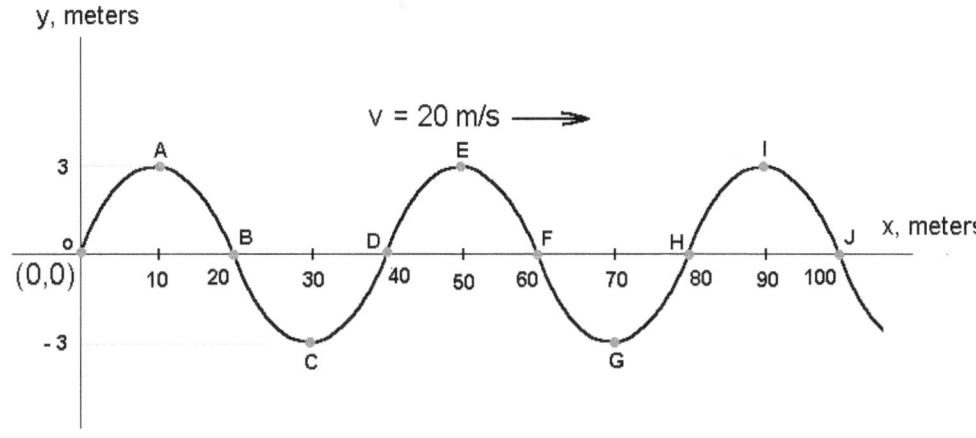

(b) Determine the frequency. *[0.5 Hz]*

(c) Determine the time period. *[2.0 s]*

(d) Determine the amplitude of the wave. *[3 m]*

(e) At this instant, state for each of the following points, whether it is moving up, down, or stationary: O, A, B, C, and D   *[O↓,A stationary ,B ↑,C stationary ,D↓]*

(f) Of the points labeled in the diagram, which points are in the same phase as the point B.
[F and J]

(g) What is the displacement of the point B in (i) ¼ T and (ii) ½ T
[(i): 3 m, (ii): 0 m]

(h) What is the distance traveled by the point B in (I) ¼ T and ½ T
[(i): 3 m, (ii): 6 m]

(i) Sketch the wave at t = ¼ T.

2. A 2-m long string has a mass of 18 g. One end of this string is held fixed and a mass of 1 kg is suspended from the other end. Determine the speed of a transverse wave through this string.

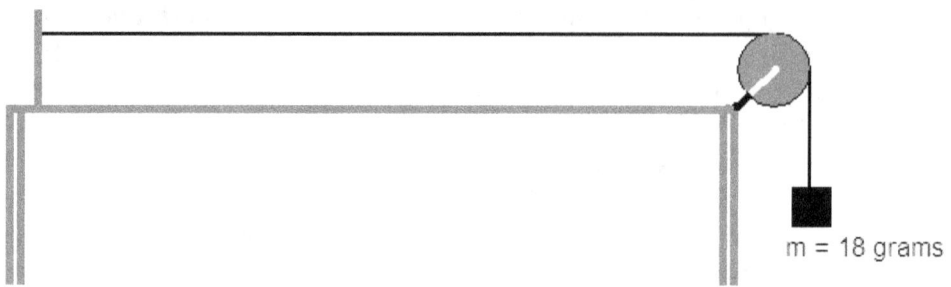

m = 18 grams

Multiple Choice

1. For a transverse wave traveling on a stretched string, the amplitude is Y, the time period is T and the wavelength is λ. In one time period, the distance moved by any particle of the string is
(A) λ (B) $\frac{1}{2} \lambda$ (C) 0   (D) 4Y   (E) 2Y

2. A transverse traveling wave on a string has a speed of 30 m/s, frequency of 0.5 Hz and amplitude of 0.3 m. The distance traveled by any particle of the string in 2 seconds is
(A) 120 m   (B) 60 m   (C) 2.0 m   (D) 1.2 m   (E) 1.0 m

3.

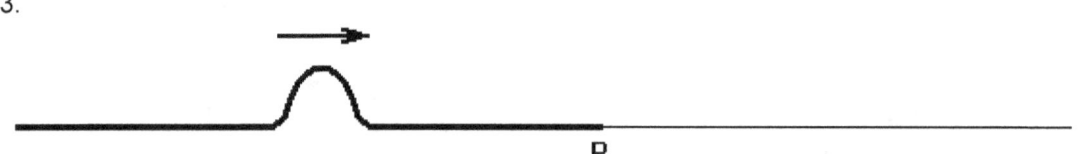

P

A composite stretched string consists of two strings of linear mass densities μ_1 and μ_2 respectively joined together at a point P. A pulse traveling along the first string reaches the point P and divides into a reflected and a transmitted pulse. Given that $\mu_1 > \mu_2$, which of the following is true about the phases of the later pulses in relation to the phase of the initial pulse

	Change of phase of the reflected pulse	Change of phase of the transmitted pulse
A	0	π
B	π	0
C	0	0
D	No reflected pulse	0
E	π	π

4.

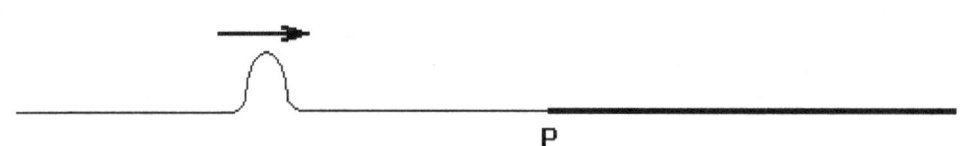

P

A composite stretched string consists of two strings of linear mass densities μ_1 and μ_2 respectively joined together at a point P. A pulse traveling along the first string reaches the point P and divides into a reflected and a transmitted pulse. Given that $\mu_1 < \mu_2$, which of the following is true about the phases of the later pulses in relation to the phase of the initial pulse.

	Change of phase of the reflected pulse	Change of phase of the transmitted pulse
A	0	π
B	π	0
C	0	0
D	No reflected pulse	0
E	π	π

Chapter 46
Standing Waves and Beats

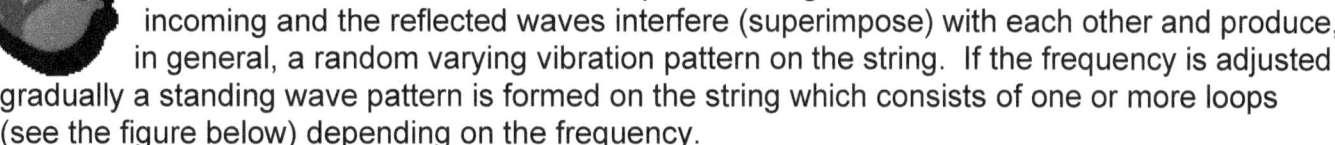

Standing waves on a string stretched between two points:

When a string stretched and under tension between two fixed points is set into vibrations the transverse wave travels to a fixed end and is reflected back with a phase change of π. The incoming and the reflected waves interfere (superimpose) with each other and produce, in general, a random varying vibration pattern on the string. If the frequency is adjusted gradually a standing wave pattern is formed on the string which consists of one or more loops (see the figure below) depending on the frequency.

The frequency needed to produce a single loop (see the figure below) is the *natural frequency, fundamental frequency,* or *first harmonic* of the string. Two or more loops are formed at higher frequencies that are multiples of the natural frequency.

The Loop: The points of zero displacements are called **nodes** and those of maximum displacement are called **antinodes**. The length of each loop equals ½ λ for the wave traveling on the string. Thus standing wave pattern can be used to easily determine the wavelength of the traveling wave on the string.

In the first mode the string vibrates with one loop; hence the wave on the string has maximum wavelength and minimum frequency in the first mode. This frequency is called first harmonic or fundamental frequency. The higher modes have multiples of the fundamental frequency.

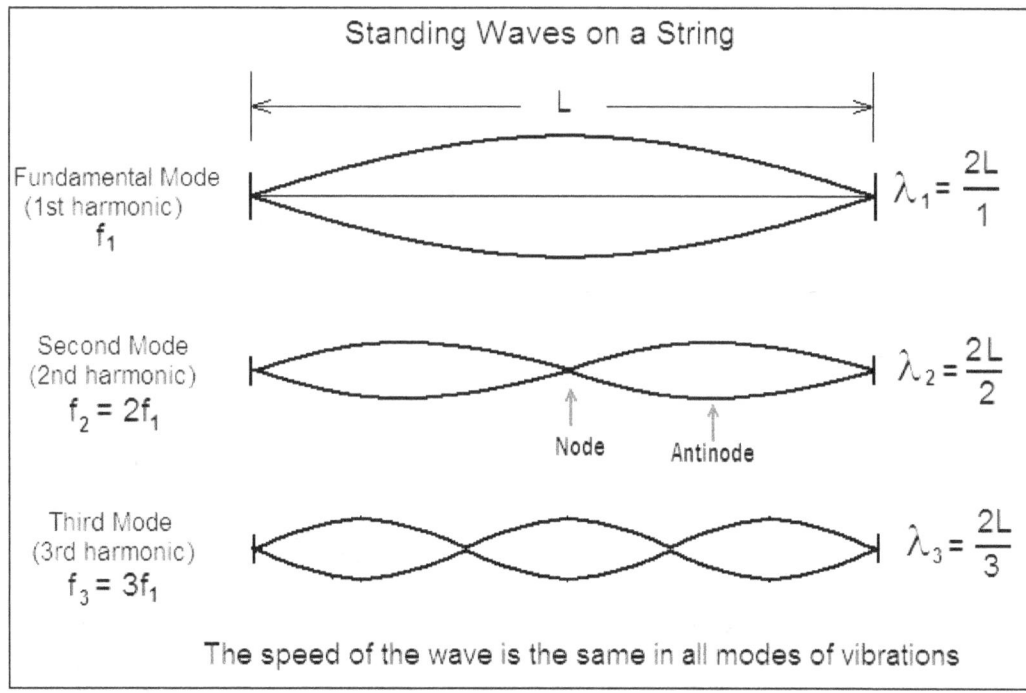

Standing Waves on a String

Fundamental Mode (1st harmonic) f_1 $\lambda_1 = \dfrac{2L}{1}$

Second Mode (2nd harmonic) $f_2 = 2f_1$ $\lambda_2 = \dfrac{2L}{2}$

Node   Antinode

Third Mode (3rd harmonic) $f_3 = 3f_1$ $\lambda_3 = \dfrac{2L}{3}$

The speed of the wave is the same in all modes of vibrations

NOTE:
- *The wavelength λ = 2 x length of one loop*

- *The speed of the wave on a string is* $v = \sqrt{\dfrac{T}{\mu}}$

Standing sound waves in pipe-organs (Resonating air column):

Open Pipe: For the open pipes, both the ends are open. When a standing wave is set up in the open pipe, there are antinodes at both the open ends. Unlike the closed pipe all the harmonics are present in the open pipe.

1ST MODE: (Fundamental harmonic or 1st harmonic)
$$\lambda_1 = 2L, \qquad f_1 = v / \lambda_1$$

2ND MODE: (2nd harmonic)
$$\lambda_2 = 2/2\,(L), \quad f_2 = v / \lambda_2 = 2\,f_1$$

3RD MODE: (3rd harmonic)
$$\lambda_3 = 2/3\,(L), \quad f_3 = v / \lambda_3 = 3\,f_1 \ldots$$

In general the resonant frequencies for the open pipe are given by
$$f_n = nv/2L, \quad n = 1, 2, 3, 4, 5, \ldots$$

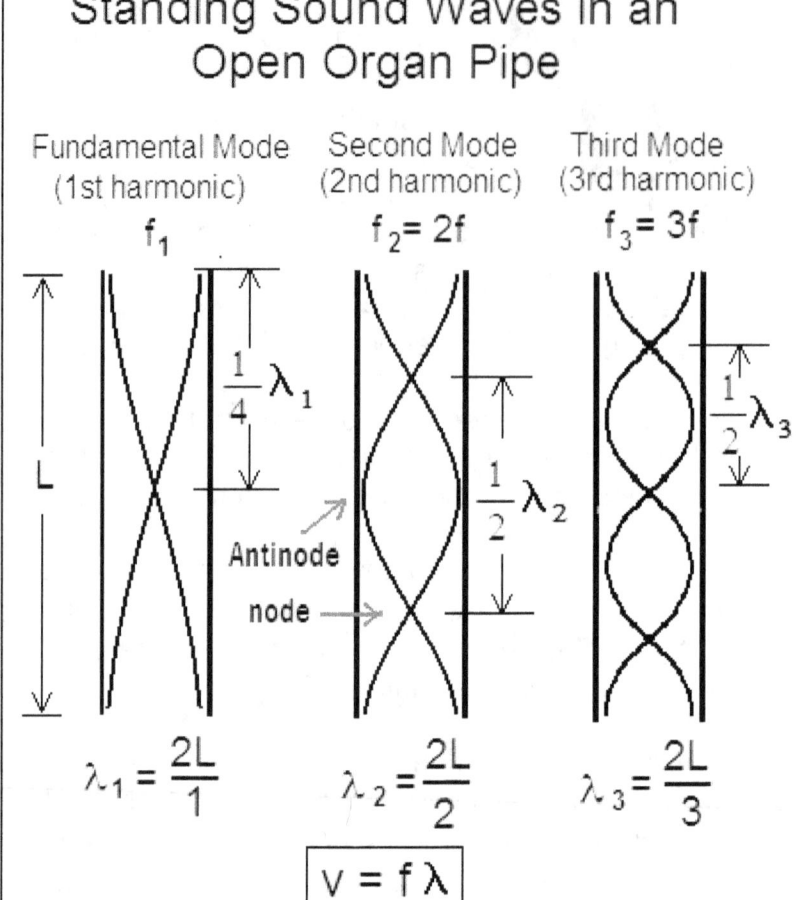

Standing Sound Waves in an Open Organ Pipe

Fundamental Mode (1st harmonic) f_1

Second Mode (2nd harmonic) $f_2 = 2f$

Third Mode (3rd harmonic) $f_3 = 3f$

$\frac{1}{4}\lambda_1$

$\frac{1}{2}\lambda_2$

$\frac{1}{2}\lambda_3$

Antinode

node

$$\lambda_1 = \frac{2L}{1} \qquad \lambda_2 = \frac{2L}{2} \qquad \lambda_3 = \frac{2L}{3}$$

$$v = f\lambda$$

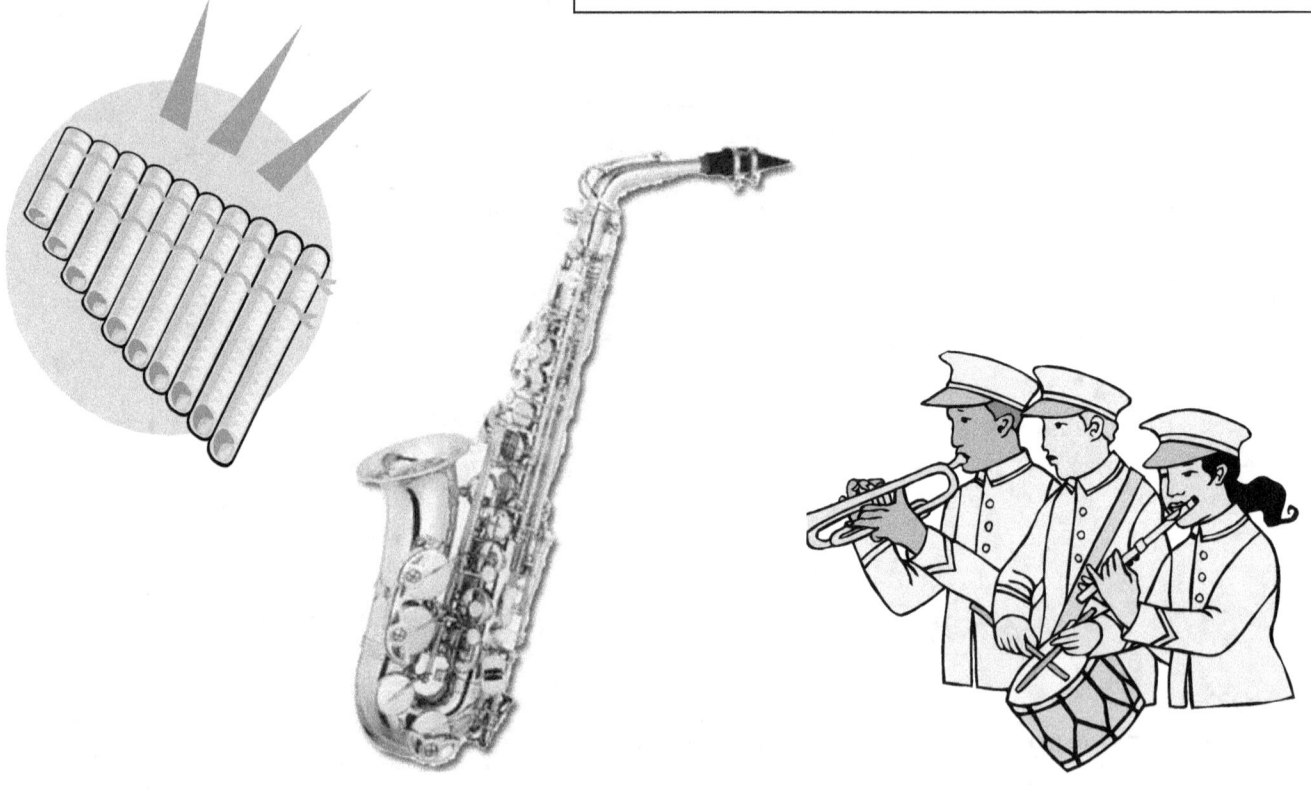

Closed Pipe: A pipe of length **L** closed at one end. For the standing sound waves created in the closed pipe, there is always a node at the closed end and an antinode at the open end. Let the speed of sound be **v**.

For the closed pipe only the odd harmonics are present; even harmonics are missing.
(Note: The speed of sound does not depend upon what mode the organ pipe is resonating)

1ST MODE: (Fundamental harmonic or 1st harmonic)
$$\lambda_1 = 4L, \qquad f_1 = v / \lambda_1$$

2ND MODE: (3rd harmonic)
$$\lambda_3 = 4/3\ (L), \qquad f_3 = v / \lambda_3 = 3 f_1$$

3RD MODE: (5th harmonic)
$$\lambda_5 = 4/5\ (L), \qquad f_5 = v / \lambda_5 = 5 f_1$$
...

In general the resonant frequencies for the closed pipe are given by
$$f_n = nv/4L, \quad n = 1, 3, 5, 7, 9, ...$$

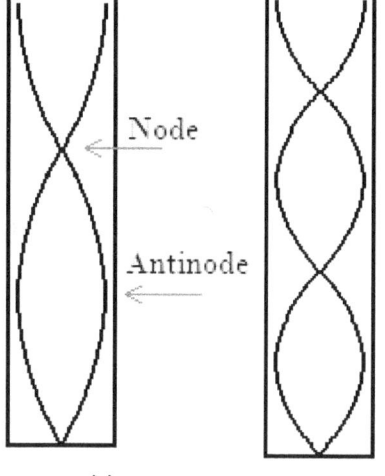

Standing sound waves in a pipe closed at one end

Fundamental Mode (1st harmonic) f_1 — 2nd Mode (3rd harmonic) $f_3 = 3f_1$ — 3rd Mode (5th harmonic) $f_5 = 5f_1$

Node — Antinode

$$\lambda_1 = \frac{4L}{1} \qquad \lambda_3 = \frac{4L}{3} \qquad \lambda_5 = \frac{4L}{5}$$

$$V = f\lambda$$

End correction: The anti node is formed a distance 0.3 D above the rim of the tube of diameter D. If D is significant in relation to L then replace L in the above equations by $L' = L + 0.3D$

NOTE
1. For any standing wave, λ = 2 x length of one loop
2. Speed of a wave does not depend on modes of vibrations

Some interesting facts & figures
- Sound travels about 4 times faster in water than in air.
- Sound travels about 15 times faster in steel than in air.
- Under ordinary conditions, the speed of sound increase by 0.6 m/s for every 1 °C rise starting at 331 m/s at 0 °C
- Cats have over one hundred vocal sounds, while dogs only have about ten.
- The blue whale can produce sounds up to 188 decibels. This is the loudest sound produced by a living animal and has been detected as far away as 530 miles.
- The crackling of a whip is due to its tip moving at supersonic speed breaking the sound barrier. Similar effect can be created with a handkerchief.
- Divide the time between lightning and thunder by and you have distance of the thunder cloud in miles.

PROBLEMS:

1. In a standing wave experiment, a string is fixed at one end and passes over a pulley 2 m from the fixed end. The tension is created in the string by suspending a mass of 1 kg from the free end. A 3-m sample of the same string has a mass of 15 g.

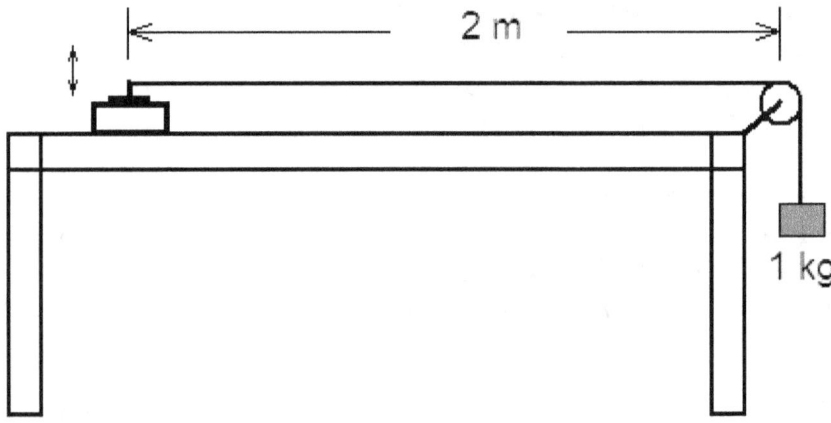

(a) Determine the fundamental frequency of the string.   *[11.1 Hz]*

(b) Determine the wavelength and frequency of the standing wave in the 4th mode of vibration.
[1 m, 44.3 Hz]

2. A string is stretched between two fixed points, 1.5 m apart. The linear mass density for the string is 0.008 kg/m and the tension in the string is 5 N.
(a) Determine the speed of the transverse wave in the string.

(b) Standing wave with a frequency of 100 Hz is set up in the string. Determine the number of loops on the string.

Problems on Resonating air-column

3. A 20-cm long organ pipe is closed at one end. Determine the frequency at which it will resonate in the 4th mode (7th harmonic). Take the speed of sound to be 330 m/s.
[2889 Hz]

4. Determine the frequency at which a 20-cm long open pipe will resonate in the 4th mode. Take the speed of sound to be 330 m/s.

5. In an experiment on the standing wave of sound, a long glass tube partly immersed in water serves as a variable length closed tube. A tuning fork of frequency 107 Hz is used in this experiment. The length of the tube is adjusted so that a resonating sound is heard. When the length of the tube is increased the resonating sound disappears and reappears when the length has increased by 1.6 m. Determine the speed of sound.   *[342.4 Hz]*

BEATS:

If two sound waves with different frequencies f_1 and f_2 interfere they give rise to the phenomenon of beats. The beats which are heard as waxing and waning of sound have a frequency given by $f_B = |f_1 - f_2|$

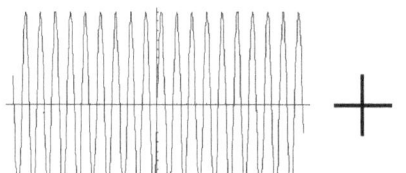 $+$ 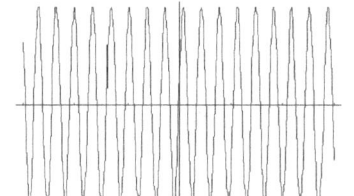 $=$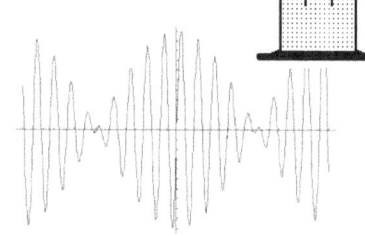

Problem on Beats

6. A tuning fork A produces beats of frequency 5 Hz with another tuning fork B of frequency 500 Hz. Determine the possible frequencies of the tuning fork A.

Multiple Choice:

1. A 0.6-m long vibrating string forms 2 loops of standing waves between two fixed points. The wavelength of the transverse wave on the string must be
(A) 0.3 m   (B) 0.6 m   (C) 1.2 m   (D) 3.33 m   (E) 6.67 m

2. A tuning fork produces 5 beats/s with a frequency of 200 Hz and 10 beats/s with a frequency of 215 Hz. The fork must have a frequency of
(A) 15 Hz   (B) 195 Hz   (C) 210 Hz   (D) 220 Hz   (E) 205 Hz

3. A stretched string is vibrating with 4 loops. If the tension in the string is increased by a factor of 4 the number of loops formed would be
(A) 1   (B) 2   (C) 4   (D) 8   (E) 16

4. A stretched string is vibrating with 4 loops. If the frequency is increased by a factor of 4 the number of loops formed would be
(A) 1   (B) 2   (C) 4   (D) 8   (E) 16

5. A tuning fork produces 1st harmonic in a <u>closed</u> pipe of length 30 cm. The same tuning fork will produce the 1st harmonic in an <u>open</u> pipe of length
(A) 15 cm   (B) 30 cm   (C) 45 cm   (D) 60 cm   (E) 120 cm

Chapter 47
Doppler Effect

Austrian physicist Christian Doppler [1803-1853], an Austrian physicist, proposed the phenomenon in 1842.

Change in frequency of sound: The frequency of sound as heard by an observer may not be same as that generated by the source. The frequency of sound may be affected by the motion of the source and/or the observer. This phenomenon is called *Doppler Effect*.

Here we consider the cases where the medium is stationary (no wind) and the source (S) and observer (O) are constrained to move along the line joining them. Let,

- v be the speed of sound in the medium (<u>not affected by the speed of the source or the observer</u>).
- v_o be the speed of the observer
- v_s be the speed of the source
- f be the actual frequency generated by the source
- f' be the frequency as heard by the observer

Then the Doppler frequency f' is given by

$$f' = f \frac{V \pm V_o}{V \pm V_s}$$

Use $v + v_o$ if the observer is moving toward the source

Use $v - v_o$ if the observer is moving away from the source

Use $v + v_s$ if the source is moving away from the observer

Use $v - v_s$ if the source is moving toward the observer

NOTE: For Doppler Effect, a source moving toward the observer at the speed of, say, 30 m/s is NOT the same as the observer moving toward the source at 30 m/s. The motion of the source changes the wavelength of sound, the motion of the observer does not.

CHANGE IN WAVELENGTH:

The wavelength of sound is changed due only to the motion of the source. If the source is moving, the new wavelength λ' reaching the listener is given by
If the

$$\lambda' = \lambda\left(\frac{V \pm V_s}{V}\right)$$

- **Use +V_s if source is moving away from the observer.**

- **Use –V_s if source is moving toward the observer.**

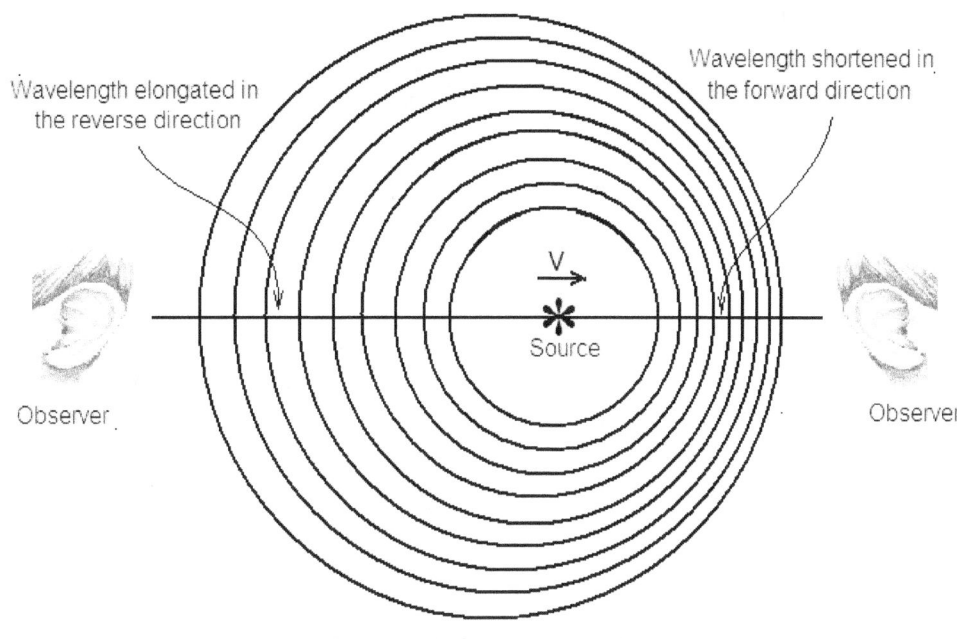

Wavelength elongated in the reverse direction

Wavelength shortened in the forward direction

$\overrightarrow{V}$

Source

Observer

Observer

Speed of sound is not affected by the speed of the source

Doppler Effect for light:

Doppler Effect also occurs for light. The wavelength of light from a star or galaxy increases when it is moving away from the earth and vice versa. It was discovered by astronomers that almost all galaxies show increase in wavelength. This effect is called 'red shift'. Scientist **Hubble** *was able to show that the red shift from almost all galaxies indicated that the universe is expanding.*

Police radar: *The Doppler Effect is exhibited by all the electromagnetic waves. Police use radio waves from their radar gun to detect the speed of a car ahead. The radar signal reflected from the car ahead is equivalent to the signal coming from that car and hence the reflected signal shows 'red shift'. The amount of red shift is converted to the speed of the car ahead by electronics.*

Some interesting facts & figures
• **Doppler Effect of light can be used to measure temperature of plasma. The ions emitting light are moving in all possible directions with respect to the observer. This results in both blue and red shifts and appears as broadening of spectral lines which is a function of the temperature.**
• **Echocardiogram uses Doppler Effect to study the speed and direction of blood flow and speed of cardiac tissues.**
• **Other applications in medical include ultrasonography, such as obstetric ultrasonography and neurology.**
• **Bat uses ultrasound and Doppler Effect to hone in on its pray.**

PROBLEMS:

v = 340 m/s ←

1. A source S is emitting sound of frequency 1000 Hz. The speed of sound is 340 m/s. The source is in between two stationary observers O_1 and O_2. If S moves toward O_1 and away from O_2 at 100 m/s determine

Observer O_1

Source S
f = 1000 Hz

Observer O_2

(a) the frequency and wavelength of sound detected by O_1 .   [1417 Hz]

(b) the frequency and wavelength of sound detected by O_2.   [773 Hz]

2. An observer O is in between two stationary sources of sound S_1 and S_2. Both the sources are emitting sound of frequency 1000 Hz. Now O moves toward S_1 and away from S_2, at 0.25 times the speed of sound.

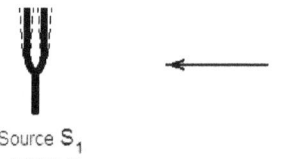

Source S_1
f = 1000 Hz

Observer O

Source S_2
f = 1000 Hz

Determine the wavelengths and frequencies of the sound from S_1 and S_2 as detected by the observer.

**3. A car is moving straight toward a cliff with a speed of 30 m/s. The driver sounds a whistle of frequency 2000 Hz. The speed of sound is 340 m/s. Determine the wavelength and the frequency of the echo from the cliff heard by the driver. *[2388 Hz, 0.155 m]*

**4. A bat is emitting ultrasonic frequency of 39 kHz as it flies toward a wall at 1/40th of the speed of sound. Determine the frequency of the echo heard by the bat.

**5. A bat moving straight toward a wall sends out ultrasonic sound waves of frequency 35.00 kHz and detects the echo of frequency 36.43 kHz. Determine the speed of the bat as a fraction of the speed v of sound. *[(1/50) v]*

Multiple Choice:

1-2. A source emits sound of frequency of 1000 Hz as it moves toward a stationary observer. The speed of sound in air is 343 m/s.
     1. The frequency heard by the observer is
     (A) 1000 Hz   (B) < 1000 Hz   (C) > 1000 Hz

     2. The wavelength of sound received by the observer is
     (A) 0.343 m   (B) < 0.343 m   (C) > 0.343 m

3-4. An observer moves toward a stationary source that is emitting a frequency of 1000 Hz. The speed of sound in air is 343 m/s.
     3. The frequency heard by the observer is
     (A) 1000 Hz   (B) < 1000 Hz   (C) > 1000 Hz

     4. The wavelength of sound received by the observer is
     (A) 0.343 m   (B) < 0.343 m   (C) > 0.343 m

5-6. An observer and a source of sound are moving <u>toward each other</u>. The source is emitting sound of frequency of 1000 Hz. The speed of sound in air is 343 m/s.
     5. The frequency heard by the observer is
     (A) 1000 Hz   (B) < 1000 Hz   (C) > 1000 Hz

     6. The wavelength of sound received by the observer is
     (A) 0.343 m   (B) < 0.343 m   (C) > 0.343 m

7-8. An observer and a source of sound are moving in the <u>same direction with the same speed with source following the observer</u>. The source is emitting sound of frequency of 1000 Hz. The speed of sound in air is 343 m/s.
     7. The frequency heard by the observer is
     (A) 1000 Hz   (B) < 1000 Hz   (C) > 1000 Hz

     8. The wavelength of sound received by the observer is

(A) 0.343 m   (B) < 0.343 m   (C) > 0.343 m

Chapter 48
Reflection and Refraction of Light

The reflection and refraction occur when a ray of light travelling in one medium meets the boundary of another medium.

Let a ray traveling in one medium be incident on the boundary of the second medium. This ray is called the *incident ray*.

A line drawn perpendicular to the boundary at the point of incidence is called the *normal.*

The incident ray generally breaks up into two rays at the point of incidence. One ray is reflected back in the same medium. It is called the *reflected ray.* The second ray enters the second medium. It is called the *refracted ray.*

Refraction

The light travels at the speed of $c = 3 \times 10^8$ m/s in vacuum and approximately so in air. When the light enters any other medium it slows down to a speed v depending on the nature of the medium.

$$n = \frac{c}{v}$$

The ratio of the speed of light in air (c) to that in another medium (v) is called the refractive

Medium	n
Vacuum	1
Air (Normal Conditions)	1.000293
Crown glass	1.5
Quartz glass	1.7
Diamond	2.42
Ice	1.31
water	1.33

index (n) of that medium. Thus

Dependence of λ on n: When a ray of light travels from one medium into another, its wavelength changes but its frequency remains unchanged. Hence

$$\frac{n_1}{n_2} = \frac{\lambda_2}{\lambda_1} = \frac{v_2}{v_1}$$

In particular, if **c** and λ_0 are the speed and wavelength of light in air (or vacuum) and **v** and λ_n are the speed and wavelength in another medium of index of refraction **n** then,

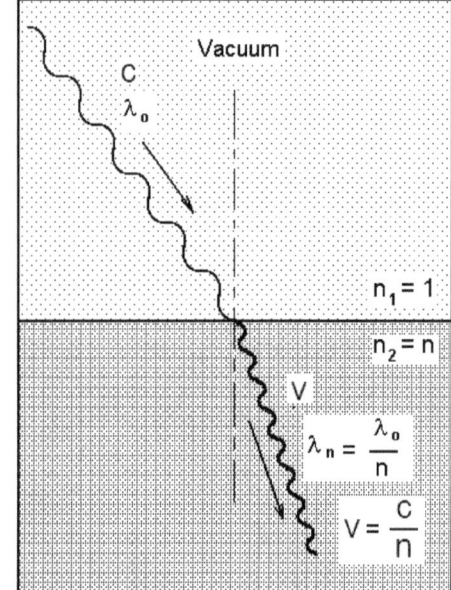

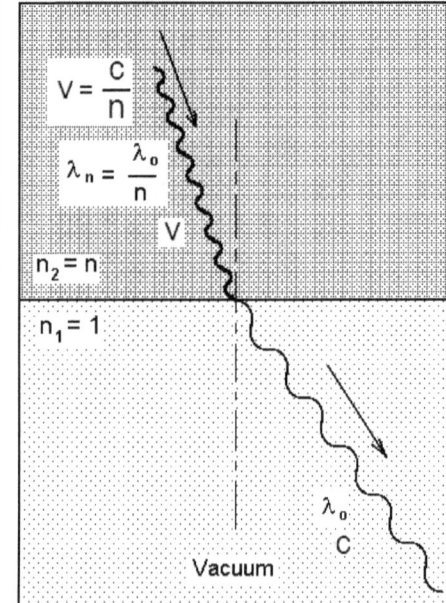

$$v = \frac{c}{n} \qquad \text{and} \qquad \lambda_n = \frac{\lambda}{n}$$

REFLECTION: The angle between the incident ray and the

14

normal is called the *angle of incidence* (θ_1) and the angle between the reflected ray and the normal is called the *angle of reflection* (θ_1'). According to the *law of reflection* $\theta_1 = \theta_1'$

REFRACTION: The angle between the incident ray and the normal is called the *angle of incidence* (θ_1) and the angle between the refracted ray and the normal is called the *angle of refraction* (θ_2).

According to the *Snell's law for refraction:*

$$n_1 \sin \theta_1 = n_2 \sin \theta_2$$

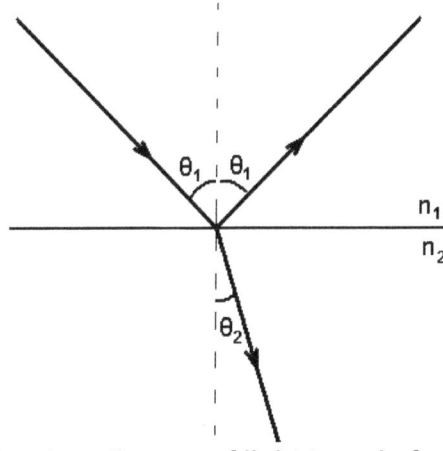

Total Internal Reflection and Critical Angle:

The phenomenon of the total internal reflection can occur only when the ray of light travels from an optically denser to an optically rarer medium.

Let a ray of light traveling in a medium be incident on another medium which is optically less dense. Starting from small values if the angle of incidence is gradually increased the angle of refraction increases faster toward 90°.

The intensity of the incident ray is divided between the refracted ray and the internally reflected ray. As the angles of incidence and refraction increase the internally reflected ray becomes more and more intense and refracted ray gets dimmer.

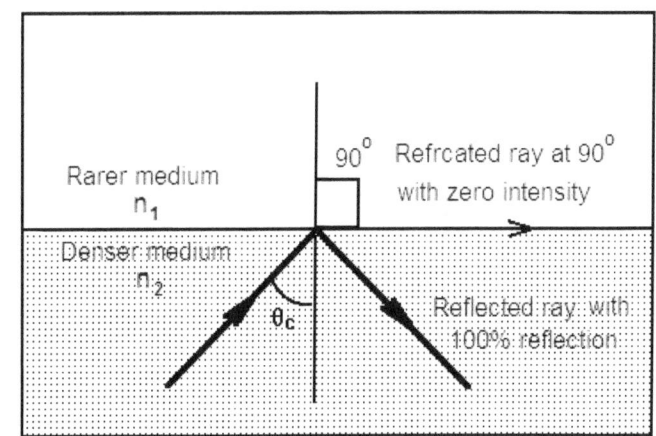

At a certain angle of incidence when the angle of refraction becomes 90° the internal reflection carries away all the energy of the incident ray and the refracted ray emerges with zero intensity. This angle is called the *critical angle* (θ_c) and the phenomenon is known as the *total internal reflection*. For all angles at and above the critical angle, the ray is totally internally reflected.

If a ray travels from a denser medium (n_2) to a rarer medium (n_1), the critical angle can be obtained using the Snell's law. Thus

$$\sin \theta_c = \frac{n_1}{n_2}$$

Reversibility of a Ray of Light: If a ray of light travels a certain path in one or more media, it

215

will retrace the same path if it starts at the other end in the opposite direction.

Dependence of n on the color of light: The value of the index of refraction for materials like glass, water, and plastics, varies slightly with the wavelength (color) of light. As is evident from the diagram of the prism below, index of refraction increases from red toward violet; i.e. $n_V > n_R$

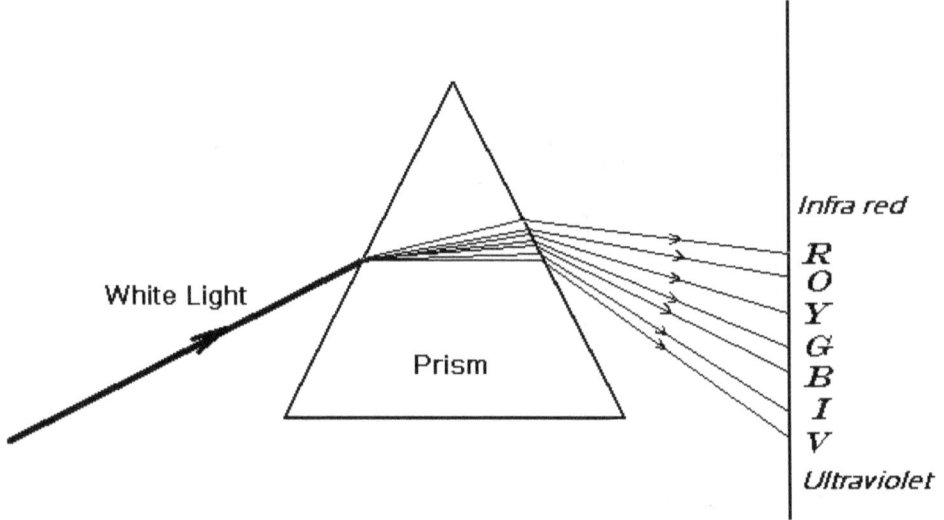

Number of images formed by two plane mirrors (Kaleidoscope):

If two plane mirrors are placed at an angle **θ** with each other the number of images formed by the mirrors is

$$N = \frac{360}{\theta} - 1$$

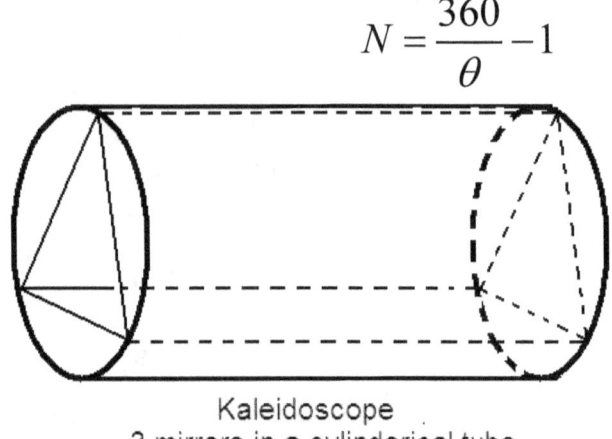

Kaleidoscope
3 mirrors in a cylinderical tube

Some interesting facts & figures
• Most gemstones contain several elements. But diamond.is all carbon.
• Diamonds are the hardest substance known to man.
• Chimps are the only animals that can recognize themselves in a mirror.
• The common goldfish is the only animal that can see both infra-red and ultra-violet light.
• Super Glue was invented by accident. The researcher was trying to make optical coating materials, and would test their properties by putting them between two prisms and shining light through them. When he tried the cyano-acrylate, he couldn't get the prisms apart.
• Chimps are the only animals that can recognize themselves in a mirror.

PROBLEMS:

1. How many images are formed with the object between two mirrors, angle between which is (a) 0° (parallel mirrors)   (b) 90°   (c) 60°   (d) 45°

[∞, 3, 5, 7]

2. A ray of light is incident at angle of incidence of 20° on the surface of a plane mirror. It is reflected to another plane mirror, which is held at angle of 70° to the first mirror. Determine the angle between the incident ray and the ray after the second reflection.

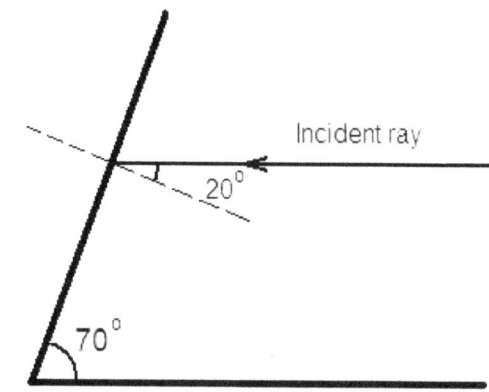

3. The speed of light in a medium is 1.8×10^8 m/s. Determine the refractive index of the medium.

[1.67]

4. Determine the critical angle at the interface between diamond (n = 2.4) and water (n = 1.33)

5. A ray of light travelling in water hits the surface of a block of ice at an angle of incidence of 30°. Determine the angle of refraction in ice.
(n(water) = 1.33, n(ice) = 1.31)

[30.5°]

6. Determine the angle from the vertical at which a diver from inside the lake must look to see the setting sun.(n_{water} = 1.33)

7. A beam of light has the wavelength of 600 nm in air.
(a) if it enters glass ($n_g = 1.5$), determine its wavelength in glass.   *[400 nm]*

(b) if the light first passes through water ($n_w = 1.33$) before it enters glass will the light have a different wavelength in glass now? Calculate and confirm.
[No. It will be same 400 nm]

**8. A ray of light is incident at angle of $65°$ on one of the faces of a $60° - 60° - 60°$ glass prism (n=1.5).
(a) Determine the angle of emergence (angle between the emergent ray and the normal)

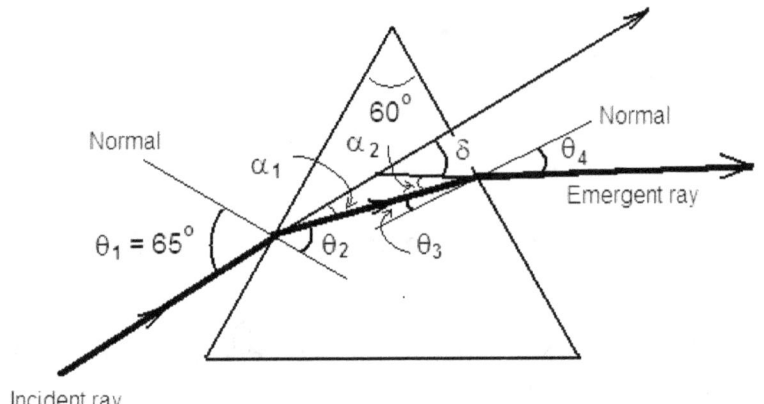

The angle of deviation $\delta = \theta_1 + \theta_4 - \theta_3 - \theta_2$

(b) Determine the angle of deviation (angle between the incident ray and the emergent ray).

(c) If the angle of incidence is now changed, determine its minimum value at which the ray can exit out of prism from the other surface.

218

9. For the normal incident ray in each of the diagrams below, sketch the rays entering and leaving the prism. Also determine the angle of incidence and the angle of refraction as the ray exits the prism.

[(a): 48.6°, (b): 19.5°]

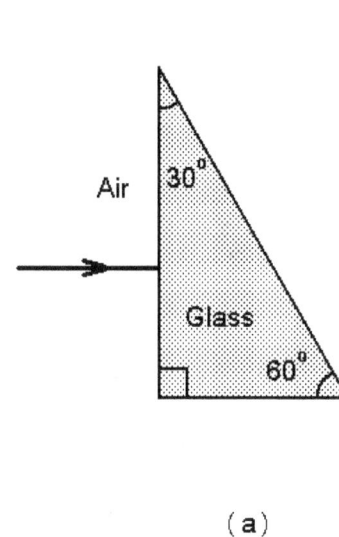

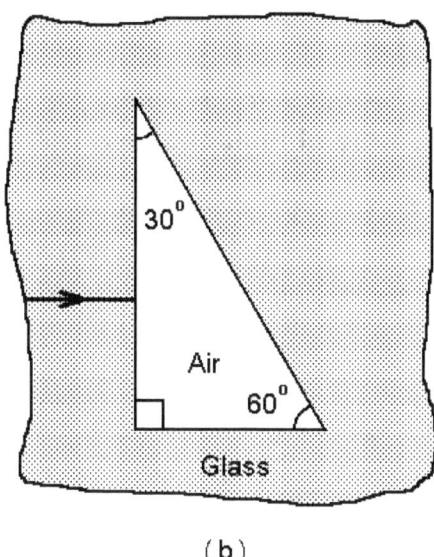

(a) (b)

Multiple Choice:

1. A narrow beam of light having blue (B), green (G), and red (R) components is dispersed by a glass prism in air. Which of the following gives the correct order of increasing deviation for the three colors by the prism?
(A) B G R (B) B R G (C) G R B (D) R G B (E) G B R

2. A beam light travels from one medium into another medium without any deviation. This implies that
(A) The first medium must be vacuum. (B) The second medium must be vacuum.
(C) Both mediums are in liquid form. (D) The angle of incidence is $0°$
(E) The angle of incidence is $90°$

3. A ray of light travels from a given medium into air ($n_{air} = 1$). It is determined that for any angle of incidence greater than α no ray emerges into air. The index of refraction of the given medium must be
(A) $\sin\alpha$ (B) $1/\sin\alpha$ (C) $\sin^{-1}(\alpha)$ (D) $1/\sin^{-1}(\alpha)$ (E) $\sin(\pi/2 - \alpha)$

Chapter 49
Image Formation by Mirrors

The image formation by plane or spherical mirrors is based on the Law of Reflection, namely
Angle of incidence = Angle of reflection

Real and virtual objects: A point is a **real object** if the rays start from that point before they hit a mirror or a lens. A point is a **virtual object** if the rays hit a mirror or a lens before they meet at that point.

Real and virtual Images: A point is a **real image** if the rays meet at that point after reflection from a mirror or after transmission through a lens. A point is **virtual image** if the rays diverge after reflection from a mirror or after transmission through a lens and appear to be coming from that point.

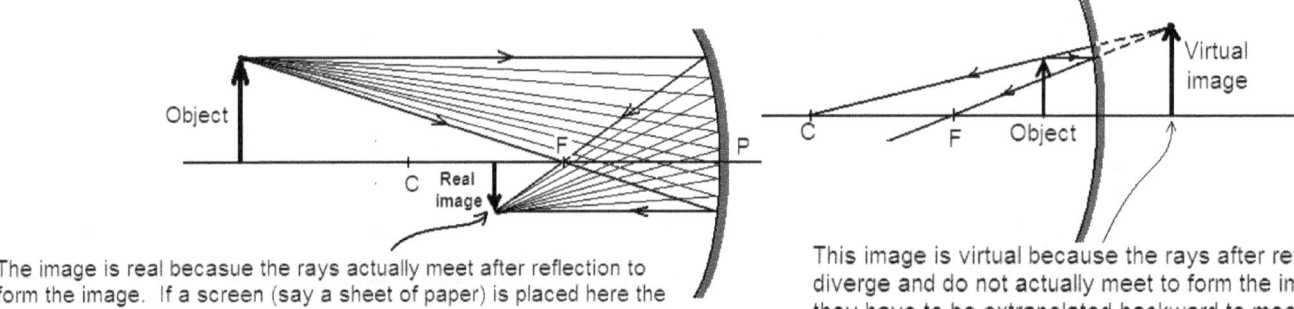

The image is real becasue the rays actually meet after reflection to form the image. If a screen (say a sheet of paper) is placed here the image appears on it if the object is bright enough.

This image is virtual because the rays after reflection diverge and do not actually meet to form the image. In they have to be extrapolated backward to meet and the image appears to be formed behind the mirror. This ima cannot becaptured on a screen.

Image formation by a plane mirror: For a real object, the image formed by a plane mirror has the following properties:

1. The image distance equals the object distance.
2. The image is same size as the object (magnification is unity).
3. The image is virtual.
4. The image is upright.

Focal Length of a concave mirror: $f = \frac{1}{2}R$

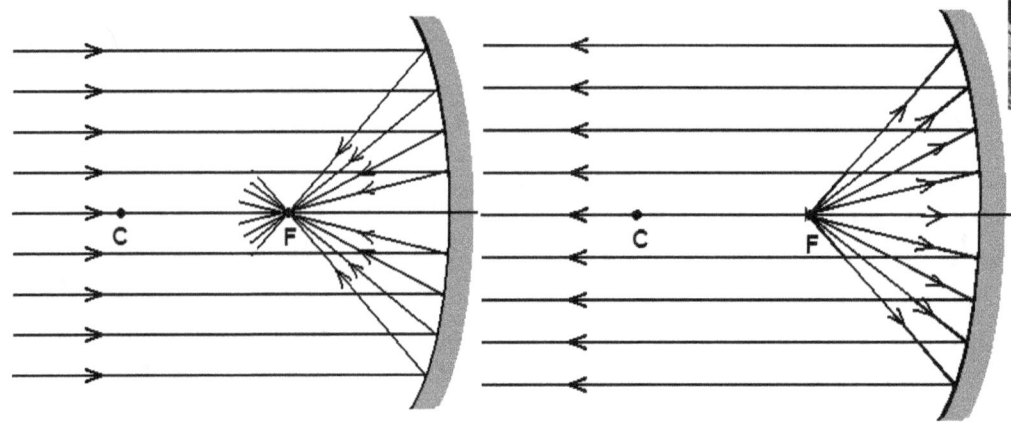

Focal length of a convex mirror: f = ½ R

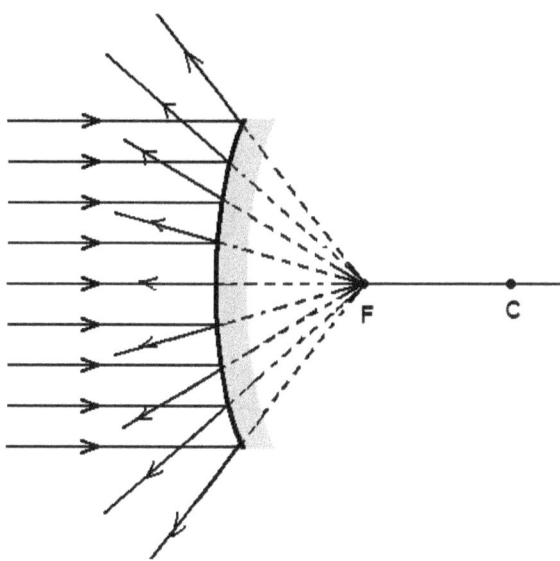

The three principal rays: Using three principal rays one can obtain image of an object for a spherical mirror. This eliminates the need to measure angles for reflected rays.

The principal rays start from the top of the given object. These rays then meet (in forward direction or extrapolated backward) after reflection from the mirror to give the position of the image.

Principal Rays for Concave Mirror

RAY 1: Draw one ray parallel to the principal axis. This ray will pass through the focus F after reflection.

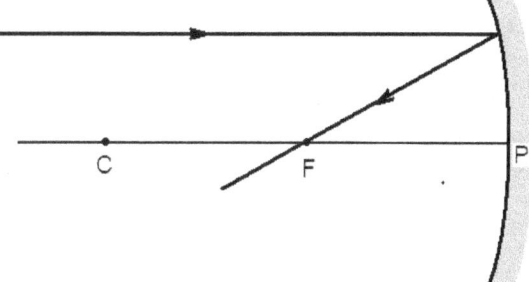

RAY 2: Draw the second ray through the focus F. This will be reflected parallel to the principal axis.

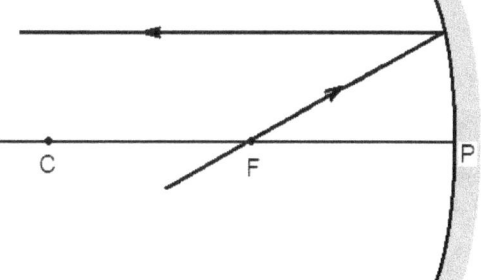

RAY 3: Draw the third ray through the center C. This will be reflected back along itself in the opposite direction.

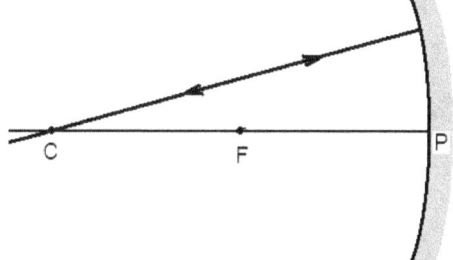

RAY 1: Draw a ray parallel to the principal axis. After reflection, it will pass through the focus F if extrapolated backward.

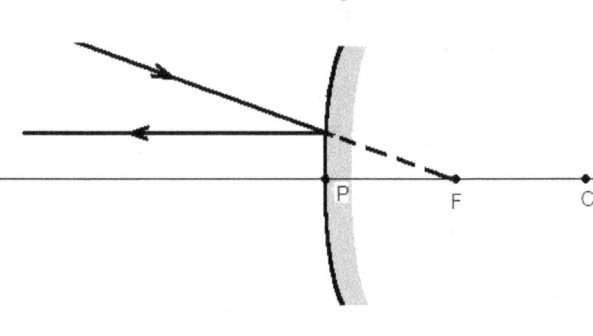

RAY 2: Draw the second ray toward the focus. This ray will meet the mirror before it can reach the focus. Upon reflection it will run parallel to the principal axis.

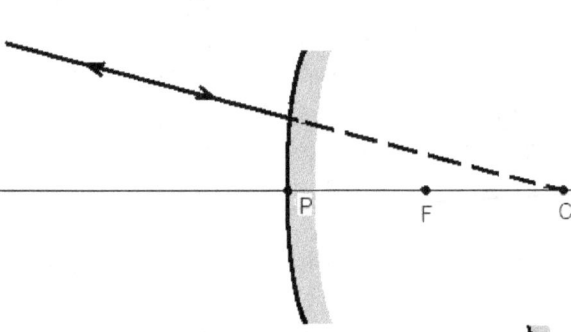

RAY 3: Draw the third ray toward the center C. It will meet the mirror before it reaches C. Upon reflection it will follow its original path in the opposite direction. The incident ray is normal to the surface.

IMAGE FORMATION BY SPHERICAL MIRRORS:

1. Ray Tracing Method

R -- radius of the spherical mirror
f   -- focal length of the mirror
s_o -- object distance
s_i -- image distance
m – magnification

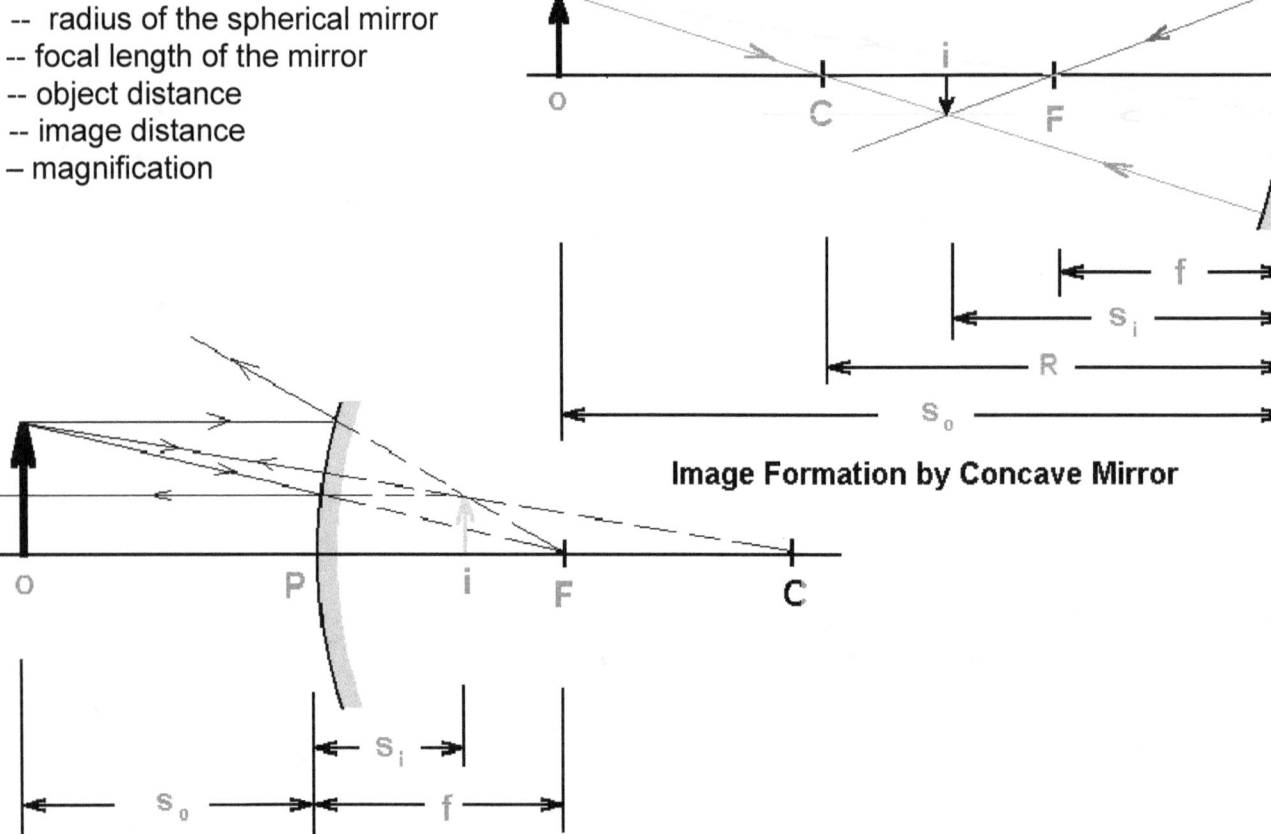

Image Formation by Concave Mirror

Image Formation by Convex Mirror

2. Analytical Method: The equations to be used for both, convex and concave, mirrors

$$f = \tfrac{1}{2}R$$

$$1/s_o + 1/s_i = 1/f$$

$$m = h_i/h_o = -s_i/s_o$$

SIGN CONVENTION:
- The object distance is + for the real objects and - for the virtual objects
- The image distance is + for the real images and - for the virtual images
- The focal length is + for the concave mirrors and - for the convex mirrors

Identifying the nature of an image:

S_i + → Real image

S_i - → Virtual image

m + → Upright image

m - → Inverted image

$|m| > 1$ → Enlarged image

$|m| < 1$ → Diminished image

CONCAVE MIRROR				
$S_o > R$ (beyond C)	$S_o = R$ (at C)	$R < S_o < f$ (Between C and F)	$S_o = f$ (at F)	$S_o < f$ (Between the mirror and F)
$|m| < 1$	$m = -1$	$|m| > 1$	No Image	$m > 1$
Real	Real	Real		Virtual
Inverted	Inverted	Inverted		Upright
Diminished	Same size	Enlarged		Enlarged

CONVEX MIRROR
For all values of S_o
m < 1
Virtual
Upright
Diminished

Example: A concave mirror of radius 60 cm produces a 25 cm tall inverted image for a 100 cm tall object.
(a) Calculate the magnification
  $m = h_i/h_o = -25/100 = -0.25$

(b) Calculate the object and image distances.
  $m = -s_i/s_o$ → $-s_i/s_o = -25/100 = -.25$ → $s_i = 0.25 s_o$
  $1/s_o + 1/s_i = 1/f$ → $1/s_o + 1/(0.25 s_o) = 1/30$ → $5/s_o = 1/30$ → $s_o = 150$ cm
  → $s_i = 0.25 s_o = 37.5$ cm
(c) Is the image real or virtual?
  Real because s_i is positive.

Some interesting facts & figures
• **Corner reflector also known as retro reflector consists of three mutually perpendicular mirrors. This system reflects returns back light beam incident on it from any direction.**
• **Astronauts left corner reflector on the moon so that laser signals sent to it will be return to the sender over a long range of lunar position with respect to earth.**
• **A corner reflector can be formed for radio and radar waves by replacing mirrors with metal plates.**

Problems:

1. A concave mirror has a focal length of 30 cm. Determine the image distance, magnification, and nature of the image (real or virtual, upright or inverted, diminished or magnified) for the object placed at the following distances:

(a) 90 cm
[45 cm, -½, R/I/Dim]

(b) 30 cm
[undefined image (at + or -∞)]

(c) 18 cm
[-45 cm, +2.5, V/U/Enlarged]

(d) 15 cm
[-30 cm, +2, V/U/Enlarged]

(e) 5 cm
[-6 cm, +1.2, V/U/Enlarged]

2. A convex mirror has a focal length of 30 cm. Determine the image distance, magnification, and nature of the image (real or virtual, upright or inverted, magnified or diminished) for the object placed at the following distances:

(a) 90 cm

(b) 15 cm

(c) 5 cm

3. A concave mirror has a radius of 24 cm. Determine the object distances to get the following magnifications:

(a) m = -2 *[18 cm]*

(b) m = -1/2 *[36 cm]*

(c) m = 5 *[9.6 cm]*

Multiple Choice:

1. An object is placed at the center of curvature of a concave mirror. For the image thus obtained the magnification m is
(A) + ½   (B) – ½ (C) + 1   (D) – 1   (E) infinity

2. Out of convex, concave, and plane mirrors, which mirror(s) will form virtual image for a real object for any object distance?
(A) concave mirror only   (B) convex mirror only   (C) Plane mirror only
(D) concave and convex mirrors   (E) plane and convex mirror

3. A mirror produces an upright, virtual, and same-size image for an object placed in front of it for all distances. The mirror must be
(A) Plane only   (B) Convex only   (C) Concave only
(D) Either plane or convex   (E) None of these

4. At which of the following object positions does a <u>concave</u> mirror form an inverted image?
(A) Between F and the mirror only
(B) Between F and 2F only
(C) Beyond 2F only
(D) Anywhere beyond F
(E) At all positions

5. At which of the following object positions does a <u>concave</u> mirror form a virtual image?
(A) Between F and the mirror only
(B) Between F and 2F only
(C) Beyond 2F only
(D) Anywhere beyond F
(E) At all positions

6. At which of the following object distances does a <u>convex</u> mirror form a virtual image?
(A) less than f
(B) Between f and 2f only
(C) Beyond 2f only
(D) Anywhere beyond f
(E) At all distances

Chapter 50
Image Formation by Thin Lenses

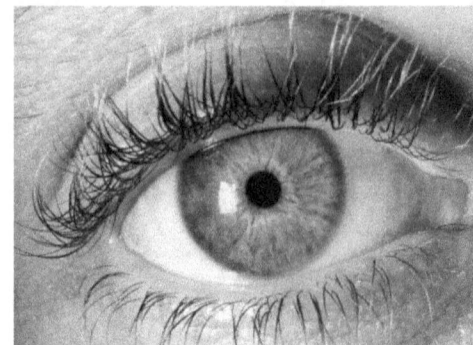

The image formation by thin lenses is based on the Law of Refraction (Snell's Law).

Converging and Diverging Lenses:

A *converging lens* will converge a beam of parallel rays of light incident upon it, to a point. A lens, which is thicker in the center and thinner toward the edge, is a converging lens (when placed in a rarer medium).

A *diverging lens* will diverge a beam of parallel rays of light incident upon it so that the rays meet at a point (focus) when extended backwards. A lens, which is thinner in the center and thicker toward the edge, is a diverging lens (when placed in a rarer medium).

Focal Length of a converging Lens:

Focal Length of a diverging Lens:

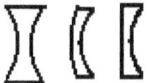

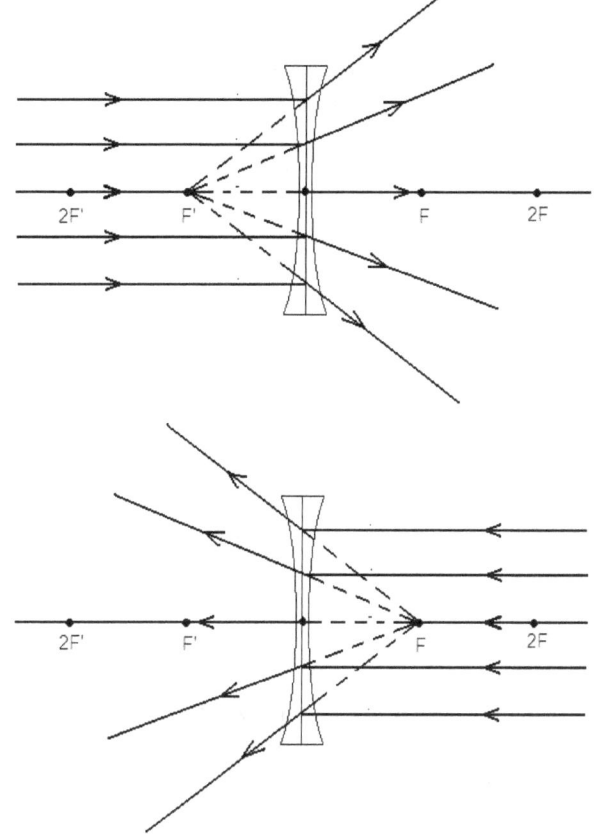

Converging Lens – Image formation with three principal rays:

The three principal rays: These rays start from the top of the given object. These rays then pass through the lens and then they may
- converge and meet to form a real image on the other side of the lens
- diverge and meet when extrapolated backward to form a virtual image.

RAY 1: Draw one ray parallel to the principal axis. This ray will pass through the focus F.

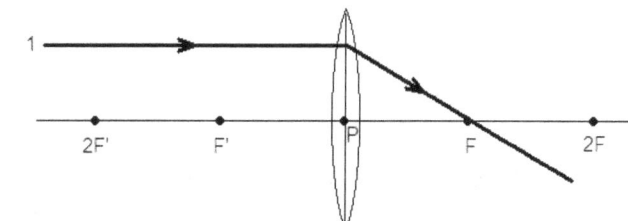

RAY 2: Draw the second ray through the focus F. This will be emerge parallel to the principal axis.

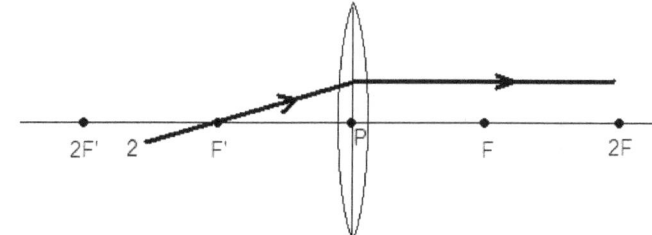

RAY 3: Draw the third ray through the center P. This will emerge undeviated.

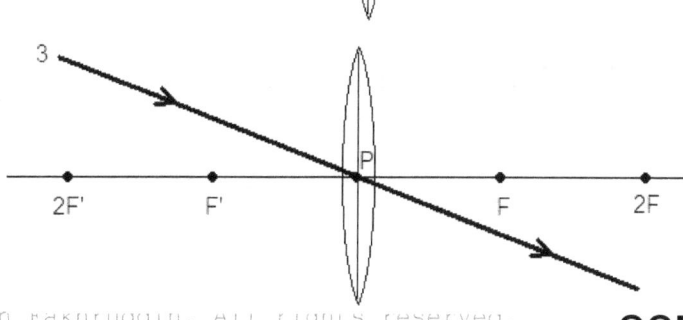

Example:

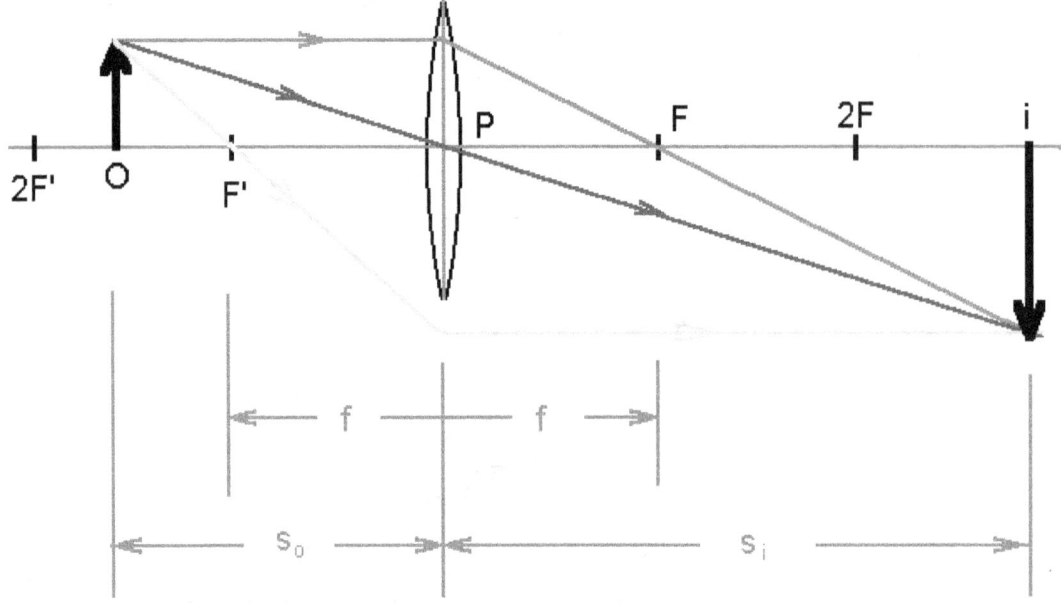

Diverging Lens – Image formation with three principal rays:
These rays start from the top of the given object. These rays pass through the lens and then they diverge and meet when extrapolated backward to form a virtual image (assuming the object is always real).

RAY 1: Draw one ray parallel to the principal axis. This ray will emerge in a way that it will pass through the focus F when extrapolated backward.

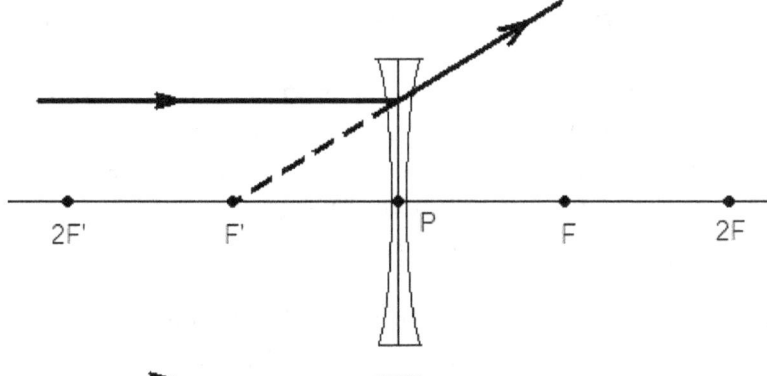

RAY 2: Draw the second ray directed toward the focus F. This will emerge parallel to the principal axis.

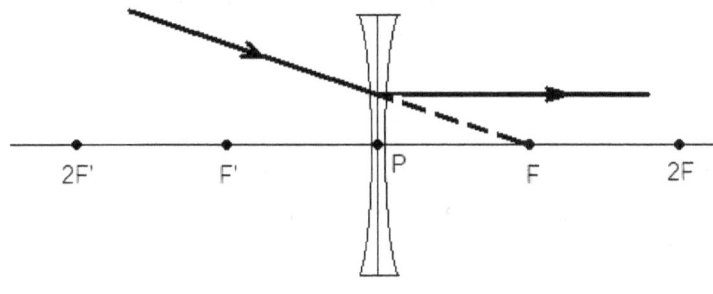

RAY 3: Draw the third ray through the center P. This will emerge undeviated.

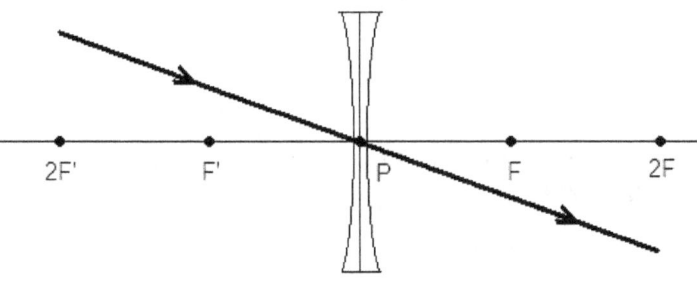

Example:

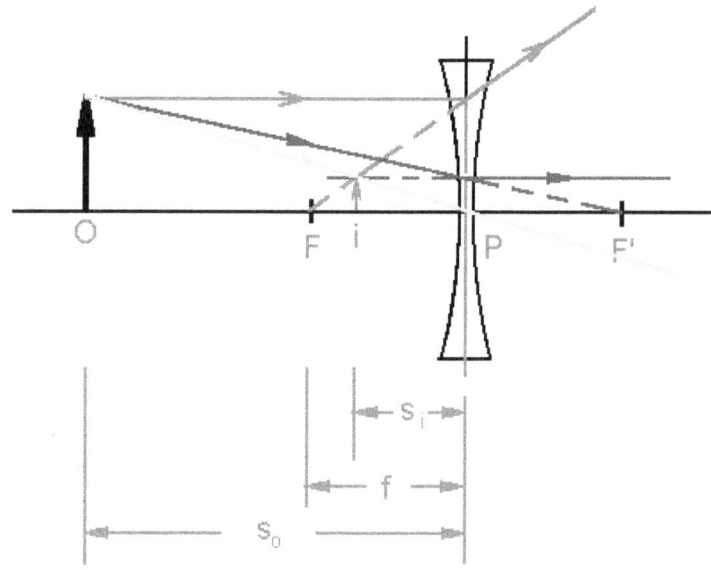

Image formation by lenses: Analytical Method

f -- focal length of the lens
s_o -- object distance
s_i -- image distance
m -- magnification

SIGN CONVENTION:
- The object distance is + for the real objects and - for the virtual objects
- The image distance is + for the real images and - for the virtual images
- The focal length is + for the converging lens and - for the diverging lens

Note: We will consider only real objects. Hence s_o is always positive

With these sign conventions one can use the following equations for both kinds of lenses:
$$1/s_o + 1/s_i = 1/f$$

$$m = h_i/h_o = -s_i/s_o$$

Identifying the nature of an image:

S_i + ➔ Real image
S_i - ➔ Virtual image

m + ➔ Upright image
m - ➔ Inverted image

$|m| > 1$ ➔ Enlarged image

$|m| < 1$ ➔ Diminished image

Example: The focal length of a converging lens is 30 cm.

(a) At what distance should an object be placed so that the image formed is upright and 3 times the height of the object?

$m = h_i/h_o = -s_i/s_o$ → $3 = -s_i/s_o$ → $s_i = -3s_o$

$1/s_o + 1/s_i = 1/f$ → $1/s_o + 1/(-3s_o) = 1/30$ → $1/s_o - 1/3(1/s_o) = 1/30$

Let $1/s_o = x$, hence $x - (1/3)x = 1/30$ → $2/3 (x) = 1/30$ → $x = (3/2)(1/30) = 3/60 = 1/20$

⇨ $1/s_o = 1/20$ → $s_o = 20$ cm

(b) Is this image real or virtual?

$s_i = -3s_o = -60$ cm. Hence s_i is negative which → image is virtual.

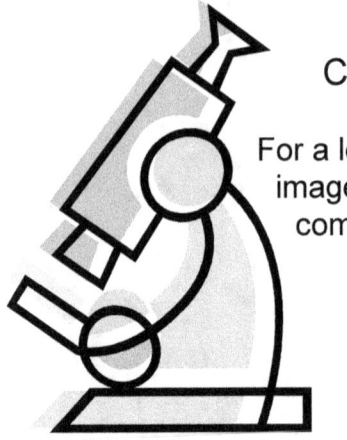

COMBINATION OF LENSES AND MIRRORS

For a lens-lens or a lens-mirror combination the intermediate image formed by the first component acts likes an object for the next component.

- Solve the problem for the first lens and obtain the image formed. This image serves as the object for the second lens.
- For the second lens, the object distance is the distance of the first image from the second lens. Solve the problem for the second lens. The image obtained is the final image for the combination.
- The net magnification is given by

$$m = m_1 \times m_2$$

where m_1 and m_2 are the magnifications due to the two lenses separately.

Examples:

1.

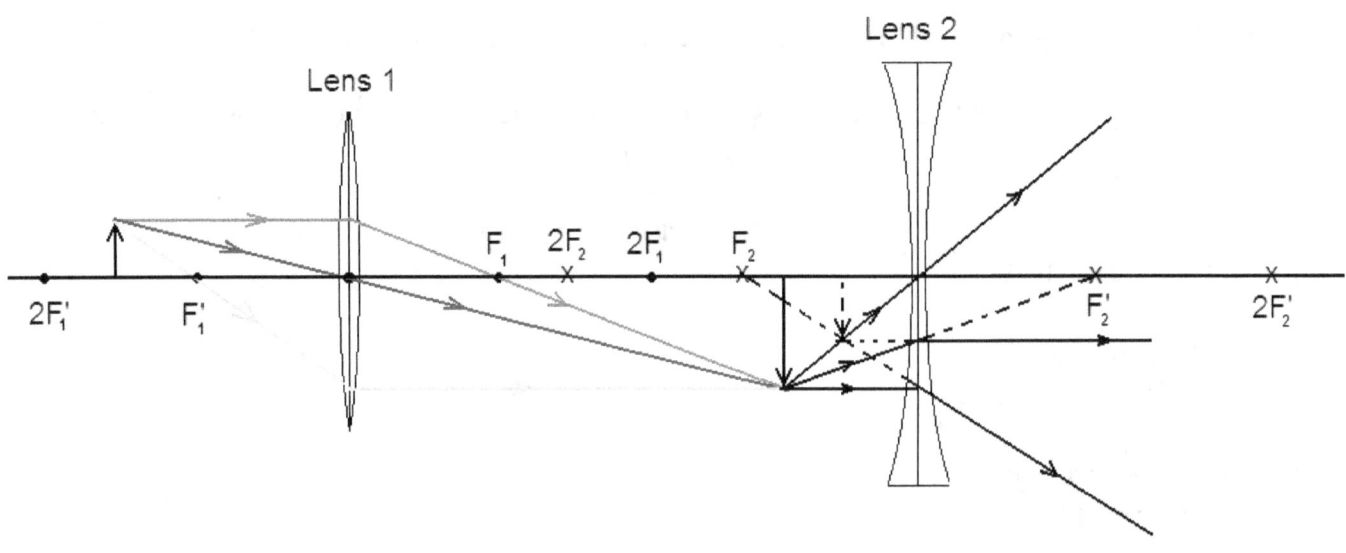

2. Lens 1 and Lens 2 are converging lenses of focal lengths $f_1 = 45$ cm and $f_2 = 30$ cm respectively placed 175 cm apart. An object is placed 75 cm to the left of Lens 1.

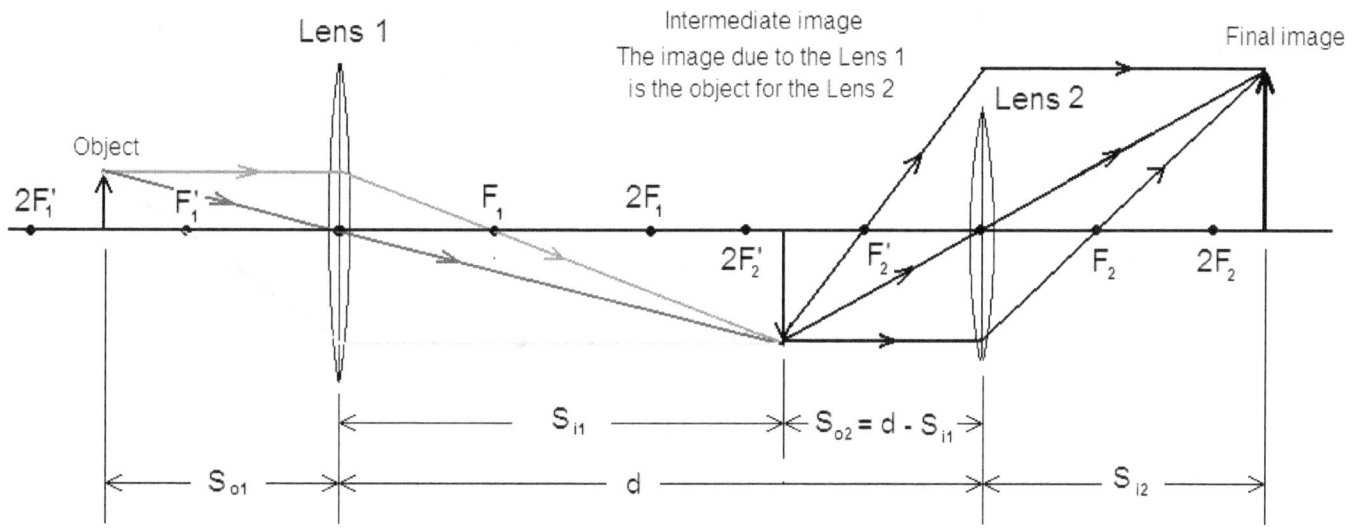

(a) Determine the position (s_{i1}) , magnification (m_1), and nature of the image formed by the Lens 1.

$1/s_{o1} + 1/s_{i1} = 1/f_1$ ➔ $1/75 + 1/s_{i1} = 1/45$ ➔ $s_{i1} = 112.5$

$m_1 = -s_{i1}/s_{o1} = -(112.5/75) = -1.5$
The image is real, inverted, and enlarged.

(b) Determine the position (s_{i2}), magnification (m_2), and the nature of the image formed by the Lens 2 treating the intermediate image as the object.

$s_{o2} = d - s_{i1} = 175 - 112.5 = 62.5$ cm
$1/s_{o2} + 1/s_{i2} = 1/f_2$ ➔ $1/62.5 + 1/s_{i2} = 1/30$ ➔ $s_{i2} = 57.69$

$m_1 = -s_{i2}/s_{o2} = -(57.69/62.5) = -0.92$
The image is real, inverted, and diminished.

(c) Calculate the overall magnification m.

$m = m_1 \times m_2 = (-1.5)(-0.92) = 1.38$.

Lens Maker's Equation: A lens in air (vacuum)

$$\frac{1}{f} = (n-1)\left(\frac{1}{R_1} + \frac{1}{R_2}\right)$$

where the lens with an index of refraction n is placed in air.

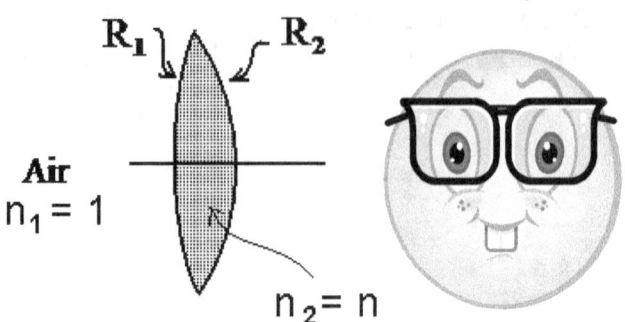

Lens in a medium different from air or vacuum:

In case the lens with index of refraction n_2 is placed in a medium of index of refraction n_1 the lens maker's equation can be written as:

$$\frac{1}{f} = \left(\frac{n_2}{n_1} - 1\right)\left(\frac{1}{R_1} + \frac{1}{R_2}\right)$$

Sign convention:
- R is **positive** if the surface is **convex** on the outside
- R is **negative** if the surface is **concave** on the outside

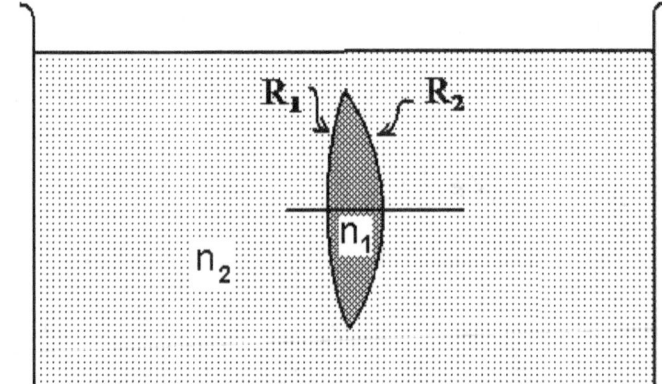

Power of a Lens:

Power (P) of a lens is reciprocal of its focal length (f). Thus, P = 1/f. To get power in SI unit of diopter use the equation

P (diopters) = 1/ f (meters)

Example: For a lens of focal length 20 cm.
Thus f = 20/100 m = 0.2 m
P = 1/f = 1/ (0.20 m) = +5 diopters

Hubble telescope orbiting the Earth

Galileo and his telescope

Some interesting facts & figures
• **Non-prescription sunglasses have zero power or an infinite focal length.**
• **Among other inventions and discoveries, Ben Franklin invented the bifocals.**
• **Hans Lippershey of Holland invented the first practical telescope of very low magnification. On 25 August 1609, Galileo demonstrated his first telescope of much greater magnification. With his telescope, Galileo discovered 3 of the four largest moons of Jupiter, phases of Venus, rings of Saturn, and lunar mountains and craters. He also correctly interpreted Kepler's discovery of sunspots; Kepler had misinterpreted the sunspot he observed as the planet Mercury.**
• **The Hubble Space Telescope weighs 12 tons (10,896 kilograms), is 13.1 meters long, has a focal length of 57.5 m and diameter of 2.4m. It cost $2.1 billion to originally build. It orbits the earth at a height of about 559 km in near circular path.**
• **Insects have compound eyes. Instead of one lens they see through two spheres with many lenses.**
• **The two large spherical eyes of a fly give an almost complete 360 degree vision.**

PROBLEMS:

1. A thin converging lens has a focal length of 25 cm. Determine the image distance, magnification, and the nature of the image formed for the following object distances:

(a) 60 cm   *[42.9 cm, -0.71, R/I/Dim]*

(b) 50 cm   *[50 cm, -1, R/I/Same]*

(c) 30 cm   *[150 cm, -3, R/I/Enl]*

(d) 25 cm   *[± ∞, undefined]*

(e) 15 cm   *[-37.5, +2.5, V/U/Enl]*

2. A thin diverging lens has a focal length of 25 cm. Determine the image distance, magnification, and the nature of the image formed for the following object distances:

(a) 50 cm   (b) 30 cm   (c) 25 cm   (d) 15 cm

3. A bi-convex lens made of glass has index of refraction n = 1.5 and the radii of curvature 15 cm and 25 cm. Determine the focal length of this lens if it is placed

(a) in air   *[18.9 cm]*

(b) in water n_w= 1.33   *[73.3 cm]*

(c) in a medium with n = 1.78   *[-59.6 cm]*

4. A convex-o-concave lens has index of refraction n = 1.5. The convex surface has a radius of 30 cm and the concave surface has a radius of 20 cm. Determine the focal length of this lens in air.

5. A convex-o-concave lens has index of refraction n = 1.5. The convex surface has a radius of 20 cm and the concave surface has a radius of 30 cm. Determine the focal length of this lens in air.   *[120 cm]*

6. A plano-convex lens has its convex surface of radius 15 cm. The index of refraction of the material of the lens is 1.7. Determine the focal length of the lens.

7. A plano-concave lens has its concave surface of radius 15 cm. The index of refraction of the material of the lens is 1.7. Determine the focal length of the lens.   *[-21.4 cm]*

8. An object is kept 90 cm in front of a screen. Determine two positions between the object and the screen at which a lens of focal length +20 cm will produce sharp images on the screen. Also determine the magnifications m_1 and m_2 in the two cases and show that **m_1 x m_2= 1**

9. An object of height 3 cm is placed 15 cm in front of a converging lens. The image produced is real, inverted, and 10 cm high. Determine the focal length of the lens.
[11.5 cm]

PROBLEMS (Combination):

10. A lens-lens combination consists of two identical lenses each of focal length +30 cm and a distance 120 cm apart. An object is placed 60 cm in front of the left lens. Determine the distance, magnification, and the nature of the intermediate and the final images formed. Draw a ray diagram for the same.

11. In a lens-lens combination the left lens has a focal length $f_1 = 30$ cm and the right lens has a focal length of $f_2 = -10$ cm. The distance between the lenses is 65 cm. An object is placed 75 cm to the left of the converging lens. Determine the positions and magnifications of the intermediate and the final images.

[Int. image: 50 cm, -0.67; Final image: -6 cm (from the div lens), -0.27]

12. In a lens-mirror combination a lens of focal length +20 cm is placed 80 cm to the left of a plane mirror. An object is placed 50 cm to the left of the lens. Determine position and magnification of the
(a) first image formed by the lens

(b) second image formed by the plane mirror

(c) third image formed when the rays reflected from the plane mirror pass again through the lens.

13. Repeat the above problem if the plane mirror is replaced by a concave mirror of focal length 15 cm.

[(a): 33.3 cm, -0.67; (b): 22.1 cm, -0.32; (c): 30.6 cm, -0.114]

Power of a lens

14. Determine the powers of the following lenses in diapers:
(a) f = +10 cm   (b) f = -50 cm(c) f = +25 mm   (d) f = -4 m

15. Determine the focal lengths of the following lenses in centimeters:
(a) P = -.5 diopters   (b) P = + 10 diopters   (c) P = +.0125 diopters

[-200 cm; +10 cm; 8,000 cm]

Multiple Choice:

1.

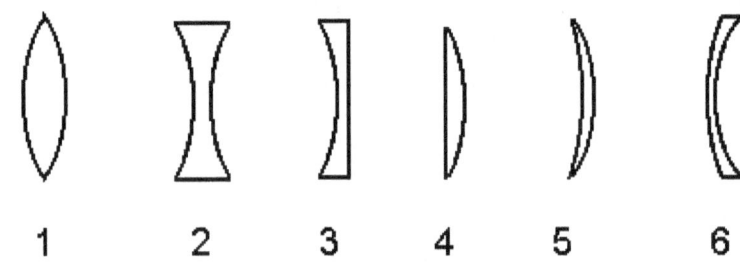

1   2   3   4   5   6

In the diagram above which lenses are the diverging types?

(A) 2, 3, 5   (B) 1, 4, 5   (C) 2, 3, 6   (D) 2   (E) 5

2. Which thin refracting device forms a virtual image for an object placed at any distance?
(A) Convex lens only   (B) Concave lens only
(C) Both convex and concave lenses   (D) Plate   (E) None of these

3. An object is placed within the focal length of a biconvex glass lens in air. Which of the following are the characteristics of the image formed?
(A) Virtual, Upright, Magnified
(B) Virtual, Upright, Diminished
(C) Real, Inverted, Same size
(D) Real, Inverted, Diminished
(E) Real, Upright, Magnified

4. An object is placed 60 cm from a lens of focal length +30 cm. The magnification of the image is
(A) +1   (B) +½   (C) infinity   (D) -½   (E) -1

5. An object is placed 60 cm from a spherical mirror of focal length +30 cm. The magnification of the image is
(A) +1   (B) +½   (C) infinity   (D) -½   (E) -1

6. Which reflecting device(s) forms a diminished virtual image for all the real object distances.
(A) Convex mirror only   (B) Concave mirror only
(C) Both convex and concave mirrors   (D) Convex and plane mirrors
(E) Only plane mirrors

Chapter 51
Interference

Principle of Superposition: When two or more wave disturbances in a medium meet at a point the resulting disturbance at that point is the *vector sum* of all disturbances.

This principle is applicable to longitudinal waves (sound) as well as transverse waves (light).

Interference between Two Waves:
Constructive Interference:

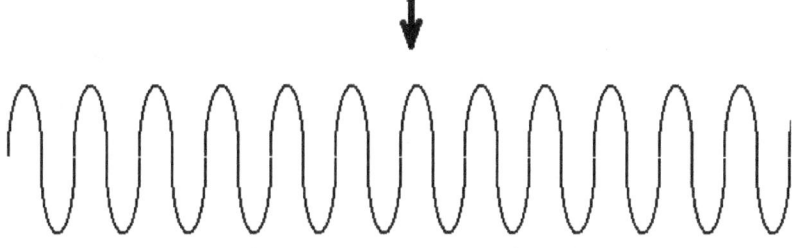

Two waves meeting in phase to cause constructive interference.
The two waves are shown slightly shifted for distinction.

If the two waves meet at a point **in phase** (crest on crest or trough on trough etc.) the resulting wave has an amplitude which is the sum of the two amplitudes.

Constructive Interference

Destructive Interference:

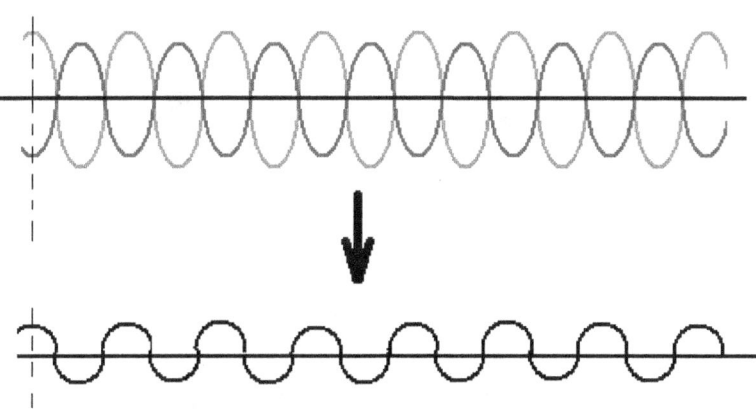

Destructive Interference

Two waves meeting out of phase. The resultant wave has amplitude equal to the difference of the two amplitudes.

If the two waves meet at a point **out of phase** (crest on trough) the amplitude of the resulting wave is the difference of the two amplitudes. If, initially, the two amplitudes were equal the resulting amplitude due to the destructive interference will be zero.

The interference phenomenon can be observed for the sound waves as well as the light waves.

Two sources giving out the waves in the same phase (or with a constant phase difference) are said to be **coherent sources**.

Whether the waves meet in phase or out of phase at a point depends on its distance from the two coherent sources from which the two waves start in the same phase.

The two waves will **interfere constructively** (be in phase) at a point if the path difference is multiple of wavelengths

$$\text{path difference} = m \lambda \; ; \; m = 0,1,2,3 \dots$$

The two waves will **interfere destructively** (be out of phase) at a point if the is integer plus 1/2 times the wavelength

$$\text{path difference} = (m - 1/2) \lambda \; ; \; m = 1,2,3 \dots$$

Beautiful colors of peacock feathers and wings of butterflies
are due to light interference

Young's Double Slit Interference: For Young's double slit experiment with monochromatic light the interference produces dark and bright fringe pattern on a screen.

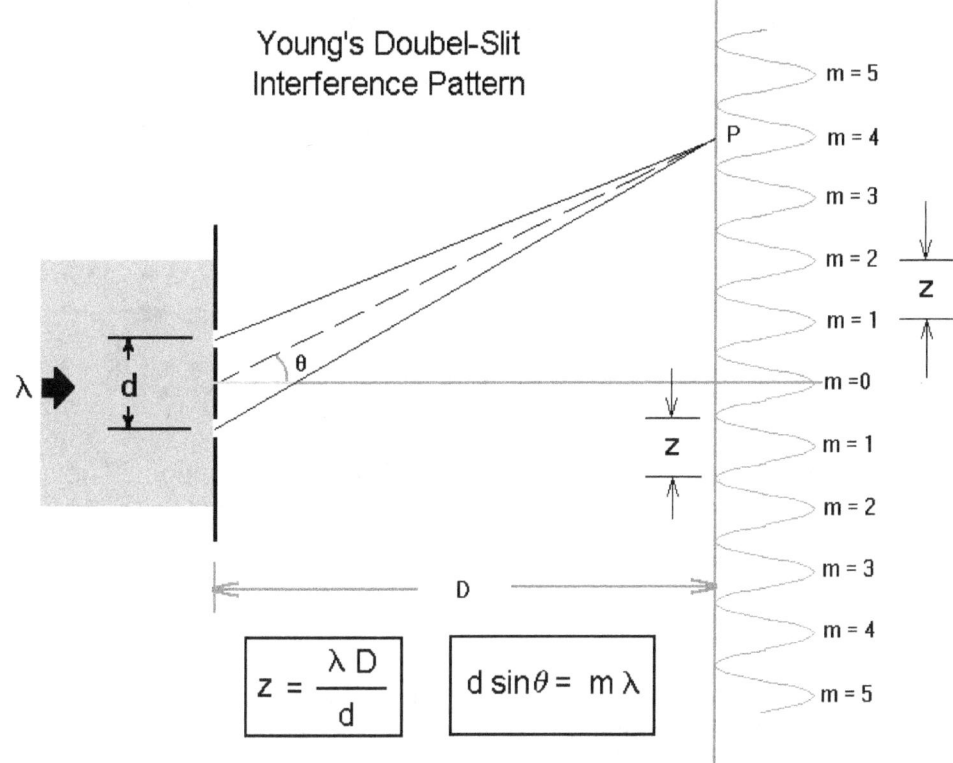

Young's Doubel-Slit
Interference Pattern

$$z = \frac{\lambda D}{d}$$ $$d \sin\theta = m\lambda$$

The Angular Positions (θ) for the Bright and Dark Fringes:

The angular positions (θ) for the interference maxima and minima are given from the equations

$$d \sin(\theta) = m\lambda, \qquad m = 0, 1, 2, \dots \text{ (bright fringes)}$$
$$d \sin(\theta) = (m - \tfrac{1}{2})\lambda, \qquad m = 1, 2, \dots \text{ (dark fringes)}$$

The Fringe Width (z):

The fringe width (z) is same for the dark and bright fringes and is given by $z = \lambda\left(\dfrac{D}{d}\right)$

where λ is the wavelength
- θ   is the angular position of a fringe
- z   is the fringe width
- d   distance between the two slits
- D   distance of the screen from the sources
- λ   is the wavelength of the monochromatic light used

Interference pigment on US currency:

Current $20 bills use interference pigments for the '20' on the lower right hand corner of the bill. The '20' appears green when viewed straight on, but looks black at other angles. The flakes can be easily seen at 30X magnification. They consist of a thin semi-opaque metal layer, typically Cr (about 5 nanometers) followed by a layer of SiO2 - the thickness is tuned to give the desired color- followed by a thick reflecting metal (usually Al) layer and then the same SiO2 and Cr layers as stated above so that the flake is symmetric.

Some interesting facts & figures
• **A noise cancellation speaker emits sound waves with the same amplitude but opposite phase to the unwanted sound and the interference of the two waves results in cancellation of the sound.**   • **The beautiful colors of many butterflies and birds such as the blue jay, hummingbirds, peacocks, and pheasants are due to interference and diffraction effects rather than pigments.**

PROBLEMS:

1. In a Young's double slit experiment, a monochromatic light of wavelength 632.8 nm is used. The slit separation is .4 mm and the screen is placed 1.5 m from the screen. Determine
(a) the fringe width. $[2.37 \times 10^{-3}\ m]$

(b) the angular position of the 10th bright fringe. $[0.91°]$

(c) the linear position of the 10th bright fringe from the center. $[2.37\ cm]$

(d) the distance between the 5th bright fringe on one side and the 4th dark fringe on the other side of the center of the screen. $[2.0\ cm]$

2. In a Young's double slit experiment, d = 0.55 mm, D = 2.0 m, and the fringe width z = 2 mm. Determine the wavelength of the light used.

3. In a Young's double slit experiment the fringe width is 3 mm when the light of wavelength 450 nm is used. With the same set up of the equipment, if another monochromatic light of wavelength 600 nm is used determine the new fringe width.

[4 mm]

Multiple Choice

1. In a Young's double slit experiment, if the distance D between the pair of sources and the screen is increased the fringe width z
(A) increases   (B) decreases   (C) remains the same
(D) increases and then decreases   (E) decreases then increases

2. In a Young's double slit experiment, if the distance d between the two sources is increased then the fringe width z
(A) increases   (B) decreases   (C) remains the same
(D) increases and then decreases   (E) decreases then increases

3. In a Young's double slit experiment, if the distance D between the source and the screen is increased the angular width θ of any fringe
(A) increases   (B) decreases   (C) remains the same
(D) increases and then decreases   (E) decreases then increases

4. In a Young's double slit experiment, if the distance d between the two sources is increased the angular width θ of any fringe
(A) increases   (B) decreases   (C) remains the same
(D) increases and then decreases   (E) decreases then increases

5. In a Young's double slit experiment, if white light instead of monochromatic light is used the center of the central fringe will be
(A) violet   (B) red   (C) greenish yellow
(D) dark   (E) white

Chapter 52
Diffraction: Single and Double slit
Diffraction Grating

The ability of the waves to bend round the obstacles in their path is called **diffraction**. Diffraction is easily observable for the waves of larger wavelengths such as sound waves. For light, which has relatively very small wavelength diffraction can be observed under special conditions.

Single-Slit diffraction: When a wave front of light passes through a slit and falls on a screen the light bends into the shadow region. This bending can be easily observed if the width of the slit is comparable to the wavelength of the light being used. The points on the wave front passing through the slit act like coherent sources and produce a pattern on the screen called the single-slit diffraction pattern. It consists of a very bright central maximum followed, on either side, by much fainter secondary maxima of rapidly decreasing intensity. The secondary maxima are all of same width and each is half as wide as the central or principal maximum

The angular position θ of mth dark fringe is given by the equation

$w \sin \theta = m\lambda, \ m = 1, 2, \ldots$

The linear position (y) of the mth dark fringe from the central maximum is given by

$$y_m = \frac{m\lambda D}{w} \quad m = 1, 2, \ldots$$

Where, **D** is the distance (large) of the screen from the slit

w is the slit width

λ is the wavelength of the light

The fringe width (z) between any two consecutive dark fringes is given by $\quad z = \lambda \left(\dfrac{D}{w} \right)$

Hence, each secondary maximum is z-wide and the central or principal maximum is 2z-wide.

Double-Slit diffraction: The double-slit diffraction is the superposition of the single-slit pattern and the Young's double-slit pattern.

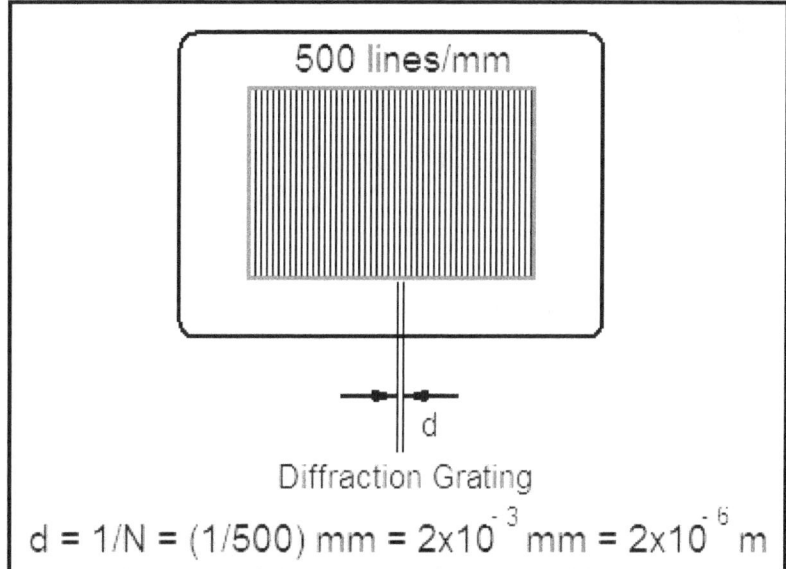

Diffraction Grating: This consists of a number of slits closely packed together

(ordinarily thousands per centimeter). The light passing through the grating is resolved into its component colors due to diffraction. Each color may have one or more orders of maxima at different angles.

$$d \sin \theta = m \lambda$$

where

d is the separation between the consecutive slits (grating element). If there are N lines/meter on the grating **d = 1/N** meters

θ is the angle from the initial direction at which the maximum occurs

n is the order of the maximum

λ is the wavelength

500 lines/mm

d

Diffraction Grating

$d = 1/N = (1/500)\ mm = 2 \times 10^{-3}\ mm = 2 \times 10^{-6}\ m$

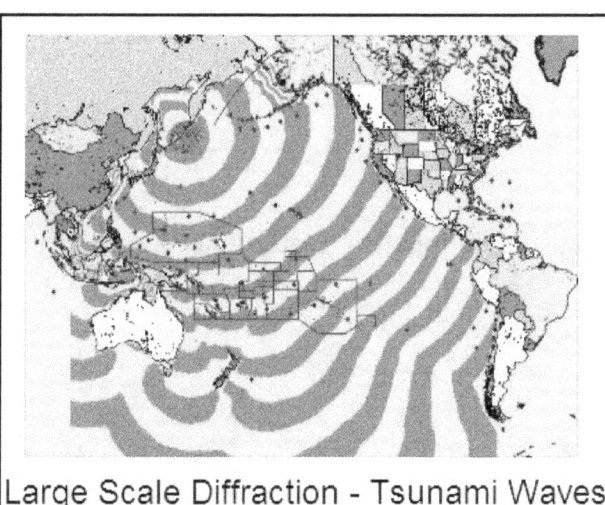

Large Scale Diffraction - Tsunami Waves

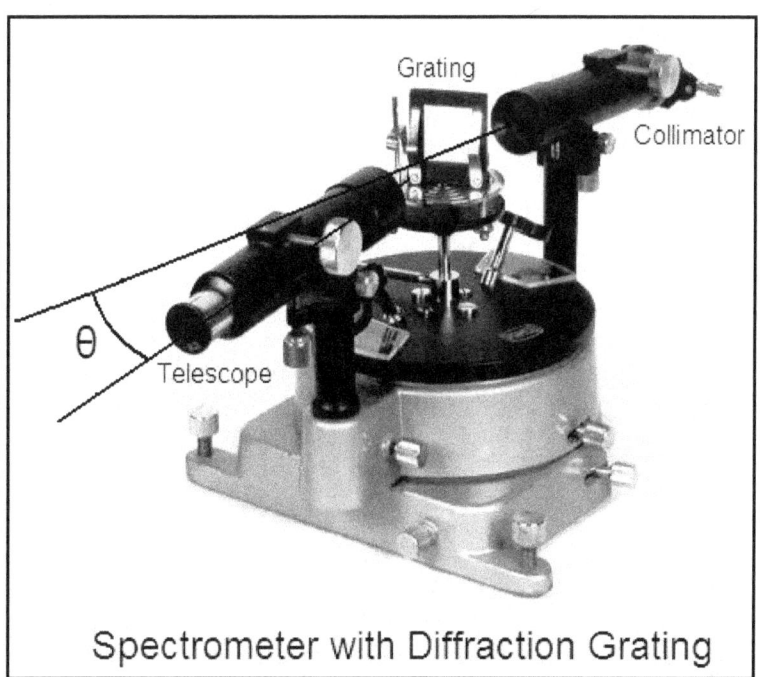

Grating

Collimator

θ

Telescope

Spectrometer with Diffraction Grating

Some interesting facts & figures

- Francesco Maria Grimaldi first made careful observation of the effects of diffraction and coined the term *diffraction*.
- Francesco Maria Grimaldi [April 2, 1618 - December 28, 1663] was an Italian Jesuit priest, mathematician, and physicist.
- A crater on the moon is named after Girmaldi.
- X-ray diffraction is used to study crystal structure.
- Electron diffraction is used to study crystal structure and also in transmission and scanning electron microscopes.

PROBLEMS:

1. In a single slit diffraction experiment: w = 0.2 mm, λ = 500 nm, and D = 1.4 m. Determine:
(a) the angular width of the principal (central) maximum.   [0.286°]

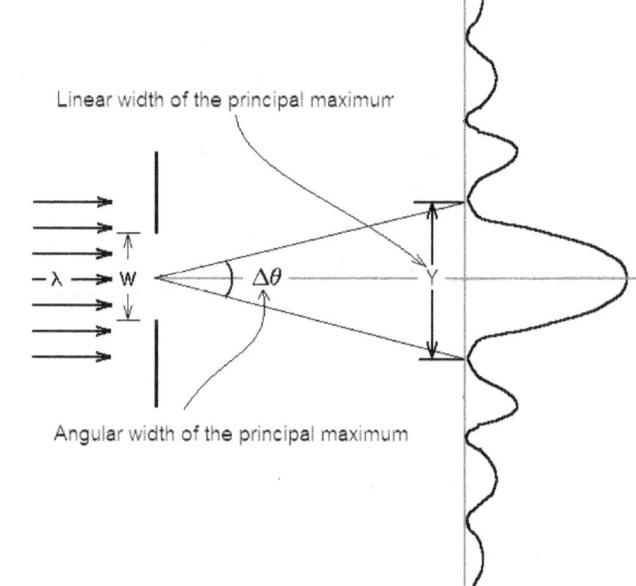

Linear width of the principal maximum

Angular width of the principal maximum

(b) the linear width of the principal (central) maximum.   [7.0x10⁻³ m]

(c) the angular position of the 10th dark fringe.   [1.43°]

(d) the linear position of the 10th dark fringe.   [3.5x10⁻² m]

2. A single slit of width 0.1 mm is used in an experiment to determine the wavelength of a given monochromatic light. The angular width of the central maximum is measured to be 0.745°. Determine the wavelength of the light.

3. A single slit has a width of 4 λ where λ is the wavelength of light used. Determine
(a) the angular width of the principal maximum.   [29.0°]

(b) the angular position of 3rd order minimum.   [48.6°]

4. A diffraction grating has 600 lines/mm. A light of wavelength 480 nm is incident on the grating.
(a) Determine the angular positions for the 1st and 2nd order maxima

(b) Determine the value of the highest order possible in this experiment.

5. A diffraction grating with 5000 lines/cm is used to obtain the spectrum of white light (400 - 700 nm). Determine the angles of dispersion for the spectrum in the first order and second order.

[9.0°, 20.8°]

Multiple Choice:

1. In a single slit diffraction experiment, the slit width is twice as much as the wavelength of monochromatic light used (w = 2λ). The angular width of the central maximum must be
(A) 30°   (B) 60°   (C) 90°
(D) 120°   (E) 180°

2. In a single slit experiment, if the wavelength of the monochromatic light used is increased the central maximum gets
(A) broader   (B) narrower   (C) remains the same
(D) increases and then decreases   (E) decreases then increases

3. In a single slit experiment, if the slit width is increased the central maximum gets
(A) broader   (B) narrower   (C) remains the same
(D) increases and then decreases   (E) decreases then increases

4. In a single slit experiment, if the screen is taken farther away from the slit is the central maximum gets
(A) broader   (B) narrower   (C) remains the same
(D) increases and then decreases   (E) decreases then increases

5. In a single slit experiment, the angular width of the central maximum depends on which of the following quantities: slit width (w), distance between the screen and the slit (D), wavelength of light used (λ)?
(A) w only   (B) D only   (C) λ only
(D) λ and w   (E) all three of them

6. In a single slit experiment, white light instead for monochromatic light is used. The center of the principal maximum must be
(A) white   (B) black   (C) violet
(D) red   (E) pink

7. In a single slit diffraction pattern, the linear distance between the two 3rd order minima (on either side of the screen center) is 6 mm. The linear width of the central maximum must be
(A) ½ mm   (B) 1 mm   (C) 1.5 mm
(D) 2 mm   (D) 2.5 mm   (E) 3 mm

Chapter 53
Electromagnetic Spectrum - Thin Film – Polarization - Intensity

THE ELECTROMAGNETIC SPECTRUM

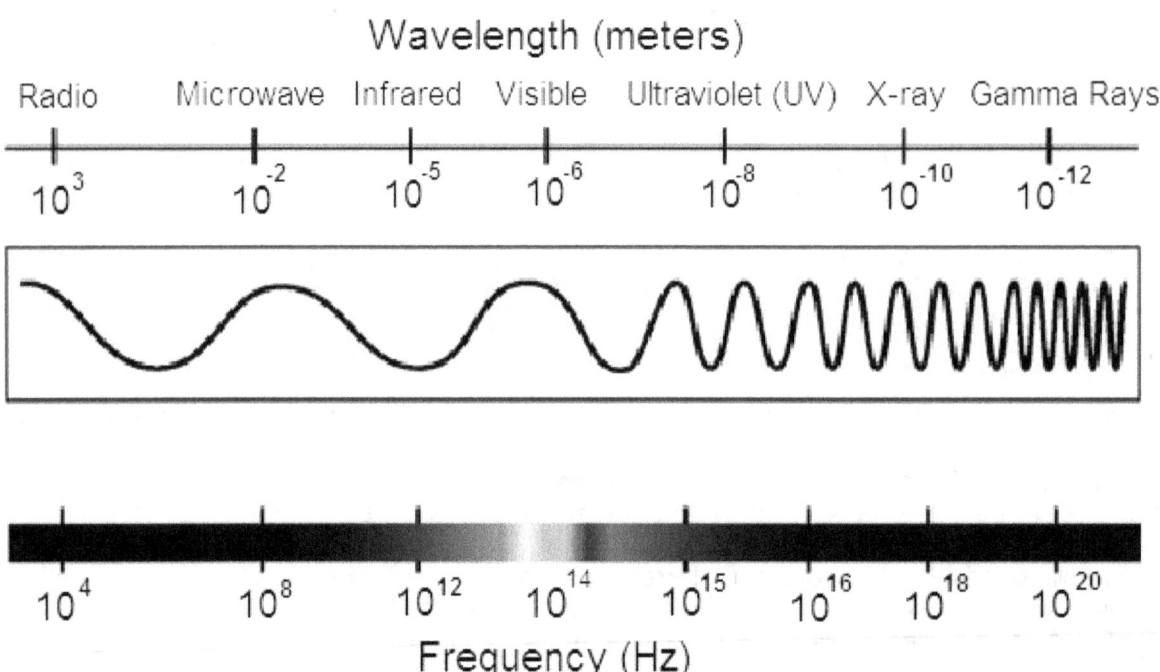

Electromagnetic Spectrum: The electromagnetic (EM) spectrum consists of the EM waves of all possible wavelengths -- normally smaller than the size of the atomic nucleus for gamma rays at one extreme to several kilometers for radio waves at the other. All the EM waves consist of synchronized transverse waves of electric and magnetic fields and travel with the same speed (300,000 km/s) through the space.

Electromagnetic Spectrum

Region	Wavelength	Frequency (Hz)	Energy (eV)
Gamma Rays	< 0.01 nm	> 3×10^{19}	> 10^5
X-Rays	0.01 nm – 1 nm	3×10^{17} - 3×10^{19}	10^3 - 10^5
Ultraviolet	1 nm – 400 nm	7.5×10^{14} - 3×10^{17}	$3 - 10^3$
Visible	400 nm - 700 nm	4.3×10^{14} - 7.5×10^{14}	2 - 3
Infrared	700 nm – 0.1 mm	3×10^{12} - 4.3×10^{14}	0.01 - 2
Microwave	0.1 mm - 0.1m	3×10^9 - 3×10^{12}	10^{-5} - 0.01
Radio	> 0.1m	< 3×10^9	< 10^{-5}

The boundaries for each region shown in the table above are for defining purposes; there is no sharp boundary between any two neighboring regions. The properties of the EM waves gradually vary with the wavelengths.

In the visible part, the wavelength decreases from red toward violet.

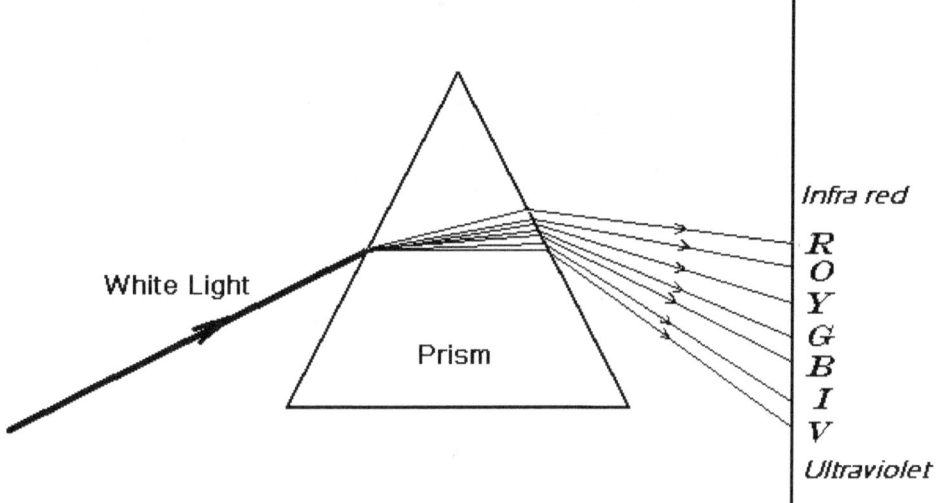

As is evident from the diagram of the prism above, the index of refraction of light varies slightly with color (wavelength). The index of refraction decreases from violet toward red.

Inverse Square Law for Light:

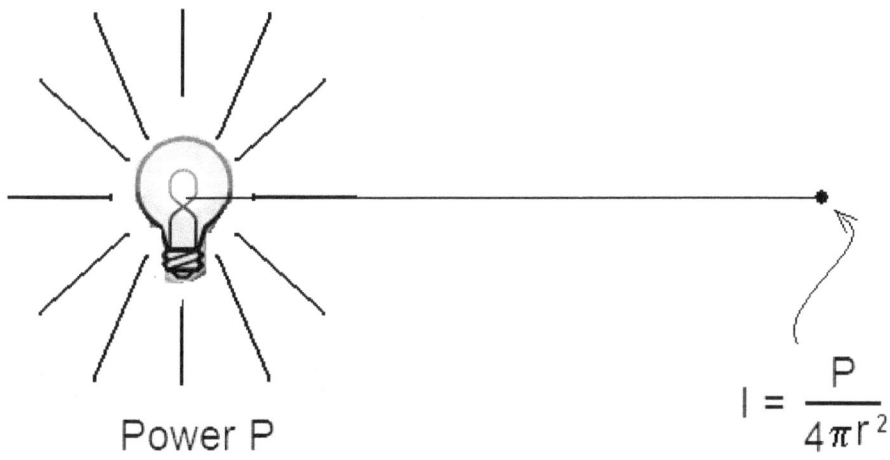

$$I = \frac{P}{4\pi r^2}$$

Let a point source emit light uniformly in all directions. If the power emitted by the source is **P** then the light intensity **I** at a point a distance **r** from the source is given by

$$I = \frac{P}{4\pi r^2}$$

If the intensity due to a point source at two distances r_1 and r_2 are I_1 and I_2 respectively then

$$\frac{I_1}{I_2} = \frac{r_2^2}{r_1^2}$$

Polarization: Polarization is the property exhibited by the _transverse waves_. Hence light and other EM waves can be polarized but sound waves (longitudinal wave) cannot be polarized. There are synthetic and naturally occurring materials, which polarize a beam of light passing through them. Light can also be polarized by reflection and scattering. For the polarized light, the field vectors **E** vibrate in a fixed plane (E-plane) and the **B** vectors vibrate in a plane (B-plane) perpendicular to the E-plane. The E-plane is called _plane of polarization_.

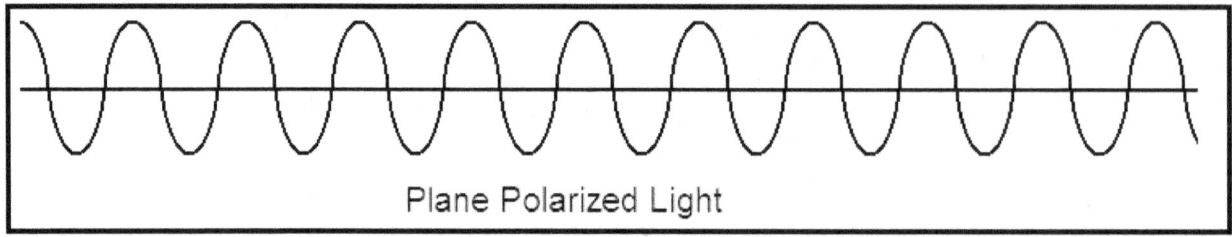

Plane Polarized Light

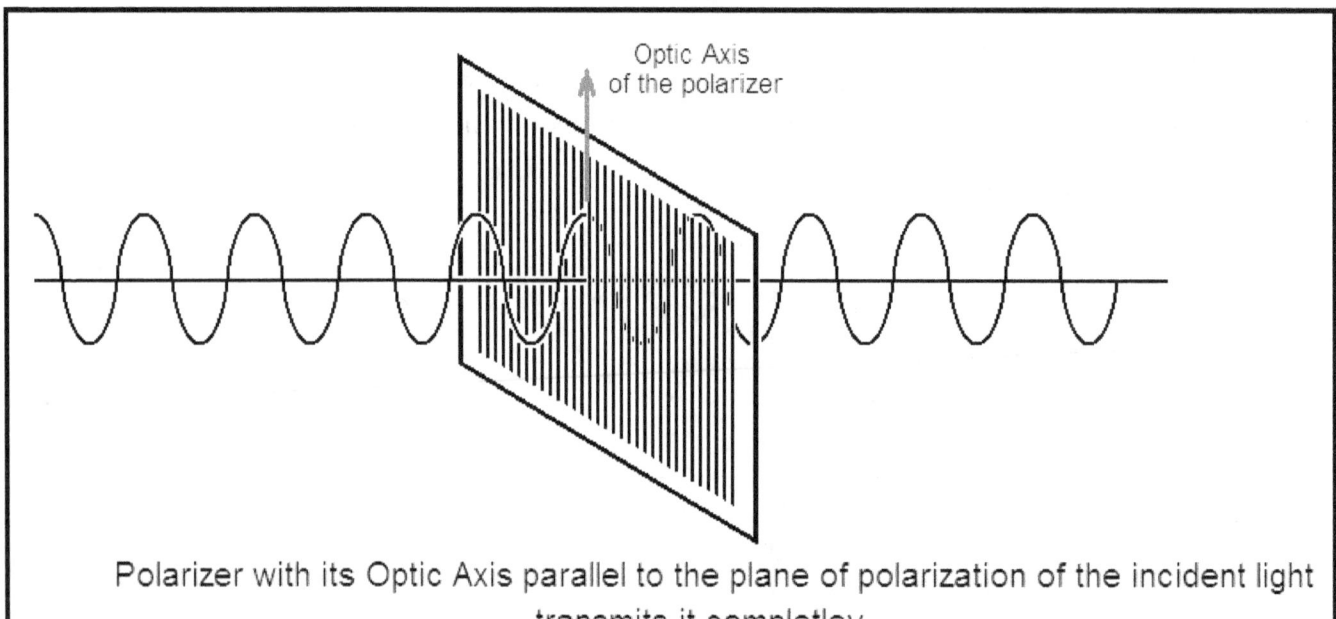

Polarizer with its Optic Axis parallel to the plane of polarization of the incident light transmits it completley

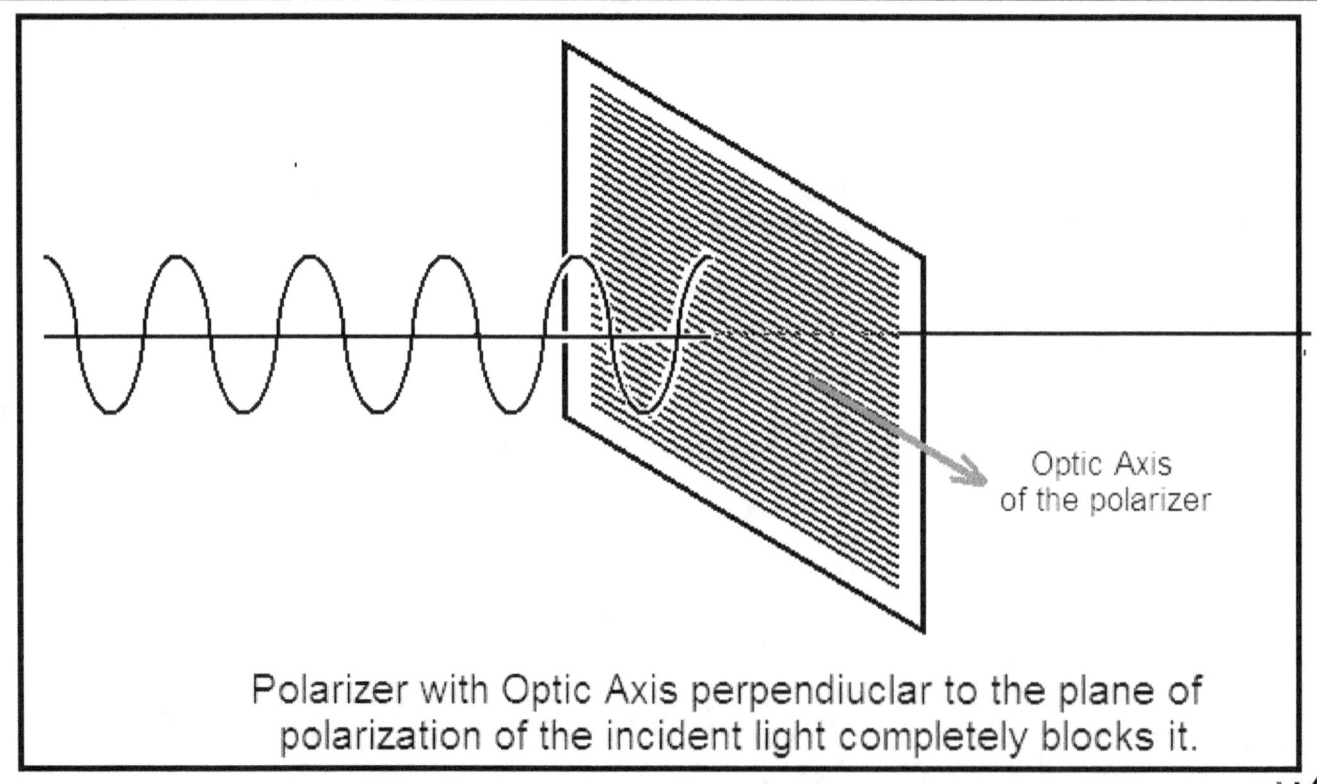

Polarizer with Optic Axis perpendiuclar to the plane of polarization of the incident light completely blocks it.

THIN FILMS:
Phase change in thin film:

Transmission:
No phase change occurs when light wave travels from one medium into another.

Reflection:
- A phase change of π occurs when light is reflected off the surface of an optically denser medium.
- No phase change occurs when light is reflected off the surface of an optically rarer medium.

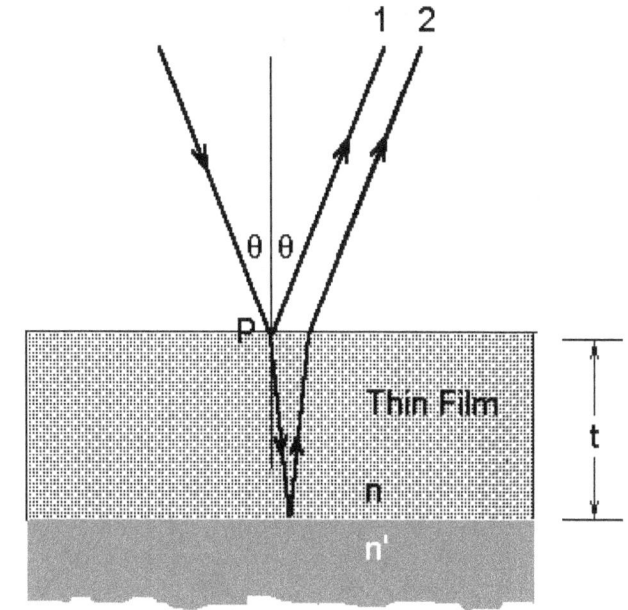

Thin Film Interference:
A thin film of thickness t and index of refraction n is deposited on the surface of another medium of index of refraction n'. As shown in the diagram, a ray of light of wavelength λ, traveling in air, is incident on the surface of the thin film at a near normal incidence, i.e. $\theta \cong 0$.

At the point P, the incident ray splits into the rays 1 and 2. The ray 1 is reflected off the surface of the thin film and the ray 2 travels into the thin film and reflects off the surface of the lower medium and emerges parallel to the ray 1. The ray 2 travels an extra optical path length of approximately 2nt.

Rays 1 and 2 then meet on a screen (retina of eye or film of a camera) and interfere. If this interference is

- **constructive** the thin film appears **highly reflecting** for the wavelength λ
- **destructive** the thin film appears **non-reflecting** for the wavelength λ

Whether the interference is constructive (for highly reflecting film) or destructive (for non-reflecting film) depends on two factors:

(i)   *The path difference 2nt*
(ii)   *The phase changes for ray 1 and 2 upon reflection*

Thin film is **highly reflecting** if
(i)   Both or none of the rays suffer a phase change of π and $2nt = m\lambda$
Or
(ii)   Only one of them suffer a phase change of π and $2nt = (m + \frac{1}{2})\lambda$

Thin film is **non-reflecting** if
(iii) Both or none of the rays suffer a phase change of π and $2nt = (m + \frac{1}{2})\lambda$
Or
(iv)   Only one of them suffer a phase change of π and $2nt = m\lambda$

Some interesting facts & figures
• Soap bubbles (films) and oil floating on water show colorful patterns due to thin film interference effects.
• The microwave was invented after a researcher walked by a radar tube and a chocolate bar melted in his pocket.
• We now know that the eyes of many invertebrates have a structure that lends itself for sensitivity to polarized light.
• The LCD screens of laptops, graphing calculators, and watches emit polarized light. Normally the plane of polarization is at $45°$ from the vertical.
• Some sunglasses use polarizer lenses. The axes of these polarizers are vertical so that they cut of the glare from road surface. This glare contains substantial polarized horizontally.

PROBLEMS:

1. Determine the frequencies for the EM waves of the following wavelengths and identify to what category of EM spectrum they belong to:

(a) 2 m   *[1.5x10^8 Hz, radio waves]*

(b) 100 nm   *[3.0x10^{15} Hz, ultraviolet]*

(c) 1 mm   *[3.0x10^{11} Hz, microwaves]*

(d) 1 x 10^{-15}m   *[3.0x10^{23} Hz, gamma rays]*

(e) 0.1 nm   *[3.0x10^{18} Hz, X-rays]*

(f) 500 nm   *[6.0x10^{14} Hz, visible]*

(g) 1.2 μm   *[2.5x10^{14} Hz, infrared]*

2. A thin coating of a material (n_m = 1.7) on glass (n_g = 1.5) is to be made non-reflecting for the 650 nm wavelength. Determine two possible smallest thicknesses of the film.

3. A thin coating of a material (n_m =1.25) is to be applied on a glass surface (n_g =1.5). Determine the minimum possible thickness of this film so that it is highly reflecting for the wavelengths 450 nm and 600 nm.   *[720 nm]*

4. A thin uniform coating of a material (n_m = 1.6) is to be applied on a glass plate   (n_g = 1.5). Determine the minimum thickness of the coating for which it will be highly reflecting for 550 nm light.

5. A point source emits 5 W of light.
(a) Determine the intensity of light at a distance of 3 m from the source.   *[0.044 W/m²]*

(b) Determine the distance at which the light intensity is 1 W/m². *[0.63 m]*

6. The light intensity at a distance of 5 m from a point source is 6 W/m². Determine the intensity at a distance of 8 m from the source.

Multiple Choice:

1. The light intensity at a distance r from a point source is I. Thus the intensity at a distance 1/3 r would be
(A) 1/3 I   (B) 1/9 I   (C) 3 I   (D) 9 I   (E) 1/6 I

2. Which of the following lists the colors blue (B), red (R), green (G), orange (O), and yellow (Y) in increasing order of wavelength?
(A) BGROY   (B) BGYOR   (C) ROYGB   (D) ORBGY   (E) RBGYO

3. If a beam of white light is dispersed by a prism which color will deviate the most?
(A) Violet   (B) Red   (C) Orange   (D) Yellow   (E) Indigo

4. A 70 nm wavelength belongs to the _____ region of the EM spectrum.
(A) Gamma rays   (B) X-rays   (C) Ultraviolet   (D) Visible   (E) Infrared

5. A 7000 nm wavelength belongs to the _____ region of the spectrum.
(A) Gamma rays   (B) X-rays   (C) Ultraviolet   (D) Visible   (E) Infrared

6. A 10^{-13} m wavelength belongs to the _____ region of the spectrum.
(A) Gamma rays   (B) X-rays   (C) Ultraviolet   (D) Visible   (E) Infrared

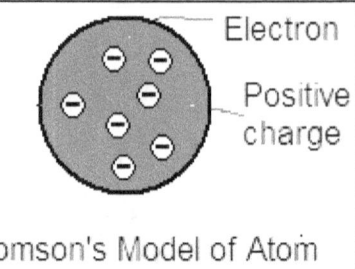

Chapter 54
Atomic Nucleus

J.J. Thomson Model:

After the discovery of electron in 1897, British physicist, **Sir Joseph John "J. J." Thomson** [18 December 1856 – 30 August 1940] proposed a model of an atom. This model later came to be known by different names such as *plum-pudding* and *watermelon* models:

Thomson's Model of Atom
(Plum Pudding Model)

In this model the pudding was the positive charge of the atom and electrons were embedded in it like plums. The total positive charge was equal in magnitude to the total negative charge of the electrons. Hence an atom was a neutral particle.

Sir J.J. Thomson

Rutherford's Planetary Model:

Ernest Rutherford (30 August 1871–19 October 1937), a New Zealand Chemist and Physicist, performed experiment to investigate the scattering of alpha particles by a thin gold foil. Based on the plum pudding model,

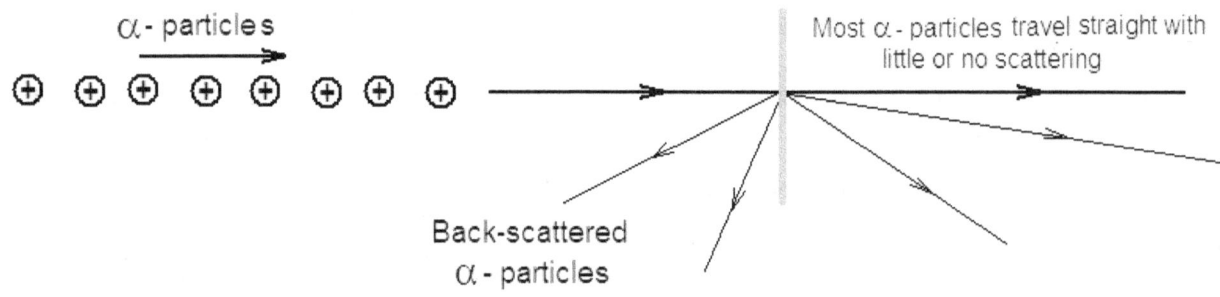

Gold -Foil Target

α- particles

Most α- particles travel straight with little or no scattering

Back-scattered
α- particles

Rutherford expected very little scattering because of large momentum for the alpha particles. However, to his surprise he discovered that some alpha particles scattered through large angles and in fact some of them had actually back-scattered. This was completely inconceivable on the basis of the plum-pudding model.

This remarkable experimental result led Rutherford to revise the atomic model. He could explain the result of his alpha scattering experiment by *nuclear model*.

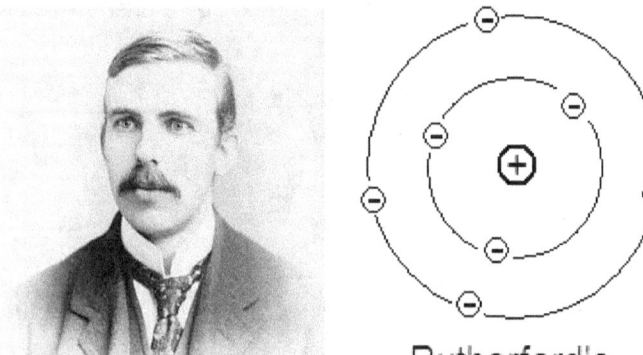

Ernest Rutherford

Rutherford's
planetary model of atom

According to the nuclear model, the positive charge of the atom and most of its mass is concentrated in a very small volume at the center of the atom. This part of the atom came to be called the *nucleus* of the atom. The electrons revolve around the nucleus in orbits similar to the planets going round the sun.

252

This model has since been further refined but basic idea of tiny atomic nucleus at the center of the outer electrons still holds true.

Rutherford Stamp depicting his picture and alpha scattering that lead to the discovery of nucleus.

Some interesting facts & figures
• In 1906, J.J. Thomson received the Nobel Prize in physics for "his theoretical and experimental investigations on the conduction of electricity by gases" • In 1906 Thomson demonstrated that hydrogen had only a single electron per atom. • Ernest Rutherford is known a father of nuclear physics for his discovery of nucleus • Rutherford received Nobel Prize in chemistry in 1908. • He coined the terms alpha and beta in 1899 for two of the three radiation due to radioactivity • In 1919 he was the first person to transmute one element into another. In 1919, he converted nitrogen into oxygen through the nuclear reaction $^{14}N + \alpha \rightarrow {}^{17}O + p$.

Multiple Choice:

1. Which scientists is credited with the discovery of electron?
(A) Albert Einstein   (B) Count Rutherford   (C) Robert Millikan
(D) Max Planck   (E) J. J. Thomson

2. Which scientists is credited with the discovery of atomic nucleus?
(A) Albert Einstein   (B) Count Rutherford   (C) Robert Millikan
(D) Max Planck   (E) J. J. Thomson

3. Which of the following experiments lead to the discovery of the nucleus of atom?
(A) Milikan's oil-drop experiment   (B) Davisson & Germer electron scattering experiment
(C) J.J. Thomson's experiments with cathode rays   (D) Michelson-Morely experiment
(E) Rutherford's experiment on α-scattering by gold foil

4. In 1911, Rutherford discovered that the J.J. Thomson model of atom could not account for
(A) penetration of α-particles through a gold foil   (B) large angle scattering of electrons by a gold foil
(C) large angle scattering of α-particles by a gold foil   (D) penetration of electrons through a gold foil
(E) penetration of photons through a gold foil

Chapter 55
Photoelectric Effect

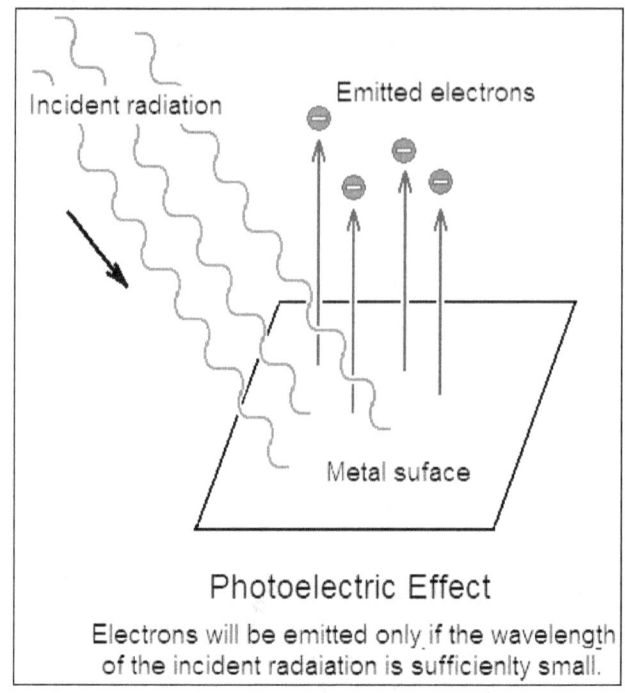

Photoelectric Effect

Electrons will be emitted only if the wavelength of the incident radaiation is sufficienlty small.

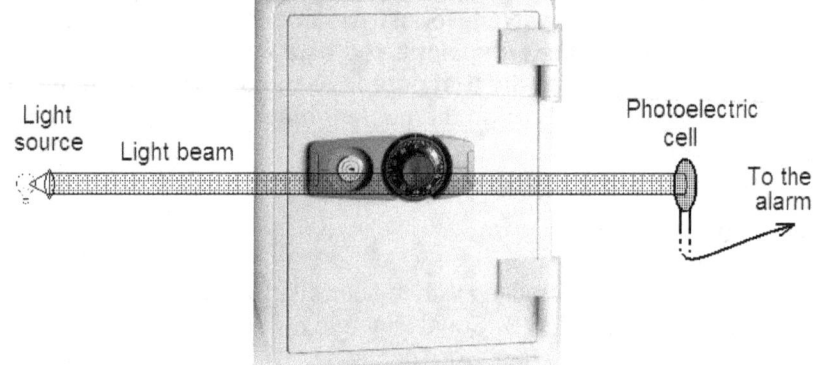

Emission of electrons from the surface of a material when light (or UV) is incident on it is called photoelectric effect.

An experiment on photoelectric effect performed using a vacuum tube with cathode as the photoelectric surface and variable voltage between the anode and the cathode reveals a number of interesting characteristics for this phenomenon, namely:

1. The emission of photoelectron is instantaneous.
2. The photoelectrons have a velocity distribution from zero to certain maximum.
3. The maximum kinetic energy (KE_{max}) of the photoelectron is independent of the intensity of the incident radiation.
4. KE_{max} depends on the frequency of the radiation.
5. If the frequency of the radiation is changed there exists a threshold frequency below which there is no emission of photoelectron however intense the radiation be.
6. The number of photoelectron (hence the photoelectric current) is proportional to the intensity of the radiation.

The wave nature of light failed to explain all of the above characteristics of the photoelectric

effect!!!
Einstein's Theory of Photoelectric Effect and Dual Nature of Light (1905):

Wave theory of light though very successful in explaining the reflection, refraction, interference, and diffraction phenomena, failed to explain the characteristics of the photoelectric effect. Einstein studied the photoelectric effect and concluded that the phenomenon could be explained by assuming that the light energies are available not in a continuous manner but in quanta called 'photon'. The energy (**E**) of the photon, according to Einstein is proportional to the frequency (**f**) of radiation.

$$E = hf, \quad h = 6.63 \times 10^{-34} \text{ J.s}$$

where **h** is the Planck's Constant.
This constant 'h' was introduced by Max Planck to explain the phenomenon of 'black body radiation' which too could not be explained by the then existing theories of physics.

Einstein proposed this revolutionary idea in 1905 along with two other landmark theories of 'special relativity, and 'Brownian motion'.

On the basis of the photon theory one can easily explain the characteristics of the photoelectric effect.

Studying Characteristics of Photoelectric Effect

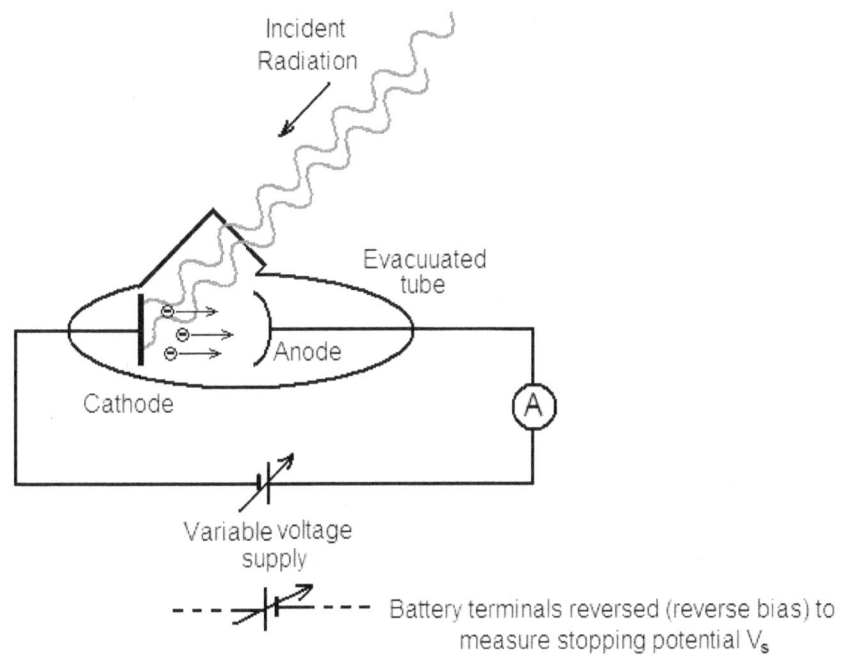

Apparatus for studying photoelectric effect

When a photon of energy **E = hf** is incident on the surface of photosensitive surface, it imparts all its energy to the electron at or below the surface, and disappears.

The electron utilizes this energy in two ways: It uses part of its energy to detach itself from the material. The deeper inside the material the electron is more the energy it needs to exit from the material. An electron at <u>the surface needs minimum</u> amount of energy to leave the material. This minimum energy required to detach the electron from the surface is called the *work function (ϕ)* of the material and depends on the nature of the material.

Thus the photoelectron are emitted with the kinetic energy which is the energy of the photon they absorbed minus the energy the used up to leave the material. The photoelectrons coming from the surface thus have the maximum kinetic energy (KE_{max}) given by the *Einstein's Photoelectric Equation*,

$$KE_{max} = hf - \phi$$

This equation also easily explains the existence of a threshold frequency (f_o) below which there will be no emission of photoelectron. At the threshold frequency a photon has energy hf_o which is just equal to the work function (ϕ) then KE_{max} is zero. Hence no electron can leave the surface at this or lower energy. Thus the threshold frequency can be obtained from the equation

$$hf_o = \phi$$

Determination of KE_{max} and v_{max} for photoelectrons:

The maximum kinetic energy (KE_{max}) can be determined by measuring the stopping voltage (V_s) that would stop the photoelectric current. As shown in the figure above, the battery terminal are reversed and the voltage is gradually increased. The photoelectrons arriving at the collector plate are slowed down. When the reverse voltage is sufficiently high, even the fastest electrons are stopped. This voltage is called stopping potential V_s. As the fastest electrons lose all their kinetic energy they gain the electric potential energy eV_s as they arrive at the anode with zero velocity. Thus

$$\tfrac{1}{2} mv_{max}^2 = KE_{max} = eV_s$$

Relationship Between Momentum (p) and Wavelength (λ) of a Photon:

Arthur Compton later showed that not only is a photon a packet of energy but it also has linear momentum given by

$$p = \frac{h}{\lambda} = \frac{hf}{c}$$

RATE OF EMISSION OF PHOTONS FROM A GIVEN SOURCE OF LIGHT:

For a given frequency, the number of photons is proportional to the intensity of light. (This explains why the number of photoelectron and not the KE_{max} of the photoelectron depend on the intensity of light.)

The number of photons/second (N/s) emitted by the light source is proportional to the power (**P**) of the source

Number of photons emitted per second = P / hf

Take power P in watts; energy hf in joules.

Some interesting facts & figures
• Einstein received Nobel Prize in physics in 1921, ""for his services to Theoretical Physics, and especially for his discovery of the law of the photoelectric effect"
• Photoelectric effect was first observed by Heinrich Hertz in 1887.

> • Many smoke detectors use photoelectric effect. The alarm is set off when either the light reaching the photocell is blocked smoke or light scattered by smoke reaches the photocell.

PROBLEMS:

1. Determine the threshold frequencies and wavelengths for the following metals

(a) silver (ϕ = 4.73 eV)   *[1.14x10^{15} Hz, 263 nm]*

(b) cadmium (ϕ = 4.07 eV)   *[9.82x10^{14} Hz, 305 nm]*

(c) sodium (ϕ = 2.28 eV)   *[5.50x10^{14} Hz, 545 nm]*

2. Determine the energy of a photon for each of the waves of the following wavelengths, in eV

(a) 2 m   (b) 1 mm   (c)   2000 nm   (d) 500 nm   (e) 100 nm   (f) 0.01 nm   (g) 1 pm

3. The work function for Ag is 4.73 eV. In a photoelectric experiment, light of wavelength 200 nm is incident on the Ag-surface.

(a) Determine the energy of a photon of the incident light (in eV).   *[6.2 eV]*

(b) Determine the threshold frequency for the Ag-surface.   *[1.14x10^{15} Hz]*

(c) Determine KE$_{max}$ for the photoelectron.   *[1.47 eV]*

(d) Determine the stopping potential in this experiment.   *[1.47 V]*

4. In a photoelectric experiment, UV light is incident on the Cs (cesium) surface for which the work function is 1.98 eV. The stopping potential was determined to be 1.15 V.

(a) Determine the KE$_{max}$ for the photoelectron

(b) Determine the wavelength of the incident light.

(c) Determine the threshold wavelength for the Cs-surface

5. In a photoelectric experiment, the threshold frequency for a surface is 7.5 x 10^{14} Hz. Light of frequency 8.5 X 10^{14} Hz is incident on the surface

(a) Determine the work function for the surface.   *[3.11 eV]*

(b) Determine the KE_{max} for the photoelectron.   *[0.414 eV]*

(c) Determine the stopping potential for these photoelectron.   *[0.414 V]*

6. A 10 mW laser emits a wavelength of 632.8 nm.
(a) Determine the number of photons/s emitted by the laser.

(b) Determine the linear momentum of each of the photon.

Multiple Choice:

1. The KE_{max} of the photoelectrons is independent of
(A) work function of the metal   (B) frequency of incident radiation
(C) intensity of incident radiation   (D) All of these
(E) None of these

2. The threshold wavelength for a metal surface depends on
(A) the nature of metal   (B) the incident wavelength
(C) incident light intensity   (D) All of these
(E) None of theses.

3. In a photoelectric experiment, if the intensity of radiation is increased
(A) the stopping potential is increased
(B) the threshold frequency is increased
(C) the KE_{max} is increased
(D) the photoelectric current is increased
(E) the work function is increased

4. In a photoelectric experiment, if the frequency of radiation is increased
(A) the stopping potential is increased
(B) the threshold frequency is increased
(C) the photoelectric current is increased
(D) the work function is increased
(E) All of the above

5. In a photoelectric experiment, the stopping potential is found to be 1.25 eV. The KE_{max} for the photoelectrons must be
(A) 1.25 V
(B) 1.25x1.6x10^{-19} V

(C) 1.25 / 1.6x10^{-19} V
(D) 1.6x10^{-19} / 1.25 V
(E) The information is insufficient to determine the answer

Chapter 56
Energy Levels
and
Emission & Absorption of Photons

Following the discovery of the atomic nucleus by Rutherford and the failure of the classical electrodynamics to support the planetary model for the hydrogen atom, Danish physicist Niels Bohr [7 October 1885 – 18 November 1962] proposed a model for the hydrogen atom using quantum ideas so that it could withstand the objections from electrodynamics as well explain the pattern of emission spectral lines of the light from the hydrogen gas discharge.

According to the Bohr's model of H-atom, there are certain discrete orbits around the nucleus, in which electron does not radiate energy even though it is accelerating. These orbits are given quantum numbers **n = 1, 2, 3 ...** in increasing order of the size of the orbits. An atom has certain energy associated with the electron in such an orbit. Hence different stationary orbits correspond to different energy levels for an atom.

The energy of the H-atom in any orbit is given by

$$E = \frac{-13.6}{n^2} eV$$

The orbit closest to the nucleus (smallest orbit) is said to be the ground state of the atom and has the lowest possible energy state. Theoretically, there are infinite number of orbits around a nucleus; as one moves to higher orbits the energy level increases.

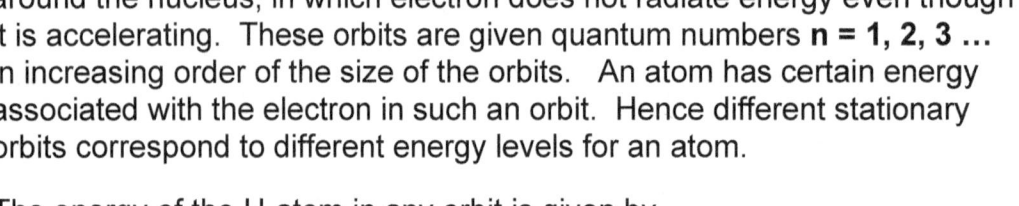

$E_{inf} = 0 \text{ eV}$ — Continuum — n = infinity

$E_5 = -0.54 \text{ eV}$ ——————— n = 5
$E_4 = -0.85 \text{ eV}$ ——————— n = 4

$E_3 = -1.51 \text{ eV}$ ——————— n = 3

$E_2 = -3.4 \text{ eV}$ ——————— n = 2

$E_1 = -13.6 \text{ eV}$ ——————— n = 1

Energy level diagram for H-atom. Only six levels are shown

Emission of radiation by an atom

When an electron is bumped to a higher orbit, the atom is said to be in an excited state. The electron later returns to the ground state directly or through one or more stationary orbits to the ground state. As the electron descends to lower energy states, the difference of energy between the two levels is radiated out as a photon. Thus,

$$hf = \left(E_i - E_f\right) \quad \text{or} \quad \frac{hc}{\lambda} = \left(E_i - E_f\right)$$

The diagram below shows an excited atom of hydrogen in which electron has been raised (by heating or passing an electric discharge through hydrogen gas) to the n=3 state.
One can calculate the three possible wavelengths emitted by the hydrogen atoms in n = 3 excited states as below.

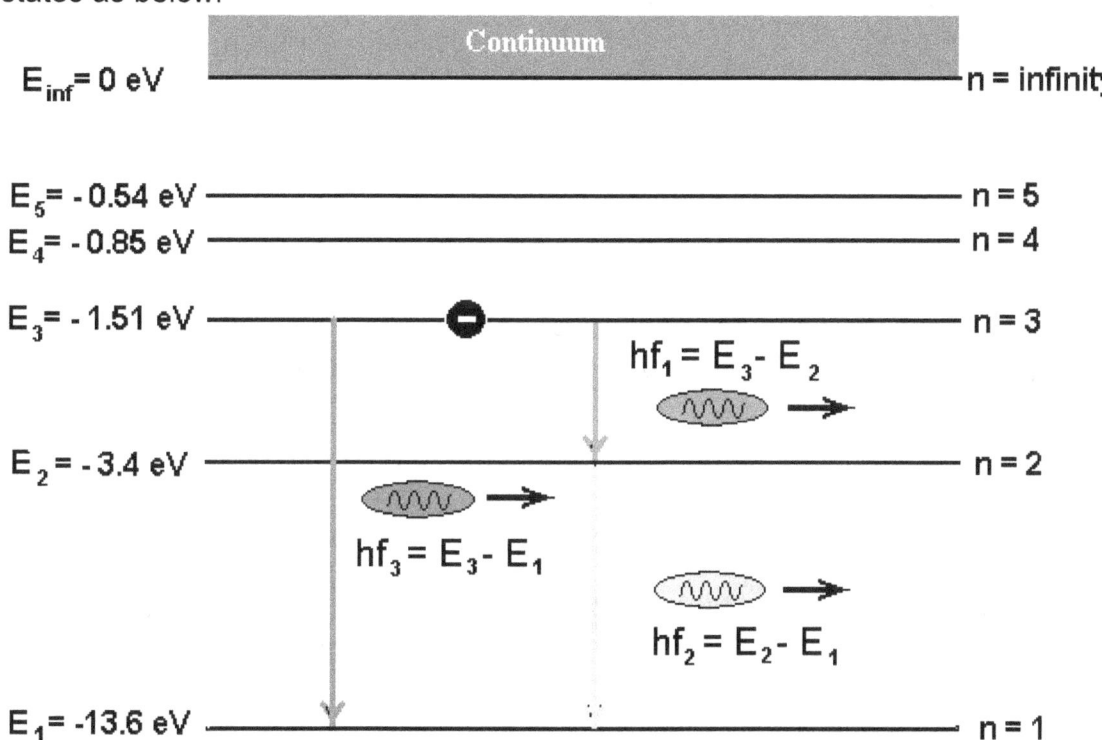

For n = 3 ➔ n = 2 transition
$hf_1 = E_3 - E_2$ ➔ $6.63 \times 10^{-34} f_1 = (-1.51) - (-3.4) = 1.89$ eV $= 3.024 \times 10^{-19}$ J
➔ $f_1 = (3.024 \times 10^{-19}) / (6.63 \times 10^{-34}) = 4.56 \times 10^{14}$ Hz ➔ $\lambda_1 = c/f_1 = (3 \times 10^8)/(4.56 \times 10^{14})$
➔ $\lambda_1 = 6.577 \times 10^{-7}$ m = 657.7 nm

For n = 2 ➔ n = 1 transition
$hf_2 = E_2 - E_1$ ➔ $6.63 \times 10^{-34} f_2 = (-3.4) - (-13.6) = 10.2$ eV $= 16.32 \times 10^{-19}$ J
➔ $f_2 = (16.32 \times 10^{-19}) / (6.63 \times 10^{-34}) = 2.46 \times 10^{15}$ Hz ➔ $\lambda_2 = c/f_2 = (3 \times 10^8)/(4.56 \times 10^{15})$
➔ $\lambda_2 = 1.219 \times 10^{-7}$ m = 121.9 nm

For n = 3 ➔ n = 1 transition
$Hf_3 = E_3 - E_1$ ➔ $6.63 \times 10^{-34} f_3 = (-1.51) - (-13.6) = 12.09$ eV $= 19.344 \times 10^{-19}$ J
➔ $f_3 = (19.344 \times 10^{-19}) / (6.63 \times 10^{-34}) = 2.918 \times 10^{15}$ Hz ➔ $\lambda_3 = c/f_3 = (3 \times 10^8)/(2.918 \times 10^{15})$
➔ $\lambda_3 = 1.028 \times 10^{-7}$ m = 102.8 nm

Absorption of radiation by an atom

A reverse process of emission of photons is the absorbtion of an incident photon by an atom. For absorption to take place the, the energy of the photon must be exactly equal to the energy difference between the two levels through which an electron goes up. If the energy of the photon is <u>more or less</u> than the energy difference between the two levels the photon will pass through the gas without getting absorbed.

Some interesting facts & figures
• **Niels Bohr received Nobel Prize in Physics in 1922 "for his services in the investigation of the structure of atoms and of the radiation emanating from them"** • **He provided the liquid-drop model for atomic nucleus.** • **His son Aage Bohr too won Nobel Prize (shared with other two physicists) in Physics "for the discovery of the connection between collective motion and particle motion in atomic nuclei and the development of the theory of the structure of the atomic nucleus based on this connection".**

PROBLEMS:

1. Determine the energy of the H-atom in n=1, 2, and 7 states.

[-13.6 eV, -3.4 eV, -0.28 eV]

2. Determine the energy, and the wavelength of a photon that would excite the H- atom from n=2 state to the n=7 state.

3. Determine the energy of the photon emitted when the electron in the H-atom jumps from n = 7 state to the ground state (n=1). *[2.13x10^{-18} J]*

4. The energy level diagram for a quantum mechanical system is shown below. An electron is in the n = 3 excited state. Calculate all possible wavelengths emitted as the photon returns to the ground state.

E_4= 11 eV ———————————— n = 4

E_3= 8 eV ———⊖——————— n = 3

E_2= 5 eV ———————————— n = 2

E_1= 2 eV ———————————— n = 1

5. The energy level diagram for a quantum mechanical system is shown below. All the atoms are in ground states. Radiation containing all wavelengths corresponding to the photon energies between 1 to 10 eV is passed through the gas. Which of the incident photons will be absorbed by the gas? *[1 eV, 2 eV, 3 eV, 5 eV, 6 eV]*

$E_4 = 8$ eV ———————————————— n = 4

$E_3 = 7$ eV ———————————————— n = 3

$E_2 = 5$ eV ———————————————— n = 2

$E_1 = 2$ eV ————(–)———————— n = 1

Multiple Choice:

1. An electron in an atom jumps from energy level E_1 to a lower energy level E_2. The wavelength λ of the photon thus emitted is given by
(A) $h / [c(E_1 - E_2)]$ (B) $(E_1 - E_2) / (hc)$ (C) $hc / (E_1 - E_2)$
(D) $hc / (E_1 + E_2)$ (E) $1/[hc(E_1 - E_2)]$

2. The theory that electrons orbit the nucleus of an atom in certain discrete orbits without losing energy was first proposed by
(A) Max Planck   (B) Count Rutherford   (C) Robert Millikan
(D) Niels Bohr   (E) J.J. Thomson

3-4. For a quantum mechanical system shown, the electron is excited to the n = 4 state.

3. For which transition will the system emit the longest wavelength photon?
(A) 4 → 3   (B) 4 → 1   (C) 4 → 2   (D) 3 → 1
(E) 3 → 2

$E_4 = 8$ eV ———(–)———————— n = 4

$E_3 = 7$ eV ———————————————— n = 3

$E_2 = 5$ eV ———————————————— n = 2

4. For which transition will the system emit the shortest wavelength photon?
(A) 4 → 3   (B) 4 → 1   (C) 4 → 2   (D) 3 → 1
(E) 3 → 2

$E_1 = 2$ eV ———————————————— n = 1

Chapter 57
Wave-Particle Duality – X-ray Production – Compton Scattering

The photoelectric effect and Compton scattering established that the EM radiation act like waves as well as particles. The momentum of the photon is given by

$$p = \frac{h}{\lambda}$$

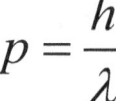

de Broglie Hypthesis:

Some years later, French physicist **Louis-Victor-Pierre-Raymond, 7th duc de Broglie**, [15 August 1892 – 19 March 1987] proposed a hypothesis that *the converse may be true, that is, the particles may behave like waves!!*

He proposed that if that is true it may be that the wavelength of the particles is given by the same formula as one used for finding the momentum of the wave. Hence he proposed that the wavelength of a particle having a momentum p would be

$$\lambda = \frac{h}{p}$$

Verification of the de Broglie Hypothesis:
His theory was soon verified by the <u>Davisson-Germer experiment</u> in which the electrons scattered by a crystal showed wave-like interference effect.

Electron Microscope: One remarkable application of de Broglie's discovery of wave nature of electron is the invention of electron microscope. For a microscope, the resolving power (ability to see details of an object) is limited by the wavelength of the light. The smallest details a microscope can show cannot be less than the wavelength of radiation used. The smallest wavelength in visible spectrum is 400 nm. Hence any detail that is smaller than 400 nm is impossible to see by the use of optical microscope. Electron Microscope uses electron beam instead of visible light. The electron beam accelerated to high energy by voltage as high as 100,000 volts, behaves like wave of a much shorter wavelength than visible light waves and hence provides much greater resolution.

Louis de Broglie

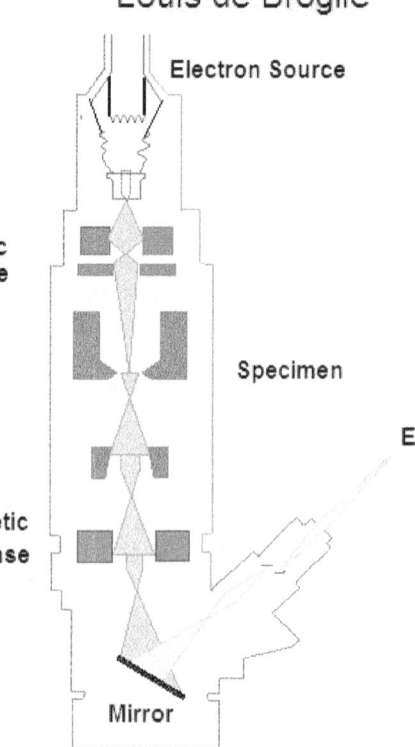

The de Broglie's wavelength for an accelerated electron as in an *Electron Microscope*:

If an electron (charge -e and mass m) is accelerated from rest, through accelerating voltage V then

$$eV = \tfrac{1}{2}mv^2 \quad \rightarrow \quad momentum...p = mv = \sqrt{2meV}$$

Thus using the deBroglie equation $\lambda = \dfrac{h}{p}$ we get, $\qquad \lambda = \dfrac{h}{\sqrt{2meV}}$

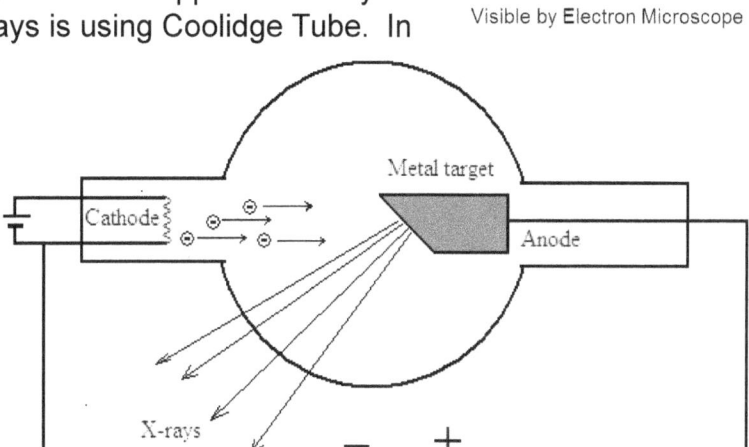

Note: The de Broglie wavelength for the electron beam in a typical electron microscope is of the order of a few picometers and resolution about 100,000 times higher than that of the most powerful optical microscope.

Blood Cells
Visible by Electron Microscope

X-ray Production: When a high-speed electron beam is topped suddenly in a target. One common method to produce X- rays is using Coolidge Tube. In the tube electrons are produced by passing current through a filament. These electrons are accelerated through a high voltage (V) hat may range from 10, 000 to a million volts. The electron beam accelerates to a high speed and hit a metal target that may also serve as the anode. The sudden deceleration of electrons causes the energy of the electrons to be liberated as X-rays. The wavelength of the x-ray this produced varies from a long wavelengths (soft x-rays) to a definite minimum value λ_{min} at the hard x-ray end. The reason for this spread of x-ray photon energies is that electron may lose some of its KE to heat while the rest appears as x-ray photon. When all the avilabale KE of the electron, KE = eV, turns into x-ray photon, this photon has maximum energy or the shortest wavelength λ_{min}. Thus,

Coolidge Tube - Production of X-rays

For the most energetic X-ray photon,

$$\boxed{hc/\lambda_{min} = eV}$$

This equation can be used to calculate λ_{min} or V.

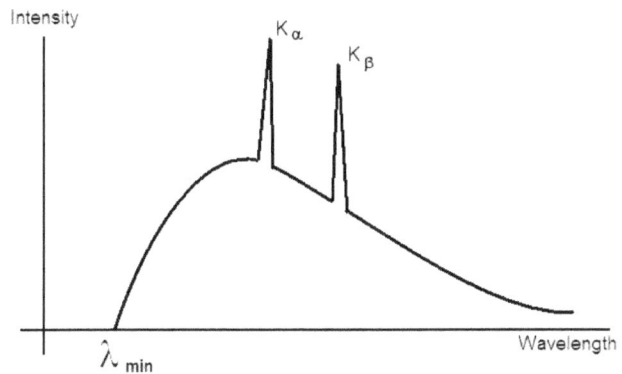

Spectrum of X-rays produced in Coolidge Tube

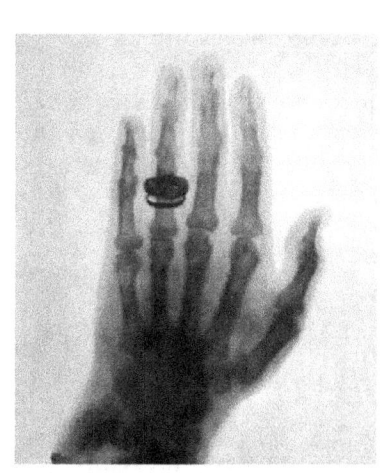

First X-ray picture
Mrs. William Roentgen's hand

Compton Scattering: Discovered by American Physicist Arthur Holly Compton [September 10, 1892 – March 15, 1962]

In **Compton scattering** an x-ray photon collides with a free electron in a metal target. The collision is just like two balls colliding on a pool table. The scattered x-ray photon has its wavelengths increased due to the loss of some energy to the electron. The change in wavelength is a function of the scattering angle θ from the initial direction.

$$\lambda' - \lambda = \frac{h}{m_o c}(1 - \cos\theta)$$

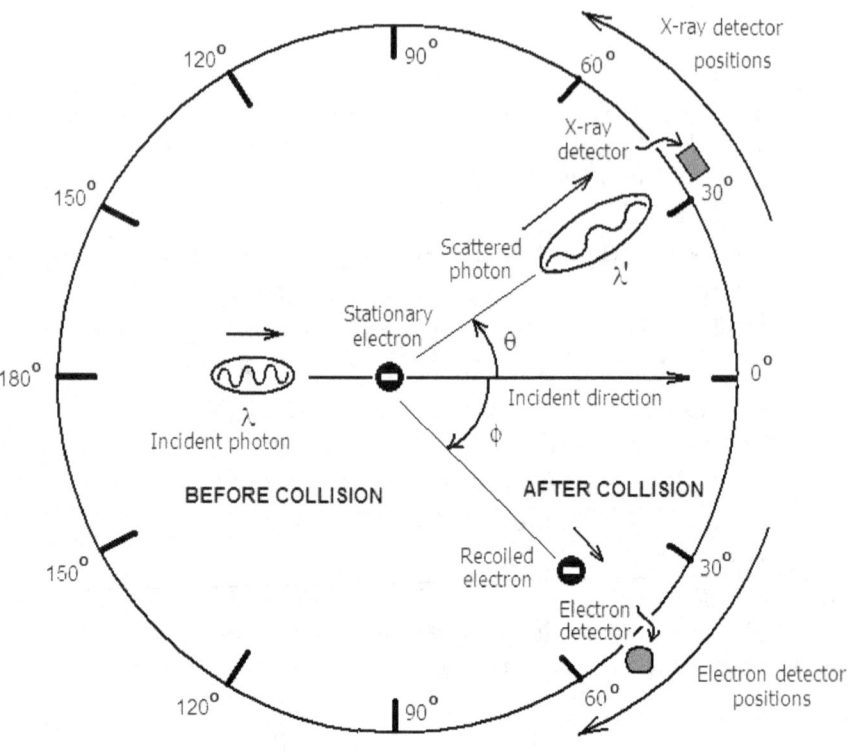

λ' is the wavelength of the scattered photon
λ is the wavelength of the incident photon
θ is scattering angle
m_o is the rest mass of electron
c is the speed of light

The quantity $h/(m_o c)$ = 0.00243 nm is called the Compton wavelength (λ_c) of electron. The Compton wavelength is indicative of the scale of the wavelength changes in the Compton scattering.

Kinetic Energy of the Recoiled Electron (KE_e)

The recoil energy of the scattered electron is equal to the loss of energy of the photon. Therefore,
   KE_e = Energy of the incident photon – Energy of the scattered photon

$$KE_e = \frac{hc}{\lambda} - \frac{hc}{\lambda'}$$

Some interesting facts & figures
• Compton earned the 1927 Nobel Prize in Physics for the discovery of Compton Effect.
• Compton scattering is of prime importance to radiobiology.
• Arthur Compton served as Chancellor of Washington University in St. Louis from 1945 to 1953.
• Louis e Broglie received Nobel Prize in Physics for discovery of wave nature of electron in 1929

PROBLEMS:

1. Determine the de Broglie wavelength for a particle of mass 0.1 microgram moving with a speed of 600 m/s.
                              $[1.11 \times 10^{-29}\ m]$

2. Determine the de Broglie wavelength for an electron accelerated through a potential difference of 12,000 V.

3. Determine the de Broglie wavelength for a proton accelerated through a potential difference of 12,000 V.   *[2.61x10^{-13} m]*

4. Determine the de Broglie wavelength for a 50-kg person walking at a speed of 1.5 m/s

5. A photon of energy 25 keV is incident on a target in a Compton scattering experiment.
(a) Determine the wavelength of the incident photon.   *[0.0497 nm]*

(b) Determine the change in wavelength for a photon scattered at an angle of 90°.
                                                                          [0.00243 nm]]

(c) Determine the percent change in the wavelength of the incident photon when scattered through 90°.   *[4.9%]*

(d) Determine the recoil energy of the scattered electron.   *[1.152 keV]*

Multiple Choice:

1. A photon of frequency f is scattered by a stationary electron in a Compton Effect experiment. The frequency of the scattered photon is f'. The kinetic energy of the recoiling electron is
(A) Zero   (B) hf   (C) hf'   (D) hf – hf'   (E) hf + hf'

2. The discovery of the Compton Effect established that
(A) Photon has energy   (B) Photon has momentum
(C) Electron has wave nature   (D) Electron has particle nature
(E) Electron has negative charge

3. According to the Compton Effect equation, at what angle will the change in the wavelength of the photon equal the Compton wavelength?
(A) 0°   (B) 45°   (C) 90°   (D) 135°   (E) 180°

4. According to the Compton Effect equation, at what angle will the change in the wavelength of the photon equal twice the Compton wavelength?
(A) 0°   (B) 45°   (C) 90°   (D) 135°   (E) 180°

3. According to the Compton Effect equation, at what angle will there be no change in the wavelength of the incident photon?
(A) 0°   (B) 45°   (C) 90°   (D) 135°   (E) 180°

Chapter 58
Nuclear Structure

An atom consists of a **nucleus** and electrons in various shells around it. The nucleus contains all the positive charge of the atom and most of its mass. Each of the electrons carries a negative charge and very little mass.

Inside the Nucleus

- What constituted the nucleus was not very well known until the discovery of the neutral particle called **neutron** by James Chadwick in 1932.
- It was then established that the nucleus contains two kinds of particles --- **protons and neutrons**.
- A **proton** has the positive elementary charge -- equal and opposite to that on an electron -- and almost same mass as the hydrogen atom.
- The **neutron** is electrically neutral and is slightly heavier than the proton.
- For a neutral atom, there are as many protons in the nucleus as there are electrons in the orbits around the nucleus.
- The **protons** and **neutrons** in a nucleus are together called as **nucleon**.

⊕ Proton   ⋒ Neutron

$Z = 3$
$A = 7$

$^{7}_{3}\text{Li}$ nucleus

Atomic Number (Z) and Mass Number (A) for a Nucleus

- The total number of protons in the nucleus is called the *atomic number* (**Z**)
- The total number of nucleon is called the mass number of the nucleus (**A**).
- Thus the number of neutrons in the nucleus is **N = A - Z**.
- Different elements have different atomic numbers.
- There are no electrons in the nucleus contrary to what was assumed before the discovery of the neutrons.

Nuclear Symbols

The symbol for a nucleus is same as that of the element to which it belongs. The chemical symbol includes two numbers as upper and lower prefixes. The lower prefix is the atomic number (Z) and the upper prefix is the mass number (A) of the nucleus as shown below:

$$^{A}_{Z}X$$

For one of the uranium isotopes, its nucleus has 92 protons and 146 neutrons. Hence its atomic number is Z = 92 and mass number is A = 92 + 146 = 238. Thus its nuclear symbol is:

$$^{238}_{92}U$$

What holds the nucleons together!

However, a problem immediately arises when it is assumed that there are protons in the nucleus. Since protons repel each other the nucleus will break apart. Then how are the protons held together against the repulsive forces in the nucleus. This lead to the discovery of the new fundamental force called **nuclear strong force**.

The **nuclear strong force** is the strongest of all the known fundamental forces. It is short range force. This acts between the nucleons only when they are closer than certain distance. This distance is of the order of the size of the nucleus. The nuclear force does not distinguish between the protons and neutrons.

Another nuclear force is called **nuclear weak force**. This force is responsible for the beta decay in which a neutron gives rise to proton and electron (beta particle).

Neutrino The Ghost Particle
In the study of the beta decay it was observed that some of the mass in the beta decay was not accounted for. Also linear momentum and angular momentum were not conserved. It was thus theorized (1934) by the physicists that there must be another particle produced in the reaction but goes undetected. Enrico Fermi named this particle **neutrino** (*little neutral one*). This particle has little or no rest mass but carries relativistic mass. This particle carries no charge and reacts very little with the matter. These properties make it very difficult to detect the neutrino. Reins and Cowman in 1953 detected neutrino and confirmed its existence.

Isotopes
The electrons around nucleus take part in chemical reactions. The nucleus is unaffected in any chemical reaction. Hence the chemical properties of an element depend on the electrons around the nucleus. However, the number of electrons and their energies depend on the number of proton on the nucleus. Hence two atoms having same number of protons and different number of neutrons will have identical chemical properties and are called **isotopes**.
Thus, Isotopes are atoms or nuclei having same atomic number (same Z) but different mass numbers (different A) hence different number of neutrons.

Examples of isotopes:
Isotopes of Hydrogen (*H*)

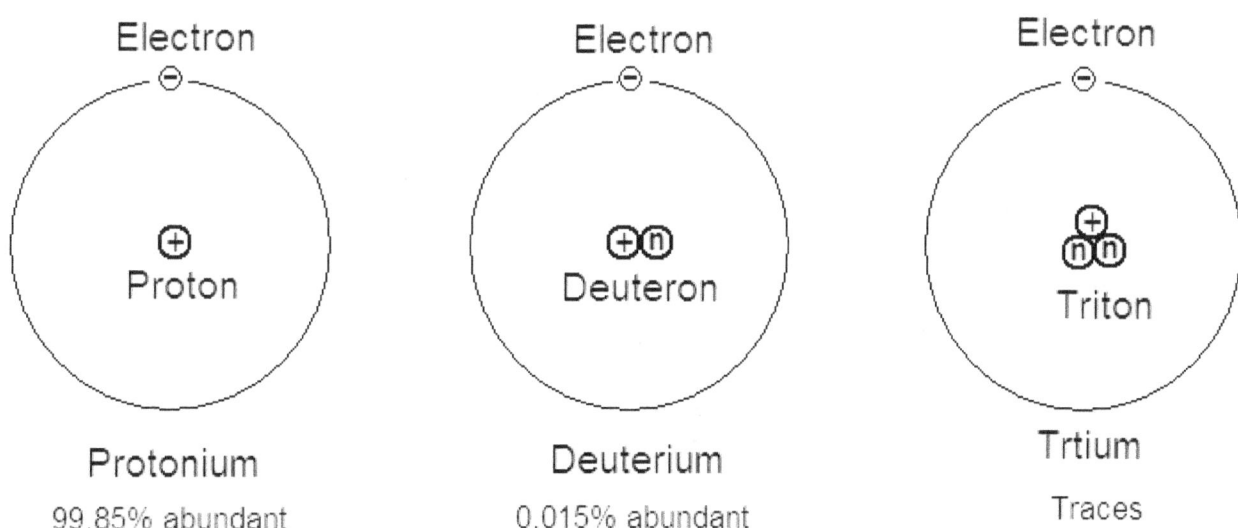

The heavier isotopes of hydrogen are also referred to as *heavy hydrogen*. Water that has heavy hydrogen in its molecules (H_2O) is called *heavy water* and has important uses in nuclear power plants. Heavy water is chemically same as ordinary water.

Isotopes of Helium (He):

$_2^4He$ (99.999% abundant); $_2^3He$; $_2^6He$

Isotopes of Phosphorus (P): $_{15}^{31}P$ (nearly 100 % abundant); $_{15}^{30}P$ (short lived)

Isotopes of Yttrium (Y): Only one isotope $_{39}^{89}Y$

Isotopes of Indium (In): $_{49}^{113}In$ (4.3% abundant) ; $_{49}^{115}In$ (95.7% abundant)

Isotopes of Uranium (U): $_{92}^{238}U$ (99.28% abundant); $_{92}^{235}U$ (0.72% abundant); $_{92}^{234}U$ (short lived) $_{92}^{232}U$ (short lived)

Some interesting facts & figures
• **Hydrogen is the most abundant element in the Universe (75%).** • **The only letter not appearing on the Periodic Table is the letter "J".** • **All the matter that makes up the human race could fit in a sugar cube if all the empty atomic space is removed.** • **More than 50 trillion neutrinos coming from the sun pass through the human body every second.**

Multiple Choice:

1. Which of the following particles are present in the nucleus of an atom?
(A) Electrons only   (B) Protons only   (C) Neutrons only
(D) Protons and neutrons only   (E) Electrons, protons, and neutrons

2. Which force holds the nucleons together in a nucleus?
(A) Gravitational force   (B) Electric force   (C) Magnetic force
(D) Nuclear strong force   (E) Nuclear weak force

3. Which force causes beta decay?
(A) Gravitational force   (B) Electric force   (C) Magnetic force
(D) Nuclear strong force   (E) Nuclear weak force

4. Which particle initially went undetected in beta decay?
(A) Electron   (B) Proton   (C) Neutron
 (D) Neutrino   (E) Alpha particle

5. The number of protons in the nucleus of an atom is called the atom's
(A) Atomic number   (B) Mass number   (C) Lepton number
(D) Principal quantum number   (E) Spin quantum number

Chapter 59
Nuclear Reactions

In any nuclear reactions there are a number of quantities, which are conserved. Two of these quantities are **mass number** (i.e. number of nucleon) and **charge.**

Total of mass numbers (**A**) and total of atomic number (**Z**) should be balanced on the two sides for any nuclear reaction.

Balancing Nuclear reactions (*This is easier than chemical ones!*)
In any nuclear reaction, including radioactivity, conserve the following quantities in order to balance the equation for reaction:

1. Nuclear charge (the total of atomic numbers)
2. Number of nucleons (the total of mass numbers)

Radioactivity: This phenomenon involves spontaneous emission of three kinds of radiation known as **alpha (α), beta (β), and gamma (γ)** radiations.

Apha-decay $\left[\alpha = {}_{2}^{4}He\right]$

Emission of an α - particle from the nucleus decreases its atomic number by 2 and lowers its mass number by 4.

$$ {}_{Z}^{A}X \rightarrow {}_{Z-2}^{A-4}y + {}_{2}^{4}He $$

Beta⁻ - decay $\left[\beta^{-1} = electron \; {}_{-1}^{0}e\right]$
Emission of a β^-- particle increases the atomic number by 1 and leaves the mass number unaffected.

$$ {}_{Z}^{A}X \rightarrow {}_{Z+1}^{A}y + {}_{-1}^{0}e + \overline{\nu} \text{ (anti-neutrino)} $$

Beta⁺- decay $\left[\beta^{+1} = positron \; {}_{+1}^{0}e\right]$
Emission of a β^+- particle decreases the atomic number by 1 and leaves the mass number unaffected.

$$ {}_{Z}^{A}X \rightarrow {}_{Z-1}^{A}y + {}_{+1}^{0}e + \nu \text{ (neutrino)} $$

Gamma – decay $\left[photon = {}_{0}^{0}\gamma\right]$

Following a nuclear reaction, the daughter nucleus may be left in an excited state (too much energy). The excited nucleus may return the ground state by releasing the excess energy as a single gamma ray photon. The emission of a gamma ray photon thus leaves the mass number as well as the atomic number unaffected.

$$ {}^{A} \qquad\qquad {}^{A} $$

$$_{Z}X^{*} \rightarrow {}_{Z}X + \gamma$$

Examples of α-decay:

$$_{92}^{238}U \longrightarrow {}_{90}^{234}Th + {}_{2}^{4}He$$

$$_{88}^{226}Ra \longrightarrow {}_{86}^{222}Rn + {}_{2}^{4}He$$

$$_{84}^{215}Po \longrightarrow {}_{82}^{211}Pb + {}_{2}^{4}He$$

$$_{62}^{147}Sm \longrightarrow {}_{60}^{143}Nd + {}_{2}^{4}He$$

$$_{92}^{235}U \longrightarrow {}_{90}^{231}Th + {}_{2}^{4}He$$

$$_{84}^{210}Po \longrightarrow {}_{82}^{206}Pb + {}_{2}^{4}He$$

Examples of β^{-}-decay: $\left(\beta^{-} = {}_{-1}^{0}e, \ \bar{\nu} = \text{anti-neutrino} \right)$

$$_{6}^{14}C \longrightarrow {}_{7}^{14}N + \beta^{-} + \bar{\nu}$$

$$_{27}^{60}Co \longrightarrow {}_{28}^{60}Ni + \beta^{-} + \bar{\nu}$$

$$_{53}^{131}I \longrightarrow {}_{54}^{131}Xe + \beta^{-} + \bar{\nu}$$

$$_{36}^{89}Kr \longrightarrow {}_{37}^{89}Rb + \beta^{-} + \bar{\nu}$$

$$_{49}^{115}In \longrightarrow {}_{50}^{115}Sn + \beta^{-} + \bar{\nu}$$

Examples of β^+-decay: $\left(\beta^+ = {}^{\ \ 0}_{+1}e, \ \nu = \text{neutrino} \right)$

$$ {}^{12}_{7}N \longrightarrow {}^{12}_{6}C + \beta^+ + \nu $$

$$ {}^{55}_{27}Co \longrightarrow {}^{55}_{26}Fe + \beta^+ + \nu $$

$$ {}^{13}_{7}N \longrightarrow {}^{13}_{6}C + \beta^+ + \nu $$

Examples of γ-decay:

$$ {}^{12}_{6}C^* \longrightarrow {}^{12}_{6}C + \gamma $$

$$ {}^{65}_{28}Ni^* \longrightarrow {}^{65}_{28}Ni + \gamma $$

Artificial transmutation: It is a procedure of using nuclear reactions to change the atomic number of the nucleus there by changing one element into another. This was a dream of alchemists in the middle ages when they made futile efforts to convert iron, mercury and other elements into gold. They failed basically because they were trying to use chemical reactions which affect only the outer electrons in an atom whereas the elements can be changed only by changing the atomic number of the nucleus. The artificial transmutation has produced many new elements beyond the naturally occurring elements in the periodic table.

Some Nuclear Reactions:

$$^{1}_{1}\text{H} + ^{7}_{3}\text{Li} \longrightarrow ^{4}_{2}\text{He} + ^{4}_{2}\text{He}$$

$$^{9}_{4}\text{Be} + ^{1}_{0}\text{n} \longrightarrow ^{10}_{4}\text{Be} + \gamma$$

$$^{9}_{4}\text{Be} + \gamma \longrightarrow ^{8}_{4}\text{Be} + ^{1}_{0}\text{n}$$

$$^{9}_{4}\text{Be} + ^{2}_{1}\text{H} \longrightarrow ^{10}_{4}\text{Be} + ^{1}_{1}\text{H}$$

$$^{9}_{4}\text{Be} + ^{1}_{1}\text{H} \longrightarrow ^{4}_{2}\text{He} + ^{6}_{3}\text{Li}$$

$$^{10}_{5}\text{B} + ^{1}_{0}\text{n} \longrightarrow ^{4}_{2}\text{He} + ^{7}_{3}\text{Li}$$

$$^{10}_{5}\text{B} + ^{1}_{1}\text{H} \longrightarrow ^{4}_{2}\text{He} + ^{7}_{4}\text{Be}$$

$$^{1}_{1}\text{H} + ^{7}_{3}\text{Li} \longrightarrow ^{4}_{2}\text{He} + ^{4}_{2}\text{He}$$

$$^{12}_{6}\text{C} + ^{1}_{0}\text{n} \longrightarrow ^{13}_{6}\text{C} + \gamma$$

$$^{13}_{6}\text{C} + ^{1}_{1}\text{H} \longrightarrow ^{13}_{7}\text{N} + ^{1}_{0}\text{n}$$

$$^{18}_{8}\text{O} + ^{1}_{1}\text{H} \longrightarrow ^{18}_{9}\text{F} + ^{1}_{0}\text{n}$$

$$^{65}_{29}\text{Cu} + ^{1}_{0}\text{n} \longrightarrow ^{66}_{30}\text{Zn} + ^{0}_{-1}\text{e} + \bar{\nu}$$

$$^{6}_{3}\text{Li} + ^{2}_{1}\text{H} \longrightarrow ^{7}_{3}\text{Li} + ^{1}_{1}\text{H}$$

Some interesting facts & figures
• Dec 17, 1938 Otto Hahn discovered nuclear fission by splitting uranium
• As of 2008, weapons-grade plutonium cost around $4,000/gram (or roughly 150 times more than gold)
• Californium costs $60,000,000/gram
• Fuel for controlled nuclear fusion (primarily deuterium) exists abundantly in the Earth's ocean: about 1 in 6500 hydrogen atoms in seawater is deuterium.
• Some experts estimate that the seawater can supply enough deuterium to provide us with fusion energy for millions of years.
• Cobalt-60, a radioactive element, has applications in cancer treatment, industrial radiography, and food and blood irradiation.
• Cobalt-60 emits β^-, with half-life of 5.27 years. It decays into Ni-60 which emits two gamma ray photons.
• Cobalt-60 does not exist in nature. It is artificially produced by nuclear reactions in labs.

Multiple Choice:

1. The nucleus of $^{9}_{4}Be$ has
(A) 9 neutrons and 4 protons
(B) 4 protons and 9 neutrons
(C) 5 neutrons and 4 protons
(D) 4 neutrons and 5 protons
(E) 4 neutrons, 4 protons, and one electron

2. When a nucleus decays by emission of a gamma ray photon,
(A) its atomic number changes
(B) its mass number changes
(C) both the atomic number and the mass number change
(D) none of the two numbers change
(E) the mass number and the atomic number are switched around

3. Different isotopes of an element have the same
(A) atomic number
(B) neutron number
(C) mass number
(D) size
(E) spin

4. The most abundant isotope of hydrogen atom has
(A) no neutrons
(B) 1 neutron
(C) 2 neutrons
(D) 3 neutrons
(E) 4 neutrons

5. The nucleus triton has
(A) no neutrons
(B) 1 neutron
(C) 2 neutrons
(D) 3 neutrons
(E) 4 neutrons

Chapter 60
Nuclear Fission and Fusion

Nuclear Fission

For some elements the nuclei are not stable. One such nucleus breaks apart in two large fragments, one or more neutrons, and a large amount of energy. This phenomenon is called **nuclear fission.** The kinds of fragments produced are not unique but do follow a certain pattern. Nuclear fission can be spontaneous or induced.

For example, ^{235}U when bombarded by slow neutrons absorbs the neutron and may break up into a nucleus of ^{140}La, ^{97}Br with the emission of 2 neutrons and large amount of energy.

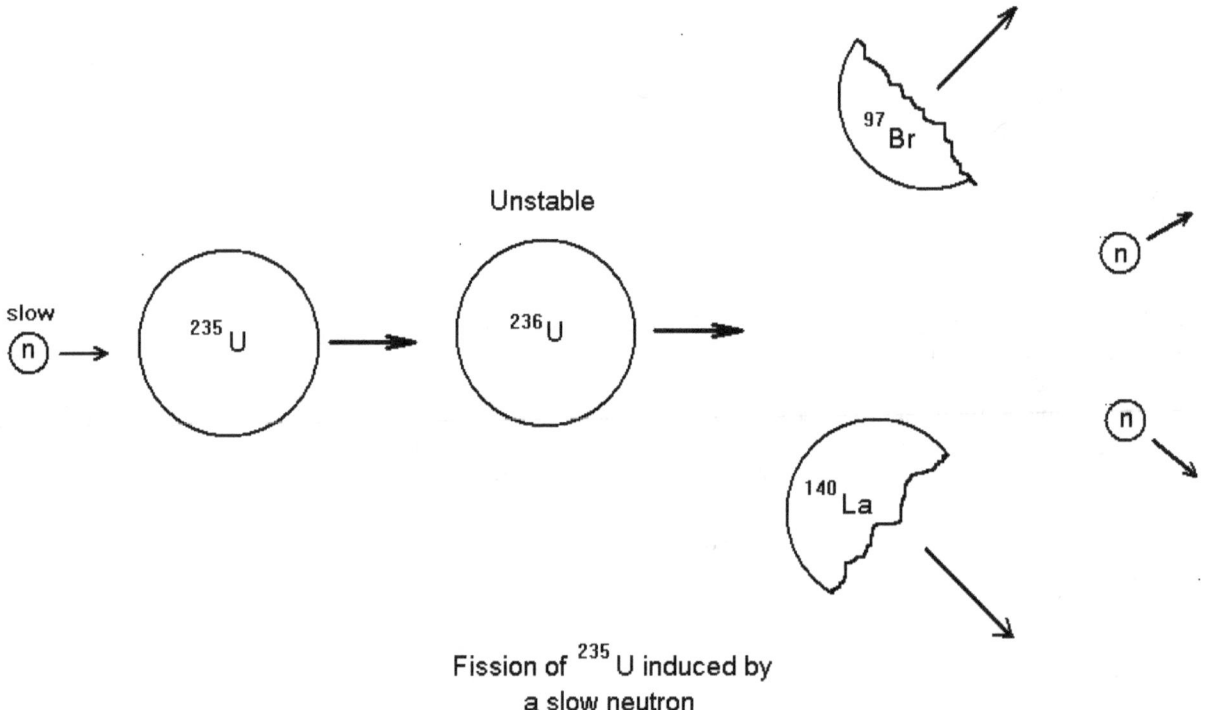

Fission of ^{235}U induced by
a slow neutron

These neutrons if slowed down may be absorbed by other ^{235}U nuclei and produce more fission. When this process becomes *self-sustaining,* it is called **chain reaction.** The *controlled* chain reaction is used in the nuclear power plants to produce electricity. In 'atomic' bomb (fission bomb) this chain reaction goes *uncontrolled* and produces explosion.

First artificial self-sustaining controlled chain reaction [December 2, 1942] was achieved by Enrico Fermi in a racquet court below the bleachers of Stagg Field at the University of Chicago.

Some fission reactions: Average energy output for ^{235}U fission = 200 MeV per event

$$^{1}_{0}n + ^{235}_{92}U \longrightarrow ^{139}_{56}Ba + ^{94}_{36}Kr + 3\,^{1}_{0}n$$

$$^{1}_{0}n + ^{235}_{92}U \longrightarrow ^{141}_{56}Ba + ^{92}_{36}Kr + 3\,^{1}_{0}n$$

$$^{1}_{0}n + ^{235}_{92}U \longrightarrow ^{140}_{54}Xe + ^{94}_{38}Sr + 2\,^{1}_{0}n$$

$$^{1}_{0}n + ^{235}_{92}U \longrightarrow ^{142}_{56}Ba + ^{92}_{36}Kr + 2\,^{1}_{0}n$$

$$^{1}_{0}n + ^{235}_{92}U \longrightarrow ^{142}_{56}Ba + ^{90}_{36}Kr + 3\,^{1}_{0}n$$

$$^{1}_{0}n + ^{238}_{92}U \longrightarrow ^{139}_{56}Ba + ^{92}_{36}Kr + 3\,^{1}_{0}n$$

$$^{1}_{0}n + ^{238}_{92}U \longrightarrow ^{140}_{57}La + ^{97}_{35}Br + 2\,^{1}_{0}n$$

Nuclear Fusion

Two nuclei are out of range for nuclear strong force
There is coulomb repulsion but no nuclear attraction

Short range for nuclear strong force

	Coulomb repulsion
	Nuclear strong force

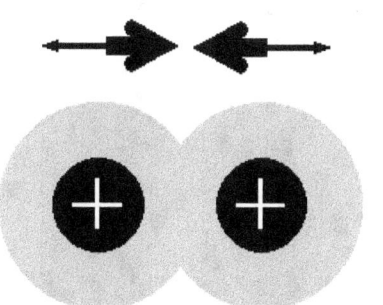

Two nuclei are close enough to be within the range of nuclear strong force.
Strong nuclear attraction fuses the nuclei with emission of energy.

In **nuclear fusion reaction,** two nuclei when brought close together may combine to form a larger nucleus and releasing large amount of energy. For example two nuclei of deuterium (heavy hydrogen) may fuse together to form an isotope of He with emission of a neutron.

To initiate the nuclear fusion reaction the nuclei have to be brought very close together so that the short-range nuclear forces could overtake. It is, however, very difficult to bring the two nuclei together because of the mutual repulsion due to their like (positive) charges. In order for them to come close enough they should be made to move toward each other with very high speeds equivalent to the temperatures of millions of degrees Kelvin.

Sun gets its energy from the nuclear fusion reactions, which was initiated at the time of evolution of the solar system. Inside the Sun's core, the temperature is 6 million Kelvin which is sufficient to sustain the fusion reaction. However, on the earth the fusion reaction created so far is of only uncontrolled nature (hydrogen bomb). This is triggered by creating high temperature with an atomic bomb (fission bomb) and allowing fusion material such as heavy hydrogen to rise to temperature of millions of Kelvin for the fusion reactions to start. But once started, the fusion reaction goes out of control and results in explosion.

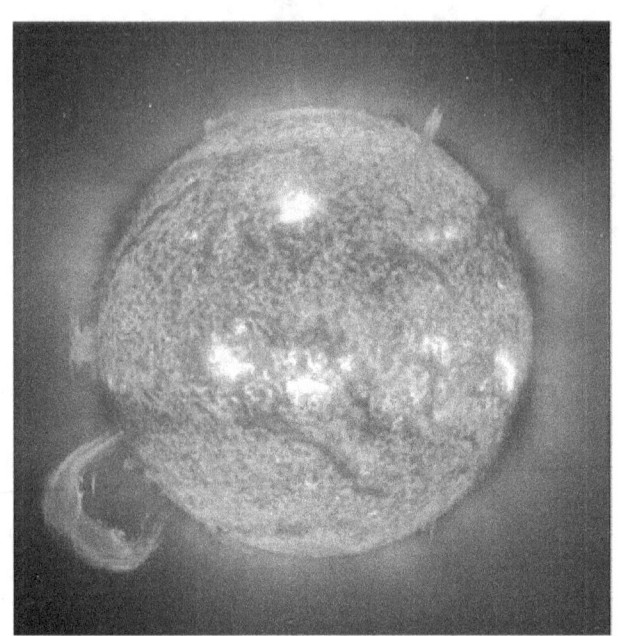

The efforts are underway to produce controlled fusion reaction to obtain clean energy (almost free of radioactive waste) for peaceful purposes.

Some fusion reactions:

$$^{1}_{1}H \rightarrow proton, \quad ^{2}_{1}H \rightarrow deuteron, \quad ^{3}_{1}H \rightarrow triton$$

$$^{1}_{1}H + ^{1}_{1}H \longrightarrow ^{2}_{1}H + ^{0}_{+1}e + \nu + \textbf{1.44 MeV}$$

$$^{1}_{1}H + ^{2}_{1}H \longrightarrow ^{3}_{2}He + \gamma + 5.49 \text{ MeV}$$

$$^{1}_{1}H + ^{3}_{2}He \longrightarrow ^{4}_{2}He + ^{0}_{+1}e + \nu$$

$$^{3}_{2}He + ^{3}_{2}He \longrightarrow ^{4}_{2}He + ^{1}_{1}H + ^{1}_{1}H + \gamma + \textbf{12.86 MeV}$$

$$ {}^{2}_{1}\text{H} + {}^{2}_{1}\text{H} \longrightarrow {}^{3}_{2}\text{He} + {}^{1}_{0}\text{n} + 3.3\,\text{MeV} $$

$$ {}^{2}_{1}\text{H} + {}^{2}_{1}\text{H} \longrightarrow {}^{3}_{1}\text{H} + {}^{1}_{1}\text{H} + 4.0\,\text{MeV} $$

$$ {}^{2}_{1} + {}^{3}_{1}\text{H} \longrightarrow {}^{4}_{2}\text{He} + {}^{1}_{0}\text{n} + 17.6\,\text{MeV} $$

H

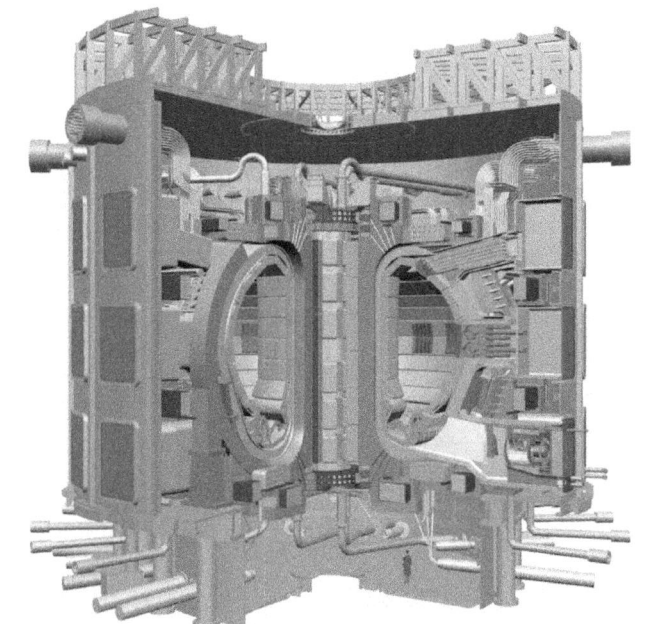

TOKAMAK Nuclear Fusion Reactor

Some important facts & figures
• **Elements heavier than iron are all produced in supernovae.**
• **We (earth and all living things) are made up of products left over by a supernova.**
• **The main source of radiation in the human body is potassium-40.**

Multiple Choice:

1. Which of the following forces poses a barrier to two nuclei fusing together?
(A) Gravitation   (B) Electric   (C) Magnetic   (D) Nuclear strong   (E) Nuclear weak

2. Which scientists first succeeded in carrying out a controlled chain reaction for nuclear fission?
(A) Niels Bohr   (B) J.J. Thomson   (C) Albert Einstein
(D) Enrico Fermi   (E) Count Rutherford

3. Which of the following is a short range force?
(A) Gravitation   (B) Electric   (C) Magnetic   (D) Nuclear strong   (E) Nuclear weak

4. Which of the following requirement pose a great difficulty in harnessing fusion energy?
(A) Need for fast neutrons   (B) Need for slow neutrons
(C) Need for temperatures in millions of kelvins
(D) Need for temperatures very close to absolute zero

(E) Need for heavy water

Chapter 61
Mass-Energy Equivalence

One of the most profound results of the Einstein's Theory of Relativity was the fact that mass could be converted into energy and vice versa.

The equation for the **mass ↔ energy** conversion is rather simple and well known:

$$E = mc^2$$

In this equation, **E** is the energy, **m** is the mass and **c** is the speed of light.

In the case of fission and fusion reactions the total mass of the product is less than the total mass of the reactant. The missing mass is called 'mass defect', **Δm**. This missing mass has been converted into energy in the fission or fusion reaction. The mass defect can be calculated by subtracting the total mass of the product particles from the total mass of reacting particle.

Thus, **Δm = mass (reacting particles) – mass(product particles)**

The energy thus evolved from an event is called Q-value of the reaction. Thus the amount of energy per event can be calculated from the equation:

$$E = \Delta m\, c^2$$

Atomic mass unit (u): This unit is defined as $(1/12)^{th}$ the mass of a ^{12}C atom. Its energy equivalence is:

$$1\, u = 1.660539 \times 10^{-27}\, kg = 931.49\, MeV$$

Particle	Symbol	Mass	Mass in amu	Energy
Electron	$_{-1}^{0}e$	9.10972×10^{-31} kg	0.0005486 u	0.511 MeV
Proton	$_{1}^{1}H$	1.672621×10^{-27} kg	1.007276 u	938.28 MeV
Neutron	$_{0}^{1}n$	1.674928×10^{-27} kg	1.008665 u	939.57 MeV
Deuteron	$_{1}^{2}H$	3.34358×10^{-27} kg	2.013553 u	1875.6 MeV
Triton	$_{1}^{3}H$	5.00736×10^{-27} kg	3.015501 u	2808.91 MeV
Alpha or He nucleus	$_{2}^{4}He$	6.64557×10^{-27} kg	4.002054 u	3727.87 MeV

Some interesting facts & figures
• Mass of C-12 atom is the standard for defining 1 u; its mass is taken as exactly 12 u.
• Atomic mass unit is exactly 1/12th the mass of C-12 atom and approximately equal to the mass of a hydrogen atom or a proton, or a neutron.
• In 2009, 15% of the world's electricity came from nuclear power.
• According to the International Atomic Energy Agency (IAEA), there were 439 nuclear power reactors in operation in the world [1] operating in 31 countries, in 2007.

Problems:

1. In nuclear reaction, 1 g of mass is converted into energy.
(a) Determine the amount of energy in joules.   *[9x10^{13} J]*

2. Two deuterons coalesce together to form a He nucleus.
   {m(deuteron) = 3.3446x10^{-27}kg , m(He nucleus) = 6.6467x10^{-27} kg}

$$^{2}_{1}H + ^{2}_{1}H = ^{4}_{2}He + \text{energy}$$

(a) Calculate the mass defect in this reaction?

(b) How much energy is evolved in this reaction? Give answers in joules and eV.

3. Determine the rest energy ($E_o = mc^2$) in eV of
(a) an electron   *[0.511 MeV]*

(b) a positron   *[0.511 MeV]*

(b) a proton   *[940 MeV]*

4. In the vicinity of a nucleus, a gamma ray photon turns into an electron-positron pair (this phenomenon is called 'pair production'). Each of the two particles carries off KE of 20 keV. Apply the conservation of mass-energy to determine the energy of the original gamma ray photon.

5. Calculate the mass defect (in u) and Q-value (in MeV) for the following nuclear reaction:

$_{1}^{2}H + _{1}^{3}H \rightarrow _{2}^{4}He + _{0}^{1}n + energy$ *[0.018335 u; 17.08 MeV]*

Given Mass $\left(_{0}^{1}n\right)$ = 1.008665 u ; Mass $\left(_{1}^{2}H\right)$ = 2.013553 u ; Mass $\left(_{1}^{3}H\right)$ = 3.015501 u ; Mass $\left(_{2}^{4}He\right)$ = 4.002054 u

6. Calculate the mass defect (in u) and Q-value (energy evolved), in MeV for the following nuclear fusion reaction: $_{1}^{2}H + _{1}^{2}H \rightarrow _{1}^{3}H + _{1}^{1}H + energy$

Given: Mass $\left(_{0}^{1}n\right)$ = 1.008665 u; Mass $\left(_{1}^{1}H\right)$ = 1.007276 u; Mass $\left(_{1}^{2}H\right)$ = 2.013553 u;; Mass $\left(_{1}^{3}H\right)$ = 3.015501 u

Multiple Choice:

1. Einstein's equation for mass-energy equivalence is given by
(A) $E = m/c^2$ (B) $E = mc^2$ (C) $E = m^2c$
(D) $E = m^2c^2$ (E) $E = c/m^2$

2. In the Einstein's equation for mass-energy equivalence 'c' stands for
(A) Speed of sound at 0 °C
(B) Speed of sound at 0 °F
(C) Speed of sound at 0 K
(D) Speed of sound in liquid He at 4 K
(E) Speed of light in vacuum

3. In a nuclear reaction, the Q-value of the reaction refers to
(A) the number of free neutrons produced
(B) the number of free neutrons missing
(C) the amount of energy produced
(D) the number of photons produced
(E) the intensity of activity from the radioactive byproducts

Chapter 62
Exam Prep: Review Problems

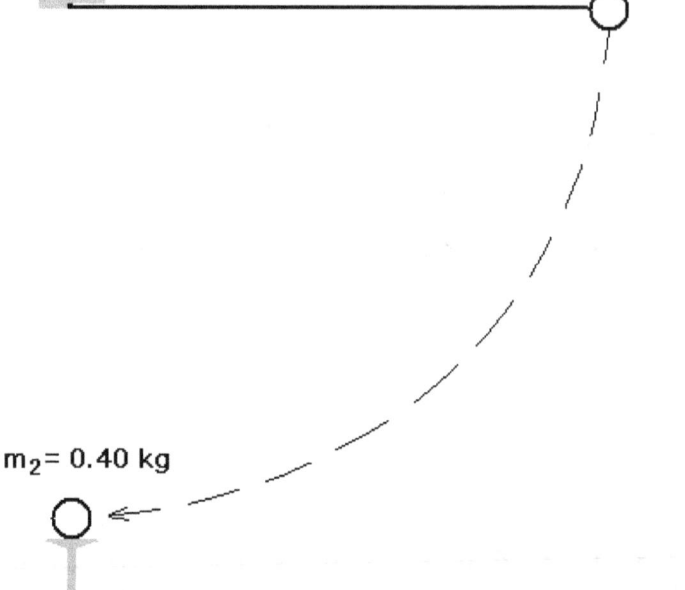

L = 1.8 m m_1 = 0.25 kg

1. A simple pendulum has a length of 1.8 m and a bob of mass m_1= 0.25 kg. As shown in the figure above, the string is initially held horizontal and then released.

(a) Calculate the speed of the bob at the lowest position.

(b) Calculate the tension in the string at the lowest position before any collision.

m_2= 0.40 kg

At the lowest position the bob collides head-on with a second small sphere of mass m_2 = 0.4 kg initially at rest. The collision is inelastic and the second sphere acquires half the kinetic energy of the first sphere.
(c) Determine the speed of m_2 after collision.

(d) Determine the speed of m_1 after collision.

(e) If the collision takes place 3.2 m above the floor calculate the horizontal displacement of m_2 after as it hits the floor.

2.
As shown in the figure here, a horizontal table is 2.4 m long. A small block of mass 4.8 kg is placed at one end of the table. The coefficient of kinetic friction between the surfaces in contact is $\mu_k = 0.14$. A constant force of 36 N directed $30°$ above the horizontal pushes the block, moving it along the length of the table. The force is removed when the block reaches the end of the table.

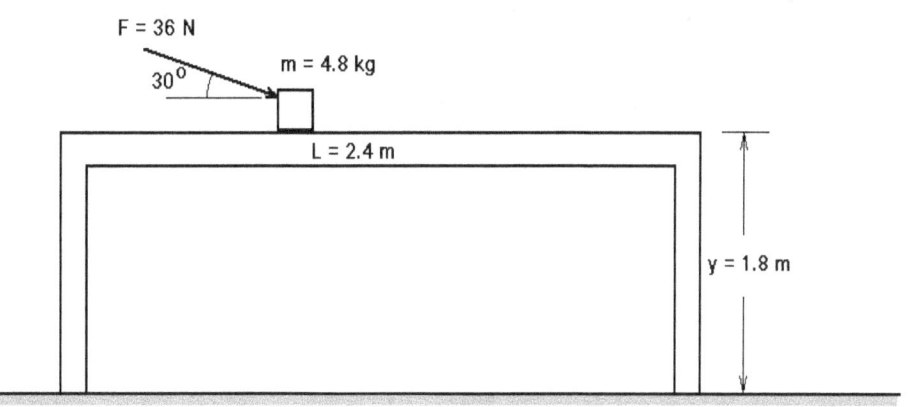

(a) Calculate the acceleration of the block along the table.

(b) Calculate the time the block will take to reach the other end of the table.

(c) Obtain an expression for the velocity of the block as a function of the displacement x from the left end of the table.

(d) Calculate the total work done on the block.

(e) If the table is 1.8 m high calculate how far from the table will the block hit the floor.

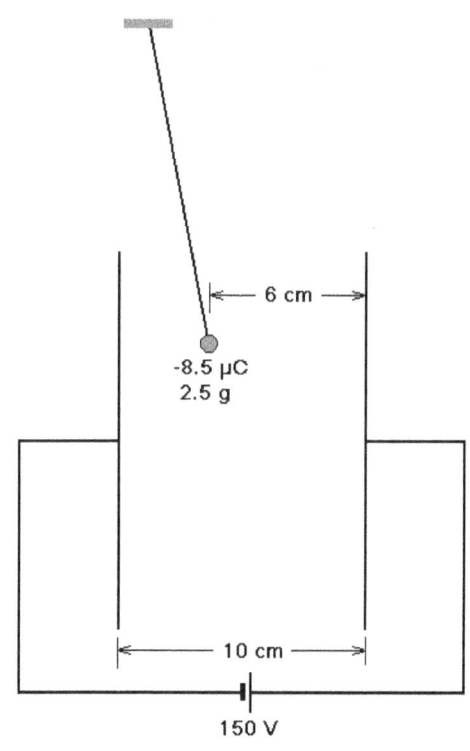

3. Two tall plates of a parallel plate capacitor are 10 cm apart and connected to a 150-V voltage supply. A small sphere of mass m=2.5 g and carrying a charge of -8.5 µC is suspended from a string between the two plates as shown in the figure above.

(a) Determine the electric field between

(b) Determine the angle made by the string with the vertical.

(c) Determine the tension in the string.

The string is now cut close to the sphere without disturbing the electric field.

(d) Determine the vertical acceleration of the sphere.

(e) Determine the horizontal acceleration of the sphere.

(f) How long will it be before the sphere strikes a plate of the capacitor?

(g) How far does the sphere fall before it strikes the plate?

(h) Trace the trajectory of the ball in the diagram until it strikes the plate.

4. A particle of mass m is launched as a projectile from a level ground with initial velocity v_o and at angle of θ above the horizontal. At the top of the trajectory, the particle explodes into two fragments of equal mass. One fragment falls vertically down. The other piece travels horizontally. Give answers to the following questions in terms of v_o, θ, and g.

(a) Determine the expected range, time of flight, and the maximum height reached for the original projectile had it not exploded.

(b) How far from the initial position will the second fragment hit the ground?

(c) What was the gain/loss of mechanical energy of the projectile due to explosion?

5. A source emits light of wavelength 650 nm

(a) Calculate the momentum of a photon of this light.

(b) Calculate the energy of a photon of this light in joules and eV..

(c) A photon of this light collides head on with an electron and bounces back. If the electron is considered to be at rest initially, determine the change in the wavelength of the photon after collision.

6. Two particles have masses $m_1 = 0.03$ kg and $m_2 = 0.08$ kg and carrying charges $q_1 = + 5$ μC and $q_2 = + 15$ μC respectively. They are held a distance of 0.25 m apart in gravity -free region by a 0.5 m long string.

(a) Calculate the tension in the string.

(b) Calculate the potential energy of the system.

The string is now cut and the particles fly apart. (The effect of the strings on the subsequent motion of the particles is negligible)

(c) Calculate the acceleration of each particle immediately after the string is cut.
As the particles fly apart

(d) What kind of path do they follow?

(e) How does the acceleration of each particle change?

(f) How does the speed of each particle change?

(g) How does the kinetic energy of each particle change?

(h) How does the potential energy of the system change?

(i) How does the total energy of the system change?

(j) Calculate the velocity of each particle after a long time.

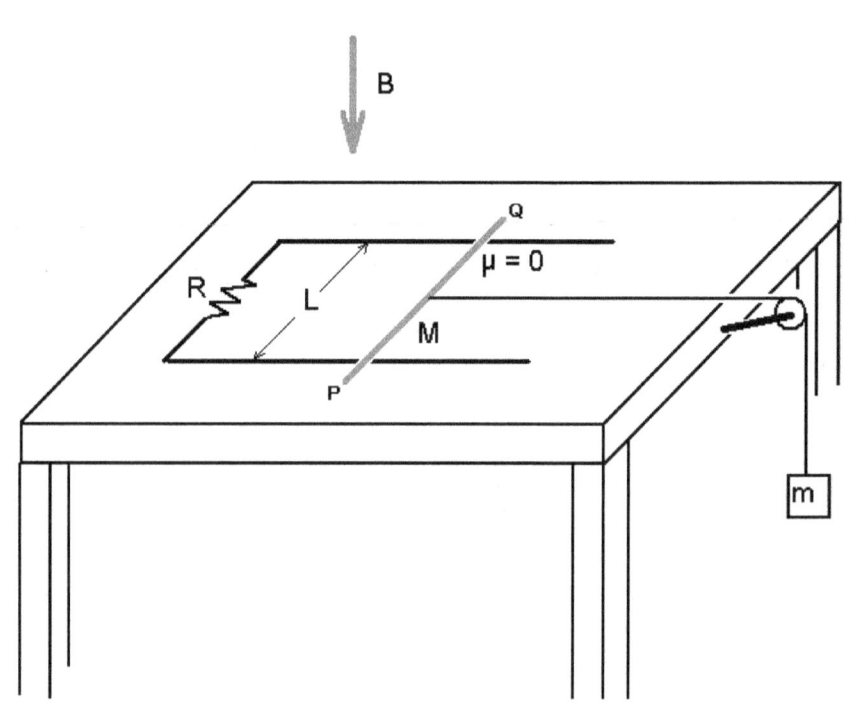

7.

As shown in the figure above, two long parallel metal rails are held fixed on a horizontal table. The rails are a distance L apart. The two ends on one side are connected to a bulb of resistance R. A metal rod PQ of mass M is kept over the rails. A uniform magnetic field B is created vertically downward. A string is attached to the center of the rod and passes over a small frictionless pulley. The other end of the string is attached to a mass m that is hanging freely. In the steady state, the rod is moving at a constant velocity v. (All frictional forces are negligible and the rails and the rod have negligible electrical resistance.)

Give your answers in terms of M, m, R, L, B, and g (not in terms of v).

(a) Determine the direction of current induced in the rod.

(b) Calculate the tension F_T in the string.

(c) Calculate the current induced in the circuit.

(d) Calculate the emf induced in the circuit.

(e) Calculate the constant velocity v.

(f) Calculate the power dissipated in the bulb.

8.
As shown in the figure above, two long parallel metal rails are held fixed on a horizontal table. The rails are a distance L apart. The two ends on one side are connected to a variable resistor R and a battery of emf V. A metal rod PQ of mass M is kept over the rails. A uniform magnetic field B is created vertically downward. A string is attached to the center of the rod and passes over a small frictionless pulley. The other end of the string is attached to a mass m that is hanging freely. The coefficient of static friction between the rod and the rails is μ_s. The rod is stationary.

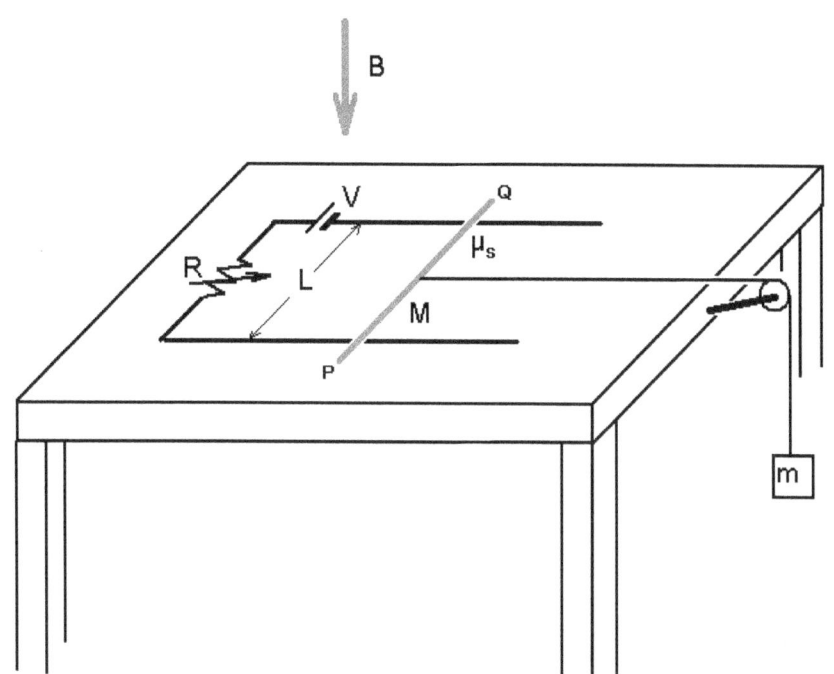

Give your answers in terms of M, m, L, B, V, R, and g.

(a) Calculate the tension in the string.

(b) Calculate the current in the rod.

(c) Calculate the force of friction acting on the rod.

(d) Calculate the value R_1 of the resistor for which the rod will begin moving to the right.

(e) Calculate the value R_2 of the resistor for which the rod will begin moving to the left.

9. In a Young's double slit experiment, the spacing between the slits is 0.2 mm, the screen is a distance of 2.4 m from the slits. The energy of a photon of the incident monochromatic light is 2.07 eV.
Calculate the angular position (θ) and the linear position (y) of the 5th bright fringe.

10. In a diffraction experiment a single slit of width 0.2 mm is used. The screen is 2.4 m from the slit. For a monochromatic light, the angular width of the central maximum is found to 0.26°.
Calculate the energy of a photon of the monochromatic light in eV.

11.
As shown in the figure below, a frictionless plane of length L is inclined at an angle α above the rough surface of a table of the same length L. A small block is released from the top of the inclined plane. Prove that the condition that the block just makes it to the end of the table is that

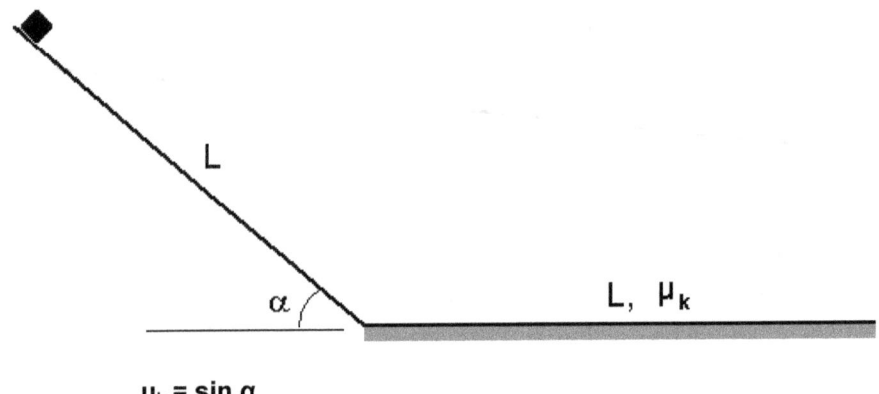

L

α

L, μk

μk = sin α

12. For a particle of mass m = 5 kg, x = 0 and v = 0 at t = 0. A variable force F is applied on the particle in the x direction. The graph below shows the variation of F with time t over a period of 50 seconds.

Calculate the speed of the particle at
(a) t = 10 seconds

(b) t = 20 seconds

(c) t = 25 seconds

(d) t = 50 seconds

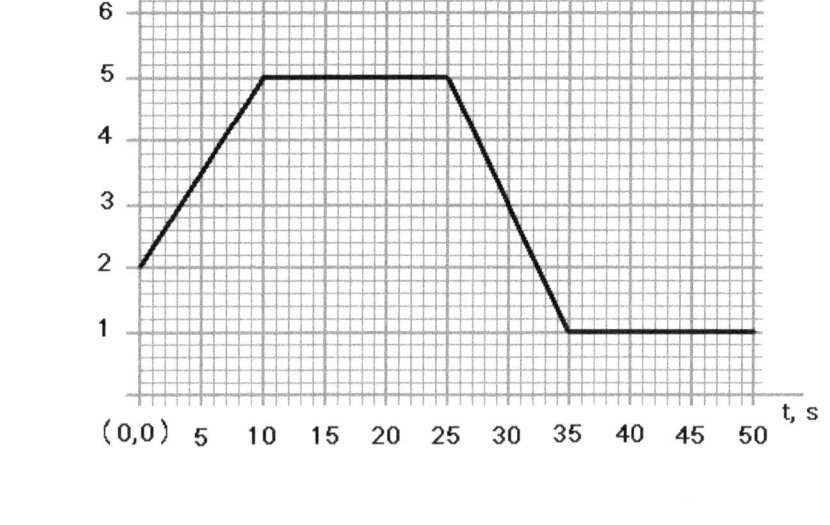

13. For a particle of mass m = 0.5 kg, x = 0 and v = 0 at t = 0. A variable force F is applied on the particle in the x direction. The graph below shows the variation of F with distance x over 50 m.
Calculate the speed of the particle at

(a) x = 10 meters

(b) x = 20 meters

(c) x = 25 meters

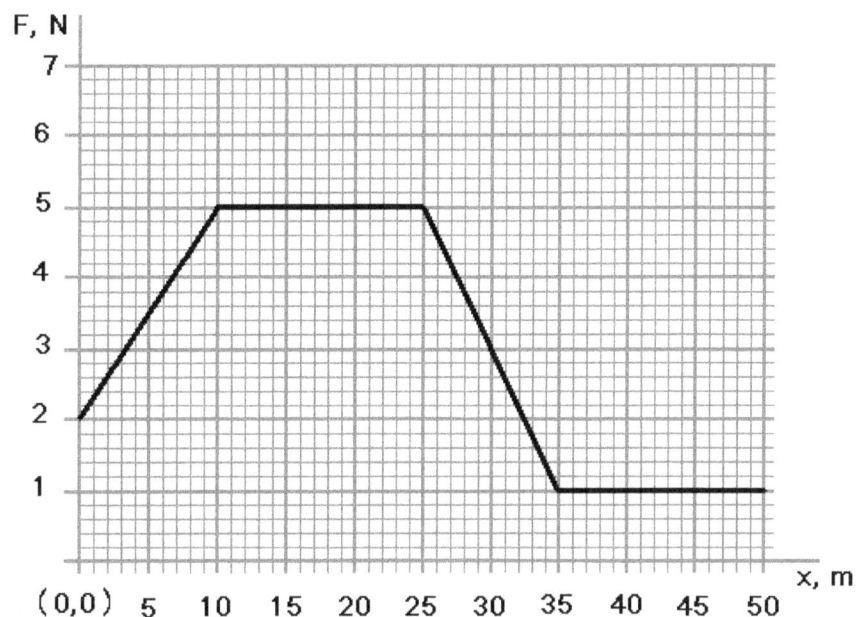

(d) x = 50 meters

14. A ball of mass 50 g is attached to the end of a thin rigid 0.4 m long rod of negligible mass. The system is whirled with frequency 1.2 Hz in a vertical circle about the other end of the rod as the center (see the figure below).

(a) Determine the force in the rod when the attached mass is at the bottom of the circle. Is this force tension or compression in the rod?

(b) Determine the force in the rod when the attached mass is at the top of the circle. Is this force tension or compression in the rod?

15. A ball of mass 50 g is attached to the end of a 1.4 m long string. The system is whirled in a vertical circle about the other end of the string as the center (see the figure below). The tension in the string when the mass is at the top of the circle is 0.8 N

(a) Calculate the speed of the ball at the top of the circle.

(b) Calculate the speed of the ball at the bottom of the circle.

(c) Calculate the tension at the bottom of the circle.

16. A block of mass 5.0 kg is attached to the end of a horizontal spring and placed on a rough ($\mu_s = 0.7$) horizontal surface of a table. When the block is at the position 0 the spring is relaxed. The spring has an elastic constant of 200 N/m and can be stretched or compressed.

Determine the range of positions along the x-axis, at which the block when placed will stay stationary, i.e. the spring can not move the block at those positions of the block.

17. Complete the following nuclear reactions. (You will need Periodic Table)

$$\beta^+ = {}^{0}_{+1}e, \quad \nu = \text{neutrino}$$

$$\beta^- = {}^{0}_{-1}e, \quad \overline{\nu} = \text{anti-neutrino}$$

$$\alpha = {}^{4}_{2}He$$

$$\gamma = \text{photon}$$

$${}^{238}_{92}U \longrightarrow {}^{234}_{90}Th + \; ?$$

290

$$^{226}_{88}\text{Ra} \longrightarrow \text{?} + \alpha$$

$$\text{?} \longrightarrow {}^{211}_{82}\text{Pb} + \alpha$$

$$\text{?} \longrightarrow {}^{143}_{60}\text{Nd} + \alpha$$

$$\text{?} \longrightarrow {}^{14}_{7}\text{N} + \beta^- + \bar{\nu}$$

$$^{60}_{27}\text{Co} \longrightarrow {}^{60}_{28}\text{Ni} + \text{?} + \text{?}$$

$$^{131}_{53}\text{I} \longrightarrow \text{?} + \beta^- + \text{?}$$

$$^{89}_{36}\text{Kr} \longrightarrow \text{?} + \beta^- + \bar{\nu}$$

$$^{55}_{27}\text{Co} \longrightarrow \text{?} + \beta^+ + \text{?}$$

$$^{12}_{7}\text{N} \longrightarrow {}^{12}_{6}\text{C} + \text{?} + \text{?}$$

Nuclear Reactions:

$$^{1}_{1}\text{H} + {}^{7}_{3}\text{Li} \longrightarrow {}^{4}_{2}\text{He} + \text{?}$$

$$^{9}_{4}\text{Be} + \text{?} \longrightarrow {}^{10}_{4}\text{Be} + \gamma$$

$$^{9}_{4}\text{Be} + \gamma \longrightarrow \text{?} + {}^{1}_{0}\text{n}$$

$$^{9}_{4}\text{Be} + \text{?} \longrightarrow {}^{10}_{4}\text{Be} + {}^{1}_{1}\text{H}$$

$$? + {}^{1}_{1}H \longrightarrow {}^{4}_{2}He + {}^{6}_{3}Li$$

$$^{10}_{5}B + {}^{1}_{0}n \longrightarrow {}^{4}_{2}He + {}^{7}_{3}Li$$

$$^{10}_{5}B + {}^{1}_{1}H \longrightarrow ? + {}^{7}_{4}Be$$

$$? + {}^{7}_{3}Li \longrightarrow {}^{4}_{2}He + {}^{4}_{2}He$$

$$^{12}_{6}C + {}^{1}_{0}n \longrightarrow {}^{13}_{6}C + ?$$

$$^{13}_{6}C + {}^{1}_{1}H \longrightarrow {}^{13}_{7}N + ?$$

$$^{18}_{8}O + {}^{1}_{1}H \longrightarrow {}^{18}_{9}F + ?$$

$$^{65}_{29}Cu + {}^{1}_{0}n \longrightarrow {}^{66}_{30}Zn + ?$$

18.
In the circuit above,
(a) Use Kirchoff's rules to calculate the direction and magnitude of the current in the circuit.

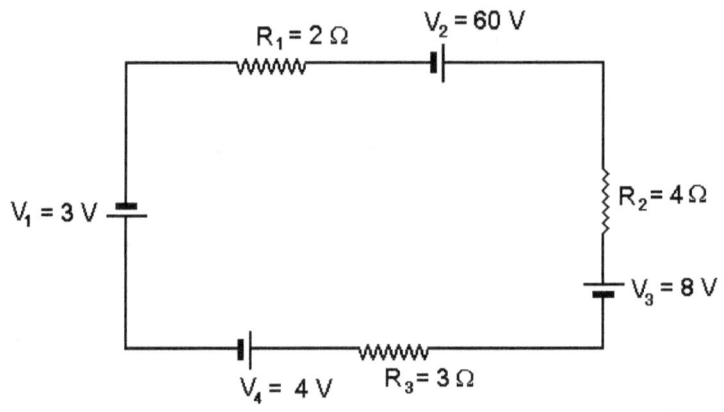

(b) Determine
    (i) which battery (batteries) are supplying power and the amount of power they are supplying

(ii) which battery (batteries) being charged and the rate at which the energy is being delivered to it.

(c) Calculate the power dissipate in the resistor R_2.

19. The lowest four energy levels of an atom in order of increasing energy are   -54.4 eV, -13.6 eV, 6.04 eV, and -3.4 eV. An electron jumps between two of these states emitting a photon of wavelength 121.9 nm..

(a) Calculate the energy of the photon in joules and eV.

(b) Determine the energy levels between which the electron jumped.

(c) Calculate the momentum of the photon.

The photon is then incident on a metal surface in a photoelectric experiment. The work function of the metal is 3.7 eV. The electron emitted has maximum possible speed.

(d) Calculate the kinetic energy of the photoelectron.

(e) Calculate the stopping potential for the photoelectron.

20. A nucleus of mass M, initially stationary, emits a photon of wavelength λ.

(a) Determine the momentum of the photon.

(b) Determine the KE of the recoiling nucleus,

21. An Atwood's machine consists of a small frictionless pulley suspended from a weighing scale. A string passes over a pulley and carries two masses m_1 and m_2 ($m_2 > m_1$) in the form of rectangular single-turn conducting coils as shown in the Fig. 1 above. The coils are allowed to move freely under gravity.

(a) Determine the acceleration of each coil.

(b) Determine the tension in the string.

(c) Determine the reading in the weighing scale in newtons.

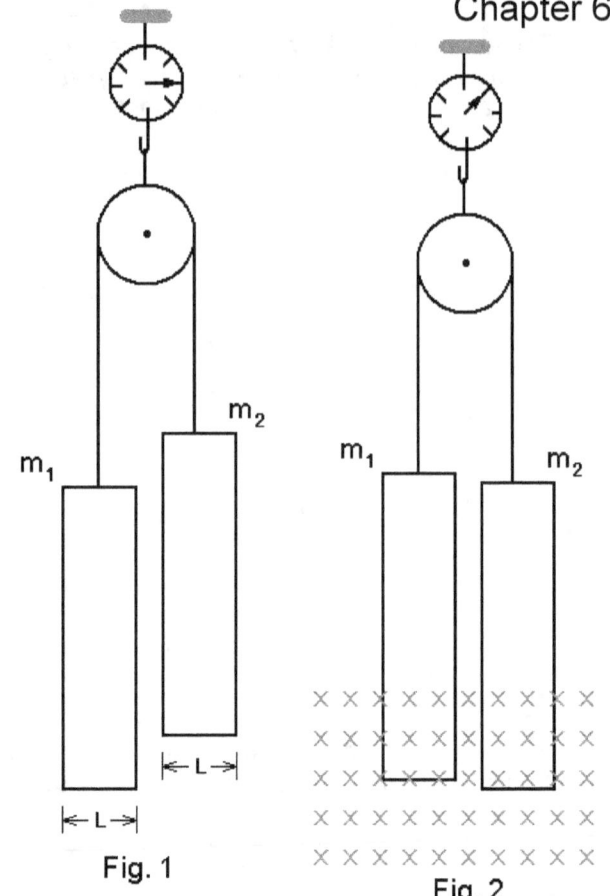

Fig. 1

Fig. 2

A uniform magnetic field B is created perpendicular to the planes of the coil (see Fig. 2). The system soon attains a constant speed.

(d) Determine the direction of magnetic forces on the two coils.

(e) Calculate the magnitude of the magnetic force acting on each coil.

(f) Calculate the magnitudes and directions of the currents in each coil.

(g) Calculate the new reading in the scale.

22.

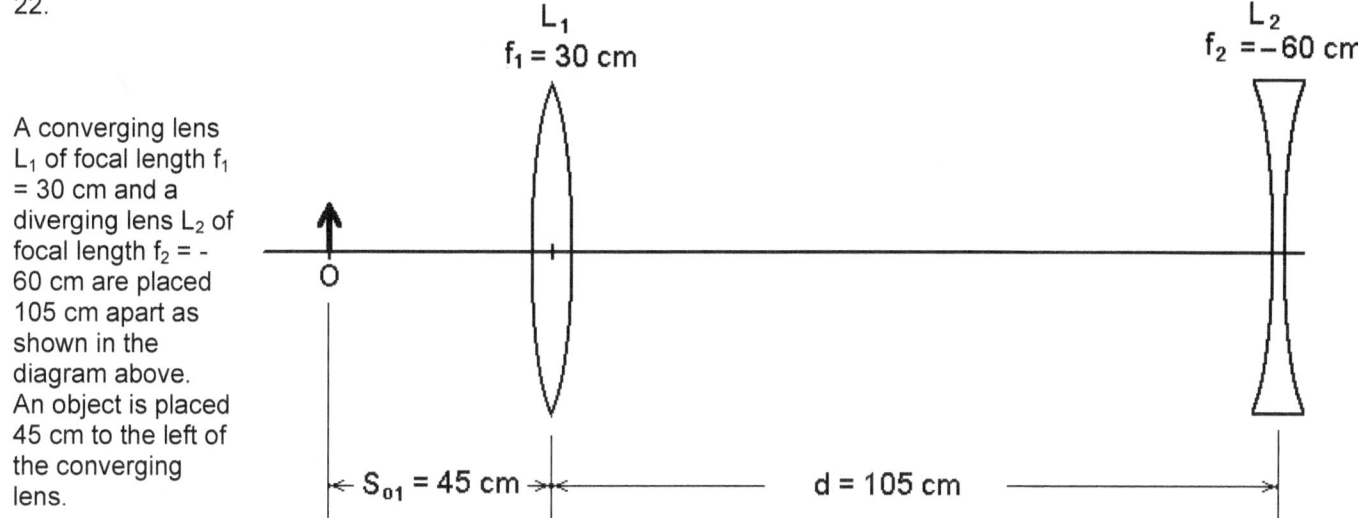

Chapter 62

L_1
$f_1 = 30$ cm

L_2
$f_2 = -60$ cm

A converging lens L_1 of focal length f_1 = 30 cm and a diverging lens L_2 of focal length f_2 = -60 cm are placed 105 cm apart as shown in the diagram above. An object is placed 45 cm to the left of the converging lens.

$S_{o1} = 45$ cm

$d = 105$ cm

Ray Tracing Method:
(a) Use three principle rays to obtain the image formed by the lens L_1.
(b) Use three principle rays to obtain the final image formed by the lens L_2.

Analytical Method:
(c) Calculate the position of the image formed by the lens L_1

(d) Determine the nature of the image formed by the lens L_1 and circle the right answers below:
Real or Virtual
Magnified, diminished, or same size
Upright or inverted

(e) Calculate the position of the final image formed by the lens L_2.

(f) Determine the nature of the final image formed by the lens L_2 and circle the right answers below:
Real or Virtual
Magnified, diminished, or same size
Upright or inverted (in relation to the object)

23.

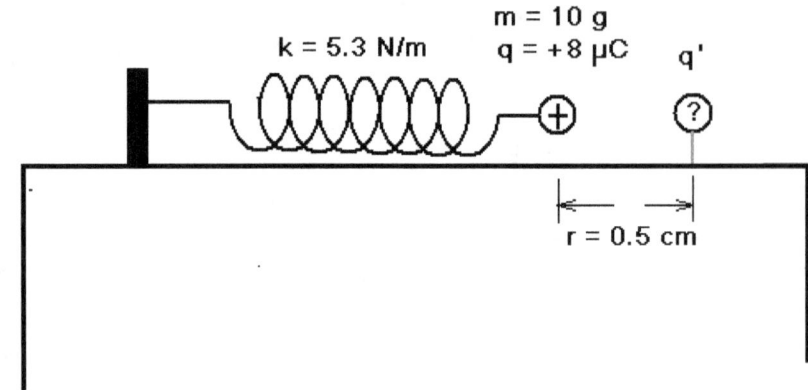

A spring is placed horizontal with one end fixed to the end of a frictionless table. The spring has elastic constant k = 5.3 N/m and caries a particle of mass m = 10 g and a charge of +8 μC at its free end. Another charged particle q' is brought near q and in steady state the spring is stretched by 3 cm when the distance between the two particles is 0.5 cm (see the figure above).

(a) Calculate the elastic storing force in the spring.

(b) Calculate the charge q'.

(c) If q' is now suddenly removed calculate the period of oscillations of the spring.

Chapter 62

24. A circuit containing two resistors, two capacitors, an ammeter, two switches and a battery is shown above. Initially the two switches are open.

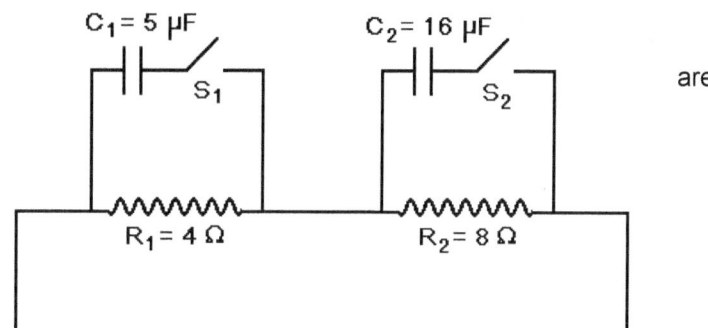

(a) Calculate the reading in the ammeter.

(b) Calculate the power supplied by the battery.

Now the switch S_1 is closed (S_2 stays open).
(c) Calculate the reading in the ammeter immediately after S_1 is closed.

(d) Calculate the reading in the ammeter long time after S_1 is closed.

Now the switch S_2 is also closed (Both S_1 and S_2 are closed).
(e) Calculate the reading in ammeter immediately after S_2 is closed.

(f) Calculate the reading in the ammeter a long time after S_2 is closed.

(g) Calculate the charges stored in the capacitors C_1 and C_2 a long time after S_2 is closed.

(h) Calculate the amount of energy stored in each capacitor.

25. The graph above is a PV-diagram for 120.28 moles of an ideal monatomic gas taken through a reversible thermodynamic cycle 1-2-3-1. For the process 3 → 1, 1.39×10^5 J of heat is removed from the gas.
The process 1→2 is isobaric.
The process 2→3 is isochoric.
The process 3→1 is isothermal.

(a) Calculate the temperatures T_1, T_2, and T_3.

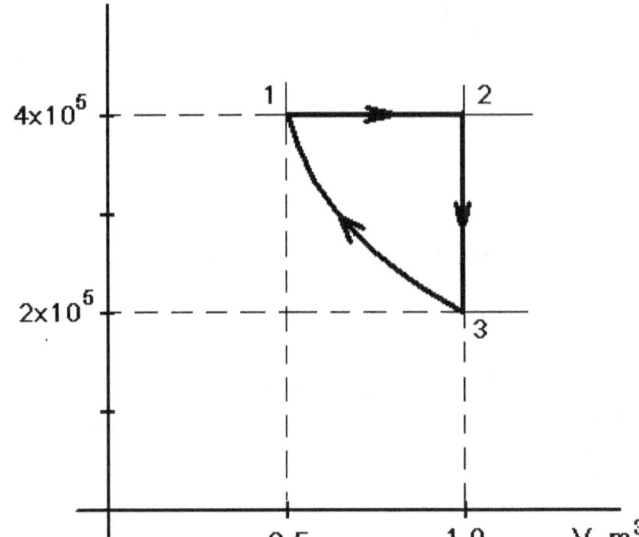

(b) Calculate P_3.

(c) Calculate the W_{12}, W_{23}, W_{31}, and W_{cycle}.

(d) Calculate ΔU_{12}, ΔU_{23}, ΔU_{31}, and ΔU_{cycle}.

(e) Calculate Q_{12}, Q_{23}, Q_{31}, and Q_{cycle}.

26. A 5-cm tall object is placed 6 cm to the left of a concave mirror of focal length 12 cm.

(a) Calculate the distance of the image formed by the mirror.

(b) Calculate the size of the image formed by the mirror.

(c) Is the image
      (i) Real or Virtual?
      (ii) Upright or inverted?
      (iii) the same size, magnified or diminished?

27. A 5-cm tall object is placed 6 cm to the left of a convex mirror of focal length 12 cm.
(a) Calculate the distance of the image formed by the mirror.

(b) Calculate the size of the image formed by the mirror.

(c) Is the image
      (i) Real or Virtual?
      (ii) Upright or inverted?
      (iii) the same size, magnified or diminished?

28.

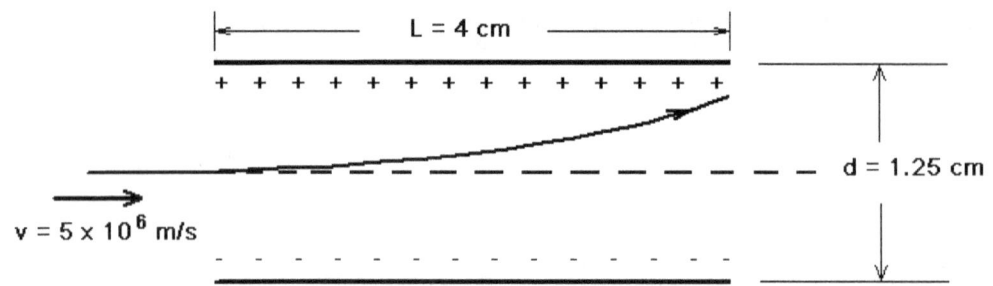

As shown in the figure above, a uniform electric field is produced between two parallel plates of length 4 cm. The plates are 1.25 cm apart and are connected to a voltage supply of 25 V.

(a) Calculate the strength of electric field between the plates.

A beam of electron moving at 5×10^6 m/s enters between the plates at right angles to the field.
(b) Calculate the acceleration of the electrons between the plates.

(c) Calculate the lateral deflection of the beam as it exits the field.

(d) A uniform magnetic field is now to be created at right angles to the electric field so that the beam passes between the plates without any deflection. Calculate the strength and direction of the magnetic field required.

29.

The figure above shows 4 long straight current carrying wires. Calculate the net magnetic field at the center of the rectangle formed by the wires.

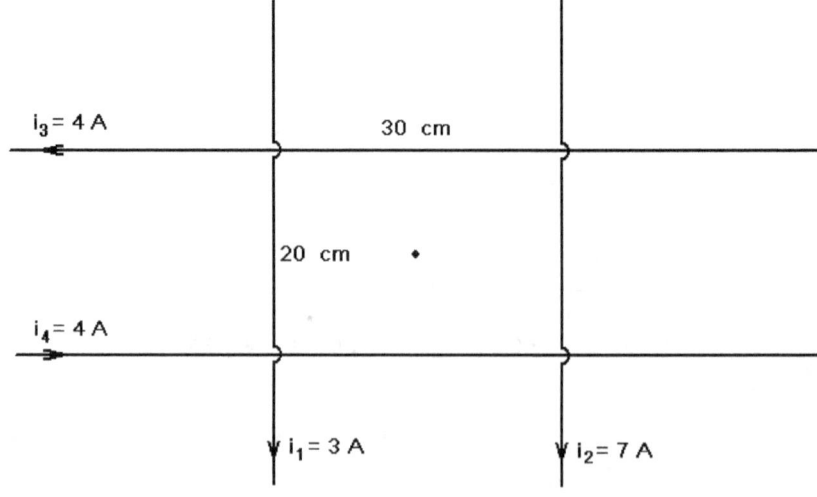

30. A block of wood floats with 25% of its volume above the water surface. If the block is placed in oil what percentage of its volume will be above the oil surface?
[ρ_{water} = 1000 kg/m^3, ρ_{oil} = 800 kg/m^3]

31. A large tank with a large cross sectional area is filled to a height of 5 m. A faucet is connected to a small hole of 0.8 cm diameter at the bottom of the tank. The top of the tank is open to the atmosphere.
(a) Determine the speed with which the water flows out when the faucet is opened.

(b) Determine the volume flow rate of the water flowing out of the faucet.

(c) Determine the mass flow rate of the water flowing out of the faucet.

(d) How long will it take to fill a 200-liter bucket of water?

32. A long straight wire carrying of 5 A is held in the plane of a 25 cm X 8 cm rectangular current loop carrying a current of 3 A as shown in the diagram below. Determine the net force acting on the current loop.

The forces on the smaller segments of the rectangle are equal and opposite and hence they cancel each other out.

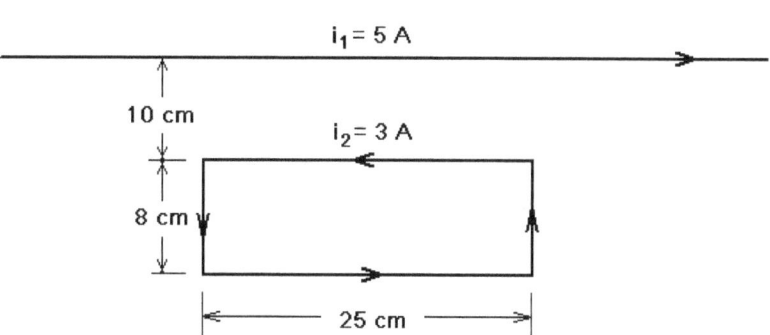

33. An electron is traveling parallel to a long straight wire and 8 cm away from it at 4.0 x 10^6 m/s. Now if a current of 3 A is turned on through the wire determine the magnitude and direction of the magnetic force acting on the electron.

34. As shown in the diagram below, a long straight wire and a circular current loop containing 8 turns lie in the same plane. Determine the radius r of the loop (in terms of d) to make the net magnetic field at its center zero.

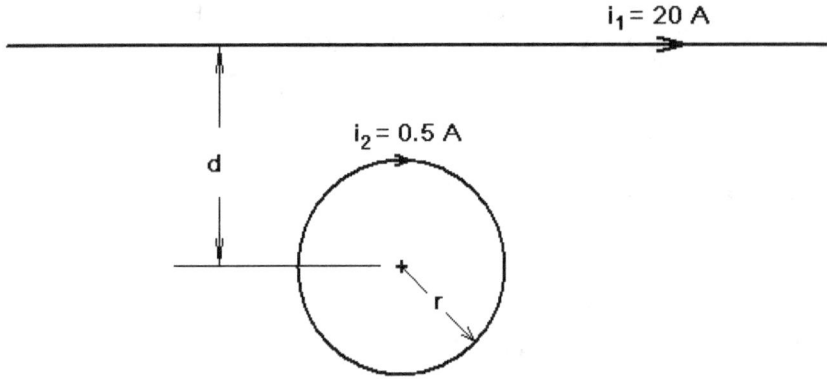

$i_1 = 20$ A

$i_2 = 0.5$ A

d

r

35. Two boys run a race on a play ground.   One runs with a speed of 5 m/s and the other with a speed of 3 m/s. They cross the finish line 8 seconds apart.   How much distance did they race?

36. A ball is thrown with a velocity of 18 m/s at 55° above a level ground. There is a 35-m tall building 10 m away from the initial position of the ball.

 Will the ball hit the building? If yes, on its wall or roof?   at what position?

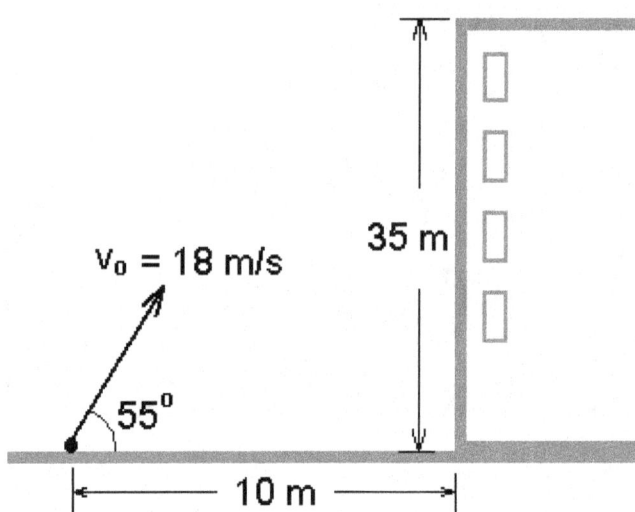

$v_0 = 18$ m/s

55^0

35 m

10 m

Multiple Choice:

1. The bob of a simple pendulum weighs W. A horizontal force is applied on a ball displacing the string through angle θ from the vertical. The tension in the string must be
(A) W
(B) Wcosθ
(C) W/cosθ
(D) Wsinθ
(E) W/sinθ

2. A concave mirror forms a real image with a magnification of -1. The object must be at
(A) the focus
(B) the center of curvature
(C) at a distance of twice the radius of curvature
(D) at a distance of one-half the focal length
(E) at infinity

3. A concave mirror (radius R and focal length f) forms an image that is upright. The object distance must be
(A) less than f
(B) equal to f
(C) equal to R
(D) equal to 2R
(E) between R and 2R

4. The image formed by a convex mirror is always
(A) Upright, virtual, and magnified
(B) Inverted, virtual, and diminished
(C) Upright, virtual, and diminished
(D) Upright, real, and same size
(E) Inverted, real, and magnified

5. A rock of mass 25 kg and volume 0.01 m^3 is suspended from a string and immersed completely in water. The tension in the string is about
(A) 0 N
(B) 150 N
(C) 250 N
(D) 400 N
(E) The answer depends on the shape of the rock.

6-7
As shown in the figure above, a block of wood of weight W is kept fully immersed in water by a string attached to the bottom of the container. The buoyancy force on the block is F_B and the tension in the string is T.

6. The tension in the string must be equal to

(A) F_B (B) W   (C) W - F_B
(D) F_B – W   (E) F_B + W

7. Which of he following best represents the freebody diagram for the block?

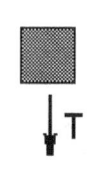 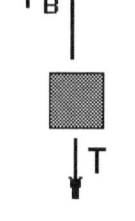 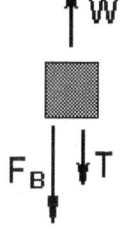  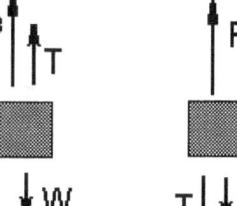

(A)   (B)   (C)   (D)   (E)

8-9

As shown in the figure above, an iron block of weight W is fully immersed in water at the end of a string. The buoyancy force on the block is F_B and the tension in the string is T

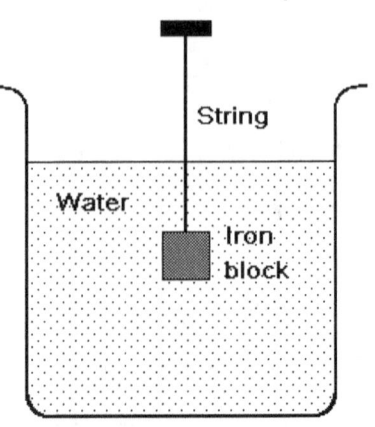

String

Water

Iron block

8. The tension in the string must be equal to

(A) F_B (B) W   (C) W - F_B (D) F_B –W   (E) F_B + W

9. Which of the following best represents the freebody diagram for the block?

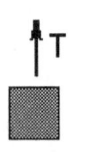
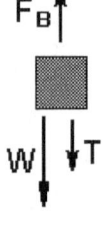
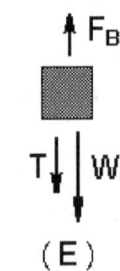

(A)   (B)   (C)   (D)   (E)

10.

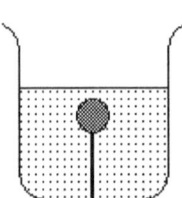

An accelerometer is constructed with beaker, water, string and a ping pong ball as shown above. Which of the following diagrams represents the apparatus being accelerated to the right?

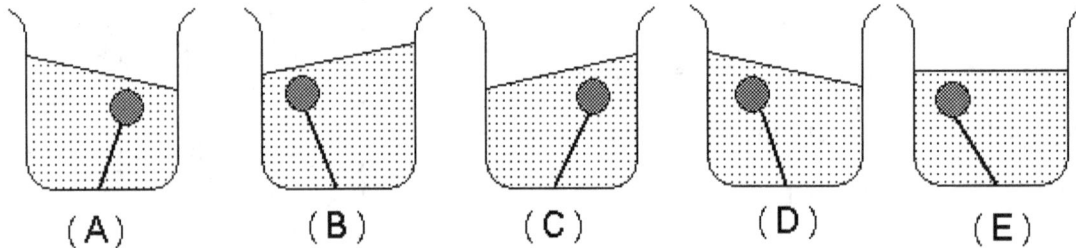

(A)   (B)   (C)   (D)   (E)

11 – 13
Two cars A and B are at x = 0 at t = 0 and are traveling along a straight road (x-axis) as shown in the graph above.

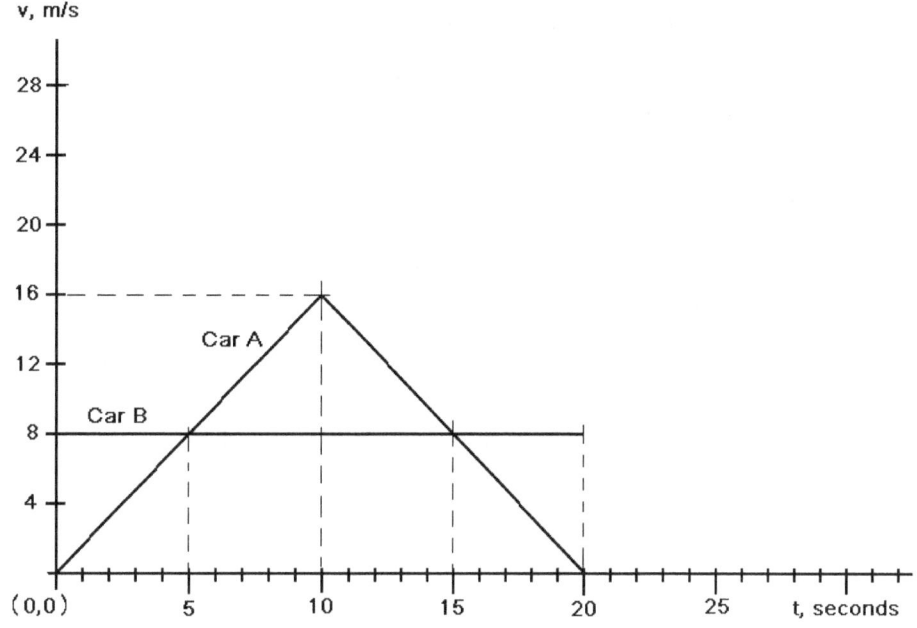

11. At what time do the cars meet again?

(A) 5 s only   (B) 15 s only   (C) 5 and 15 s only
(D) 10 s and 20 s only (E) Never

12. At what time does car A return to the origin?

(A) 5 s   (B) 10 s   (C) 15 s   (D) 20 s   (E) Never

13. At what time do the cars have the same speed going in the same direction?

(A) 0 s   (B) 10 s   (C) 20 s   (D) 5s and 15 s   (E) Never

14. A long straight wire is carrying a current into the plane of the diagrams below. Which diagram correctly represents the direction of magnetic field lines produced by the wire?

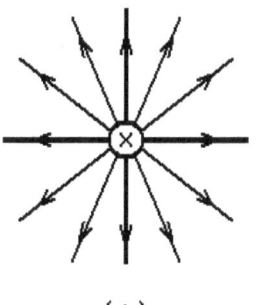

(A)

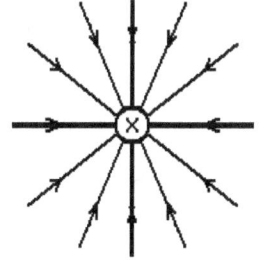

(B)

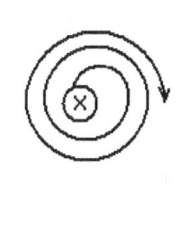

(C)

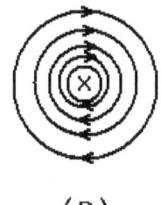

(D)   (E)

15. A wire has resistance of 1 ohm. Another wire of the same material but twice the length and twice the diameter would have a resistance of
(A) 4 ohm   (B) 2 ohm   (C) 1 ohm   (D) 0.5 ohm   (E) 0.25 ohm

16. A boat is floating in a lake in which water has density of 1.2 g/ cm^3. The boat displaces 10,000 newtons of water in that lake. The weight of the boat must be

(A) 10,000 newtons   **(B) 12,000 newtons**   **(C) about 833 newtons**
(D) 12,000,000 newtons   **(E) about 0.833 newtons**

17. Which of the following terms refer to sound frequencies above the audible range?
(A) Ultrasonic (B) Infrasonic   **(C) Supersonic**   **(D) Subsonic**

18. Which of the following terms refer to a speed less than the speed of sound?   **(A) Ultrasonic**   **(B) Infrasonic**   **(C) Supersonic (D) Subsonic**

19. Among sound waves, radio waves, gamma rays, visible light, ultra violet, and microwaves which is(are) electromagnetic waves?
(A) Sound waves only   **(B) Radio waves only**   **(C) Microwaves only**
(D) All except sound waves   **(E) All except microwaves**

20. A horse of mass M is pulling a wagon of mass m with an acceleration a. How much <u>horizontal</u> force is the ground applying on the horse?

(A) ma   **(B) Ma**   **(C) (M + m)a**   **(D) (M – m)a**   **(E) ½ (M + m)a**

21. A particle is performing SHM with amplitude of 10 cm and a period of 2 seconds. The distance traveled by the particle in 6 seconds is

(A) 30 cm   **(B) 60 cm**   **(C) 120 cm**   **(D) 180 cm**   **(E) 240 cm**

22. On a certain planet, the period of a 1-m long simple pendulum is 1 second. The period of a 4-m long pendulum on that planet would be

(A) 16 seconds   **(B) 8 seconds**   **(C) 4 seconds**   **(D) 2 seconds**
(E) 1 second

23. The period of a spring mass system on earth is 5 seconds. If this system is oscillating on another planet where gravity is 4 times the gravity on the earth the new period of the system will be

(A) 80 seconds   **(B) 20 seconds**   **(C) 10 seconds**   **(D) 5 seconds**
(E) 2.5 seconds

24. A coin placed on a horizontal turntable is revolving without slipping with the turntable. The centripetal force on the coin is
(A) Its own weight   **(B) static friction**   **(C) kinetic friction**   **(D) normal force**
(E) Centrifugal Force

25. A metal rod PQ is bent into the form of a part-rectangle as in the figure below. Now if the bent rod is heated uniformly the gap between its ends will

(A) increase   **(B) decrease**   **(C) stay the same**
(D) answer depends on the nature of the metal
(E) answer depends on the thickness of the rod

DETERMINATION OF THE LINEAR COEFFICIENT OF THERMAL EXPANSION BY OPTICAL METHOD

Hasan Fakhruddin

PURPOSE: To determine the linear coefficient of thermal expansion of the given U-shaped pieces of metals using thin slit diffraction.

APPARATUS: Slit-metal assembly, laser, screen, water bath, hot plate, thermometer, clamp etc.

DESIGN OF THE APPARATUS

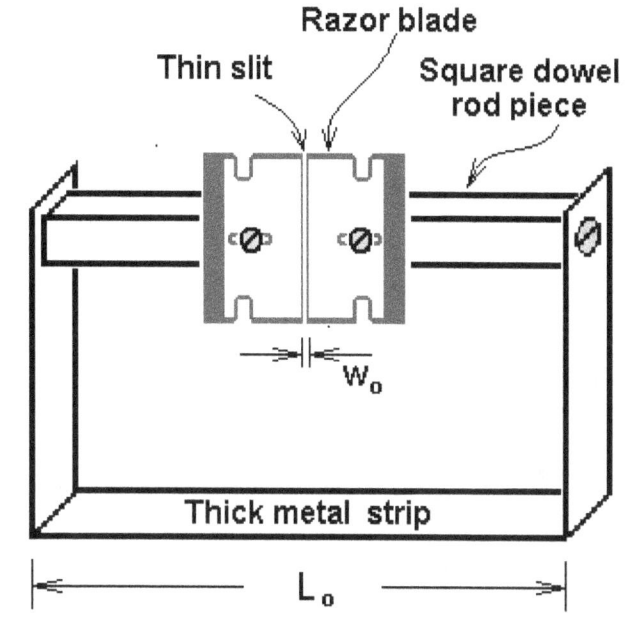

Fig. 1. A strip-slit assembly. The blades are attached to the ends of the bent strip by two dowel rod pieces. The expansion of only the horizontal portion of the strip affects the slit width.

EXPERIMENTAL SET UP

The Figure 2 shows the schematic of the usual single-slit experimental set up. A laser beam of wavelength λ passes through the single slit and produces a diffraction pattern of fringe width z_0 on a screen a distance D away. The beaker may be appropriately covered to prevent slight warming of the blades due to the rising warm air and moistures from the water bath. The water bath is slowly heated to a temperature T.
The expansion of the vertical portions of the bent strip does not affect the fringe width.

As the temperature increases from T_0 ➜ T:
The length of the horizontal section of the strip increases from L_0 ➜ L.
The slit width increases from w_0 ➜ w.
The fringe width decreases from z_0 ➜ z.

The increase in length ΔL and the linear coefficient of thermal expansion α can be determined from the Equations 1 and 2 derived below:

Fig.2. Laser beam passing through the thin slit produces a diffraction pattern on a screen. The metal strip expands as it is heated and increases the slit width thereby changing the diffraction pattern. The slight warming of the blades due to rising warm air and moistures can be minimized by covering the beaker.

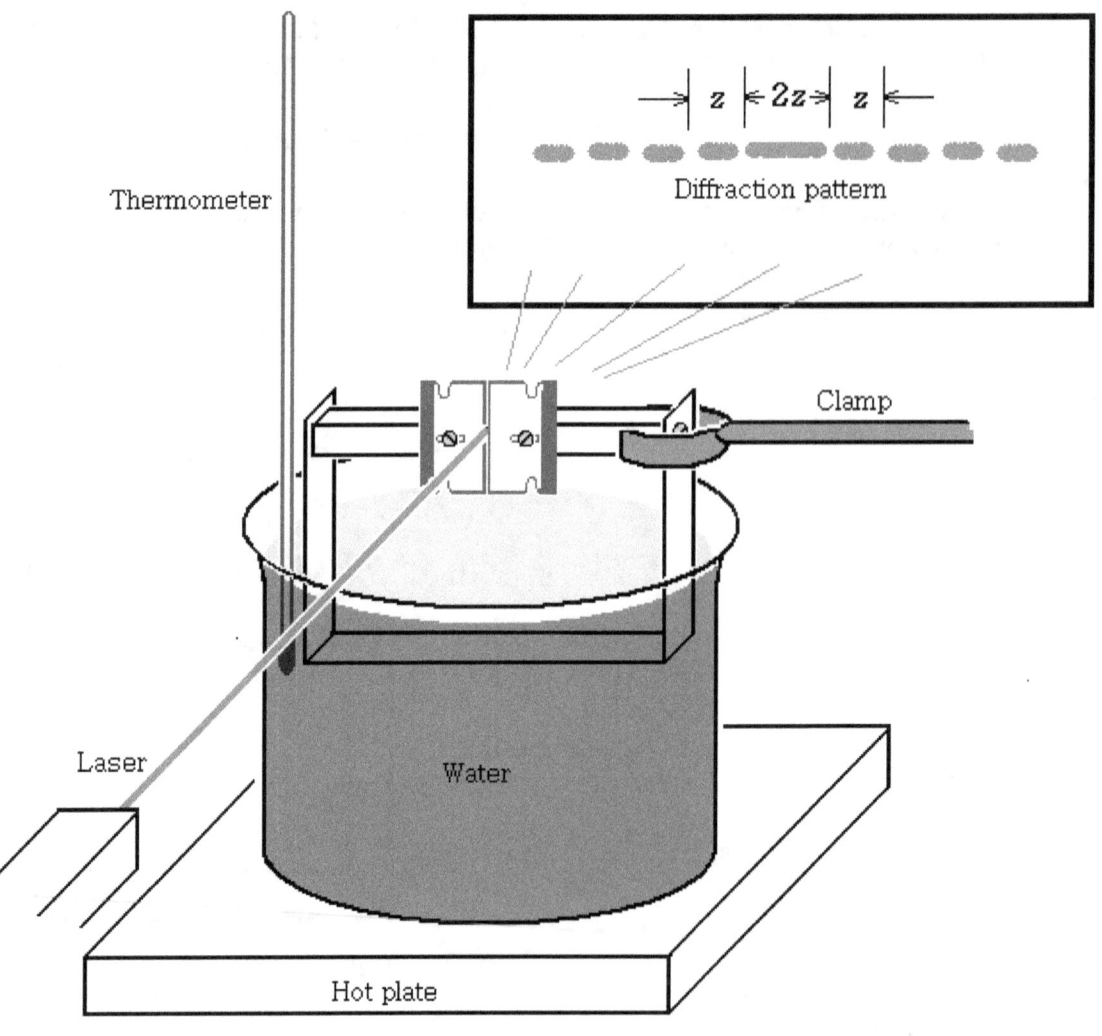

DERIVATION

Using the usual equation[2] for single slit diffraction,

$$w_o = \frac{\lambda D}{z_o}$$

$$w = \frac{\lambda D}{z}$$

$$\Delta L = w - w_o = \frac{\lambda D}{z} - \frac{\lambda D}{z_o}$$

Hence, $\Delta L = \lambda D\left(\frac{1}{z} - \frac{1}{z_o}\right)$ (1)

Now, the increase in length is given by the equation[3]

$$\Delta L = \alpha L_o (T - T_o)$$

Hence,

$$\alpha = \frac{\Delta L}{L_o (T - T_o)}$$

Using the expression for ΔL from Eq. 1, the above equation becomes

$$\alpha = \frac{\lambda D}{L_o} \frac{\left(\frac{1}{z} - \frac{1}{z_o}\right)}{(T - T_o)}$$
(2)

Thus the Eq. 2 can be used to calculate the value of the linear coefficient of thermal expansion α.

DATA
Metal 1:
$L_o =$ _____

$D =$ _____

$\lambda =$ _____

Temperature	Distance for 10 fringes	Fringe width z	1/z

Metal 1:
$L_o =$ _____
$D =$ _____
$\lambda =$ _____

Temperature	Distance for 10 fringes	Fringe width z	1/z

Metal 1:
$L_o =$ _____
$D =$ _____
$\lambda =$ _____

Temperature	Distance for 10 fringes	Fringe width z	1/z

DATA ANALYSIS

This data can be analyzed graphically as described below:

Solving the Eq. 2 for 1/z we get,

$$\frac{1}{z} = \frac{\alpha L_o}{\lambda D} T + \left(\frac{1}{z_o} - \frac{\alpha L_o T_o}{\lambda D} \right)$$

Thus a graph of 1/z vs. T will be a straight line with

$$slope = \frac{\alpha L_o}{\lambda D} \qquad (3)$$

Hence,

$$\alpha = slope \left(\frac{\lambda D}{L_o} \right)$$

SOME HELPFUL POINTERS

- Make sure that the wooden pieces do not get wet; they can change shape if they get wet and seriously interfere with the data acquisition.
- Measure z by finding distance between a large number of fringes symmetrically about the central maximum and dividing the distance by the number of fringes.
- Optimize the initial slit width. The slit width will increase during heating. Therefore it should be very narrow initially but must allow for a measurement of distance between, say 10 fringes.
- For temperatures above 100 $^\circ$C consider using of silicon oil in place of water.

SAMPLE RUN

Here is a sample run for calculation of α for an aluminum strip.

Material:
Aluminum strip (approximately 16 cm x 2 cm x 2 mm thick)
L_o = 0.0904 m (horizontal portion of the bent strip)
D = 1.70 m
λ = 632.8x10^{-9} m

Data and Graphical Analysis:

T ($^\circ$C)	Z (m)	1/Z (m^{-1})

4

17	0.0064	156
40	0.0051	196
57	0.0043	233
75	0.0037	270
90	0.0033	303

Table I. The measured values of the fringe width z and its reciprocal as a function of temperature T

This data can be analyzed graphically as described below:

Solving the Eq. 2 for 1/z we get,

$$\frac{1}{z} = \frac{\alpha L_o}{\lambda D} T + \left(\frac{1}{z_o} - \frac{\alpha L_o T_o}{\lambda D} \right)$$

Thus a graph of 1/z vs. T will be a straight line with

$$slope = \frac{\alpha L_o}{\lambda D} \qquad (3)$$

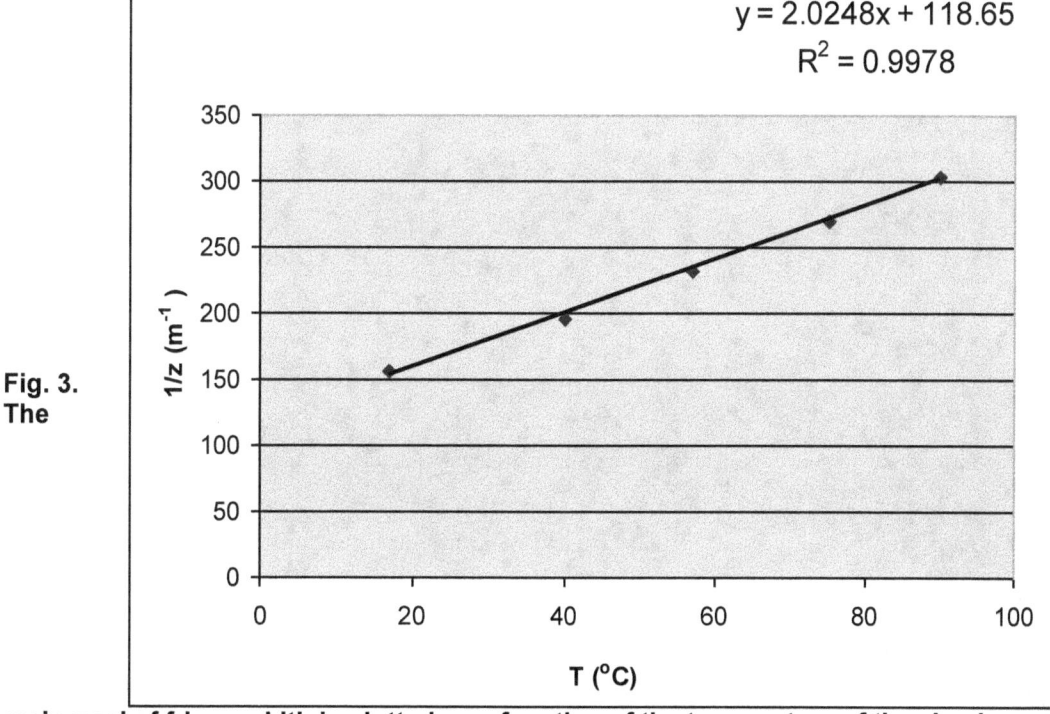

Fig. 3. The

reciprocal of fringe width is plotted as a function of the temperature of the aluminum strip. The slope of the best-fit straight line is then used to calculate α.

From the graph in the Fig. 3, slope = 2.02 m^{-1} °C^{-1}. Using this value of the slope in the Eq. 3 we get

$$\frac{\alpha L_o}{\lambda D} = 2.02 \text{ m}^{-1} \text{ °C}^{-1}$$

Substituting the values of L$_o$, λ, and D in the above equation, we get:
Experimental value of α = 24.0 x 10^{-6} °C^{-1}
Accepted value of α for aluminum is 23.1x10^{-6} °C^{-1} giving a 3.90 % error.

The Periodic Table of Elements

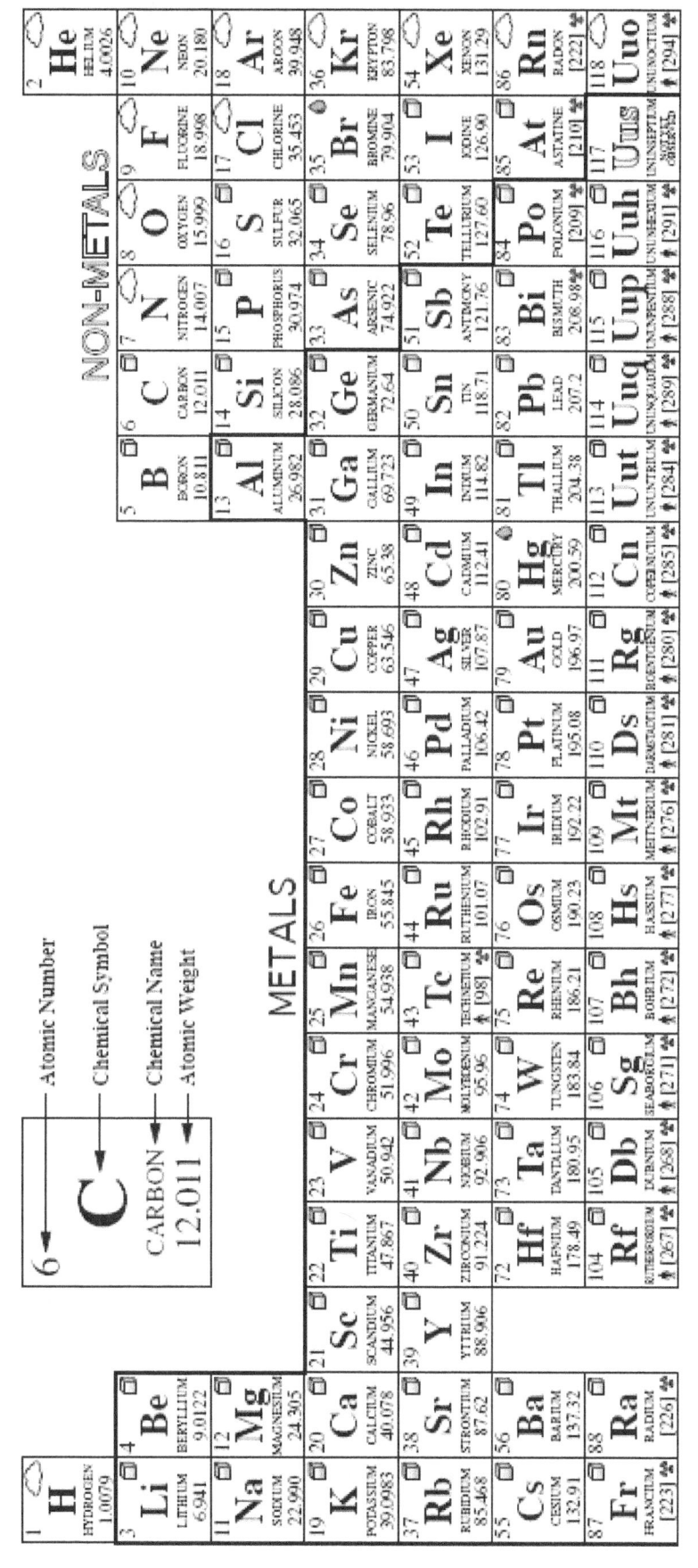